SO GREAT WAS THE SLAUGHTER

NEXUS

NEW HISTORIES OF SCIENCE, TECHNOLOGY, THE ENVIRONMENT, AGRICULTURE & MEDICINE

NEXUS is a book series devoted to the publication of high-quality scholarship in the history of the sciences and allied fields. Its broad reach encompasses science, technology, the environment, agriculture, and medicine, but also includes intersections with other types of knowledge, such as music, urban planning, or educational policy. Its essential concern is with the interface of nature and culture, broadly conceived, and it embraces an emerging intellectual constellation of new syntheses, methods, and approaches in the study of people and nature through time.

SO GREAT WAS THE SLAUGHTER

Market Hunters, Sportsmen, and Wildlife Conservation in Arkansas

BUCKLEY T. FOSTER

THE UNIVERSITY OF ALABAMA PRESS
Tuscaloosa

The University of Alabama Press
Tuscaloosa, Alabama 35487-0380
uapress.ua.edu

Typeface: Scala Pro

Cover imgage: Buffalo National River, Arkansas, by luis sandoval; stock.adobe.com

Cover design: Sandy Turner Jr.

Cataloging-in-Publication data is available from the Library of Congress.
ISBN: 978-0-8173-2224-3 (cloth)
ISBN: 978-0-8173-6187-7 (paper)
E-ISBN: 978-0-8173-9546-9

To Dad, for his outdoor instruction.
To Momma, for everything else.

Contents

Illustrations

Foreword

Hunters know the land. They know local environments. They know animal behavior; the rhythms of bird migration and the growing skittishness of deer as hunting season unfolds. They also know, Buckley T. Foster shows us in *So Great Was the Slaughter*, the whims of market forces and the rhythms of population decline. *So Great Was the Slaughter* frames the nineteenth-century origins of the Arkansas conservation movement as a clash between the increasingly rare market hunter and the new creature of the sportsman. The wealth and leisure time of Americans grew as the numbers of wildlife declined, and prompted a speciation of sorts, embodied by the sportsman, who hunted for recreation instead of food or profit. In this wonderfully thorough case study of Arkansas, Foster carefully documents how the sportsman and market hunter, though both motivated to maximize their harvests, diverged over time in their sense of responsibility, rather than rights, to the land and its inhabitants. The environmental changes, to be sure, wrought by previous overhunting by subsistence and market hunters alike, prompted cultural shifts and then political ones. At the time of this printing there is a small but steadily growing corpus of scholarly studies of hunting and its history. *So Great Was the Slaughter* tells us that the story of modern game management in Arkansas was motored by intertwined environmental, political, and economic concerns. Foster's examination of the origins of the conservation movement in Arkansas with an eye to these complicated forces at play is a robust illustration of the benefits of the interdisciplinary approach the NEXUS Series seeks to advance. We are pleased to include this contribution in the NEXUS Series.

—Alix Hui on behalf of the NEXUS editors

Preface

This journey has proved interesting. There were challenges, just as with any project. Archives closed because of the pandemic, which made research take much longer than expected. One of the early professional readers of the manuscript suggested that I shame present-day sportsmen for their choice of activities, an indictment on the world of hunting. I am a hunter and an academic, believing that one does not negate the other. During recent decades, the debate between hunters and antihunters has simmered and sometimes boiled over. I do not intend to debate the ethics of modern sport hunting. Instead, I set out to examine the death of Arkansas market hunting and the rise of the sportsman conservationist.

While traveling to many different archives, I gathered enough material for a second manuscript covering federal involvement in Arkansas conservation, including more about the creation of the Big Lake National Wildlife Refuge, specific market hunters who broke federal law, the Migratory Bird Law and Treaty, Helena federal judge Jacob Trieber, and Earnest V. Visart as a federal game warden.

Acknowledgments

As with any significant research and writing undertaking, scholars owe much to those who have assisted them. I know I will miss more than one person because I am forgetful and failed to write them down over the seven-year project. In no particular order, I am grateful to those below. Any mistakes in this manuscript are my own, despite all the assistance I received from these folks. Each of these special people made developing this manuscript easier and more enjoyable.

My family and friends. My wife, Holly, and daughter, Lily, endured my long hours and grumblings. Kenny Talley, Shane Alexander, and Grant Herndon, while stuck with me in the duck blind, have suffered hours of my ramblings about obscure Arkansas hunting and fishing information. Despite the monotony I put them through, all my friends and family encouraged me throughout the process. Thanks to my advisor, Dr. John F. Marszalek, and his wife, Jeanne, for instructing me about history and life. Thank you, "Kung Fu" Donnie Joe George, Booneville High School Arkansas history teacher, for being my friend.

At the University of Central Arkansas. Vaughn Scribner for editing and suggestions. David Welky and Don Jones provided exceptional advice. Judy Huff in the History Department office, who is the best and sweetest coworker ever! Department chair Wendy Lucas for funds to travel to the National Archives branch in Fort Worth. Stephen O'Connell in the Geography Department. Karen Pruneda, interlibrary loan coordinator at Torreyson Library, brought many resources to my door when many places had closed. The University of Central Arkansas Foundation's research grant allowed me to travel to the National Archives branch in Maryland.

Other scholars and authors. John Kyle "Big Daddy" Day for his direction and editing skills from beginning to end. I do not know if I could have ever finished this project without his assistance and encouragement. Christopher "Mort" Mortenson, David "Chicken" Schiffler, and Scott "Big Baby" Cashion for their support and advice. I am sure they are exhausted from my "hunting talk." Brooks Blevins for his advice about hunting in the Ozarks. Retired historian Lynn Morrow assisted me tremendously with information about the Missouri bootheel, Missouri archives, and early St. Louis hunting club members who came to Arkansas. Retired physician and author Wayne Capooth for details about several Arkansas hunting clubs.

At the various archives and institutions. Emily Summers, processing archivist, University of Arkansas at Little Rock, Center for Arkansas History and Culture. Brian Robertson at the Butler Center for assisting me with photographs and manuscripts. April White at the Old Independence Regional Museum. Donna Goldstein and Gerald Norwood (Henry Norwood's grandson) at the South Sebastian County Historical Society. Xyta Lucas at the Bella Vista Historical Museum. Jeff Danos, director of operations at the Eureka Springs Historical Museum. Lindsey Irvin at the Grant County Museum. Joseph Alley at the Helena Museum of Phillips County. Maria Jackson, museum program assistant at the Arkansas Post Park. Madeleine Thompson, director, Library and Archives, at the Wildlife Conservation Society. State archivist John Dougan and visual materials archivist Erika Woehlk at the Missouri State Archives. Laura R. Jolley at the Center for Missouri Studies. John C. Konzal at the State Historical Society of Missouri. Jennifer Frazier at the Missouri Department of Conservation. Tom Kanon at the Tennessee State Library and Archives. Bess Robinson at the University of Memphis. Tab Lewis and Haley Maynard at the National Archives in Maryland. Lara Szypszak at the Library of Congress. Amanda Nelson at Wesleyan University, Middletown, Connecticut. Shane K. Bernard at the McIlhenny Company Archives for assistance with correspondence between E. A. McIlhenny and Visart. April Soman and Jeffrey Williams at the Arkansas Game and Fish Commission. Michael Reinemer at the Izaak Walton League of America.

Finally, the University of Alabama Press chose two spectacular readers. They provided knowledgeable and helpful direction and suggestions. Their advice made the manuscript a much better work. The entire team at the press, especially the editor in chief, Daniel Waterman, was a dream to work with.

Introduction

Six years ago, I stumbled across a story of war. The conflict did not involve two nations but two groups of American citizens: sportsmen and market hunters.[1] This fight did not occur in the trenches of a battlefield but around a lake covering land in Missouri and Arkansas. In 1812, the New Madrid earthquake on the Arkansas, Missouri, and Tennessee borders dropped the land surface, creating several lakes in an area referred to as the sunk or sunken lands. Reelfoot Lake in Tennessee and Big Lake near Manila, Arkansas, are the most well-known. When thinking about lakes, most people picture a sizable stretch of water several feet deep. In the case of Big Lake, it was indeed sizable, covering over ten thousand acres, but it was not deep, averaging only three feet. Instead, it was more like a slough or semiswamp. Because of its location along the Mississippi Flyway and the ideal conditions it offered for migrating birds, hundreds of thousands of waterfowl visited Big Lake. Myriads of fish inhabited the lake, too. They still do.[2]

Market hunters came to Big Lake because of the game and fish, profiting from its rich bounty. In the late nineteenth century, the modern sportsman appeared. Soon after, competition between sportsmen and market hunters over wildlife grew nationally. One group wanted to profit from a natural resource, and the other tried to conserve it. Wealthy and middle-class nonresident sportsmen, primarily from Tennessee and Missouri, purchased Arkansas land and built private clubs, barring access to non–club members. Led by Tennessee millionaire, politician, state game warden, and sportsman Joseph A. Acklen, a group of wealthy sportsmen from Nashville purchased a ten-foot strip of land around Big Lake, claiming control of the ten-thousand-acre lake under riparian rights.

The Big Lake market hunters did not accept the Big Lake Shooting Club's presence and claimed they had the right to hunt and fish in the waterway. A fight ensued, with the club filing lawsuits and the market hunters burning club property and occasionally shooting at and terrorizing members and employees. After a lengthy legal and physical battle, the club donated Big Lake to the federal government, which turned it into a wildlife preserve in 1915. The Big Lake Shooting Club dissolved, and the United States Biological Survey took up the battle against the Big Lake market hunters. I wanted to write about that war and the characters in the fight over this lake because it exemplifies the conflict between marker hunters and sportsmen not only in Arkansas or the South, but nationally.

Although several studies exist about market hunting, conservation, and sportsmen, no academic book-length studies specifically covered Arkansas from roughly 1800 to 1925. The Big Lake War occurred during the middle of the American conservation movement and the lengthy fight over Arkansas game and fish laws. Because of this discovery, I realized that before I tackled the Big Lake War, I needed to study the events leading up to and surrounding this conflict. This book concentrates on hunters, conservationists, and the in-state fight for better game and fish protection in Arkansas from 1800 to 1925. I chose this period because there are a few studies concerning Arkansas hunters during the colonial period. So, my work begins when Arkansas (as part of the Louisiana Territory) became part of the United States. I chose to end my monograph in 1925 because that was ten years after the creation of the Arkansas Game and Fish Commission (AGFC). The Great Depression took hold in Arkansas, even before 1925, like much of the South. Conservation efforts shifted around this time from protection to more earnest repopulation efforts. Therefore, 1925 seemed like the proper year to end this study.[3]

The struggle at Big Lake was not only a battle between market hunters and wealthy sportsmen over a lake in northeastern Arkansas. Budding American conservationists pushed for stricter game and fish protection all over the nation during the period. Arkansas remained a rough and wild country far later than its neighbors, containing vast wildlife populations for far longer than many other states. Transportation remained primitive in Arkansas until the late nineteenth century, keeping the wildlife somewhat protected from only the most determined market hunters or natives. Arkansas, much the same game-filled wilderness that early Native Americans and colonial French and Spanish had hunted for the market, began to experience wildlife loss at an alarming rate by the late 1800s. As new technologies like the automobile, pump and semiautomatic shotguns, steamboats, and railroads appeared, farmers cleared more land for agriculture, and shooters began slaughtering Arkansas game and fish for the meat markets. More people had access to wildlife. Settlers continued to fill their pots with game and fish, just as their fathers and grandfathers had done for generations. Like other southern states, nonresident hunters poured into Arkansas, some for only a short trip, while others purchased thousands of acres to build hunting and fishing clubs. Arkansas sportsmen, along with nonhunting conservationists, led by individuals like Arkansas State Sportsmen's Association game warden Earnest V. Visart, Judge Lee Miles, Senator Junius Futrell, and Helena bird lover Mrs. Louise Stephenson, created the foundation of Arkansas's wildlife management system, including the creation of the Arkansas Game and Fish Commission.

These Arkansas conservationists, with assistance from outsiders like Tabasco heir and bird conservationist Edward A. McIlhenny, American Game Protective Association vice president Ray P. Holland, US game warden George A. Lawyer, nonresident hunting and fishing club members, and the United States Biological Survey, fought against stubborn politicians, connected game and fish merchants, and a frontier mentality that Arkansawyers had the God-given right or American privilege to kill, trap, or catch as much game and fish as they desired. The battle raged for decades. Unfortunately, by the time Arkansas legislators passed enough laws to protect the game and fish adequately, most of the wildlife was already gone. Only through restocking efforts does Arkansas have the significant game and fish populations it does today.

Earlier, Arkansas stood as a wildlife-filled land for thousands of years. When the earliest Native Americans arrived in Arkansas, they discovered a veritable paradise teeming with wildlife. They hunted and gathered to survive, eventually planting crops to supplement their diet. Later, Henri de Tonti established Arkansas Post in 1686 to trade with the Quapaw, particularly for beaver pelts. The French conducted business with other Indian Nations within a few years, like the Osage, Caddo, and Chickasaw. After furbearing animals, French, Spanish, and later American hunters traveled all over Arkansas. European and Native Americans competed in the Arkansas market hunting business for over a century. Certain wild animals equaled wealth, and Arkansas contained a lot of them.[4]

The Mississippi Flyway, one of the most prolific arteries for migratory birds, passes down the Mississippi River valley in eastern Arkansas. Another longitudinal route is the Central Flyway, located just across the western border into modern-day Oklahoma. Millions of birds move through these two corridors during their seasonal migrations. Therefore, Arkansawyers have used these waterfowl as a food source from prehistoric to modern times.[5]

Frenchman Jean-Bernard Bossu, traveling along the Mississippi River in the winter of 1750–1751, described how the Quapaws kept wild ducks in cages in their villages. He claimed that the Quapaws caught the fowl with their hands: "Toward the end of autumn a multitude [of ducks arrived]. For this hunt the Akancas places some tame waterfowl or stuffed birds on the water. The others, deceived by the lures, are not afraid to approach. Then [the] young savages swim underwater like fish, grab the waterfowl by the feet, hook them by their heads to their loin clothes and bring them back alive. Finally, they pluck their wing feathers and put them in enclosures at their dwellings so they will have them to eat in time of need when the hunt yields nothing."[6]

With the 1803 Louisiana Purchase, American settlers who continued arriving in Arkansas took advantage of the bounty and lived on a game meat and fish diet during their earliest days. A few Arkansas families continued to hunt and fish as their primary food source. Over time, others transitioned toward a more agricultural lifestyle, adding a few acres for cultivation. As time passed, families had the opportunity to clear more extensive tracts of land and produce agricultural products for the market, sometimes supplementing their table with wild game and fish. Nevertheless, some men continued to market hunt and fish as an occupation.

By the end of the Civil War, there were no more bison and beaver in Arkansas, or at least not enough to make a living. Instead, as American industrialization caused urbanization and population growth, a new need for Arkansas game and fish grew. A new type of market hunter emerged. These men did not seek hides but meat. They could not kill enough deer, bears, turkeys, and birds to feed the masses in cities like Chicago, St. Louis, and nearby Memphis. New technologies such as railroads and pump shotguns allowed professional hunters to slaughter and ship millions of animals.

Around the same time, the modern sportsman was born. These men and women, mostly men, hunted and fished for pleasure, not as a necessity. In the earliest days, observers had difficulty distinguishing between a market hunter and a sportsman. They both killed as many animals as they possibly could. The more game a market hunter killed, the more profit he made. The more game an early sportsman killed, the more he could brag, claiming himself a great hunter. The combined actions of these two groups destroyed Arkansas's game and fish.

However, as wildlife disappeared, the two groups began to differ significantly. The market hunter continued to kill, thinking his profits would dry up, killing as much as possible while it lasted. Just as mines played out and loggers cleared forests, the market hunters' resources would disappear, too, causing them to move on or change occupations. Most of the sportsmen, however, became conservationists. If they expected to continue participating in hunting and fishing as a leisure activity, they must protect the game and fish from annihilation. Therefore, market hunters and sportsmen became enemies. Sportsmen in Arkansas, whether nonresident or resident, despised market hunters. "There is nothing more destructive to wild game of any kind than hunting, netting, trapping, or snaring for market. We might as well offer bounties for the destruction of our wild game as to allow it to be sold on the market," argued one ornithologist in 1912, echoing the sentiments of thousands of sportsmen. However, professional hunters despised any restrictions on their livelihood, which brought them into conflict with conservationists.[7]

In Arkansas, during the late nineteenth century, several opposing forces appeared. Nonresidents came into Arkansas, sometimes staying for weeks or months, killing, catching, and shipping wildlife out of Arkansas. Sportsmen sent their harvests home, and market hunters shipped theirs to market. As wildlife numbers decreased, many Arkansawyers blamed the nonresidents, considered thieves, for stealing the people's property. Furthermore, habitat destruction for farm production plagued most southern states' wildlife populations. Between 1850 and 1910, Arkansas farmers cleared 7,294,724 acres of wildlife habitat for agriculture.[8]

Wishing to protect their pastime, some nonresident sportsmen bought land in Arkansas and established hunting and fishing clubs. These clubs banned locals. Many Arkansawyers resented these outsiders and fought against the clubs' presence. It was one thing for these nonresidents to come in and hunt and fish. Now, they prohibited Arkansawyers from hunting and fishing in places where traditionally and generationally they had done so for decades. Later, however, some of these nonresident clubs assisted Arkansas conservationists in pushing the Arkansas legislature to pass stricter game and fish protective laws.[9]

From 1885 to 1925, sportsmen and fellow conservationists from Arkansas and beyond fought against professional hunters, corrupt or stubborn politicians, and closed-minded citizens to establish state game and fish protection laws. As the conservationists' movement gathered steam nationally and locally, Arkansawyers established chapters in associations such as the Izaak

Walton League, the Audobon Society, and other similar organizations. These clubs created educational programs in schools and towns across Arkansas. They gave presentations and materials to countless citizens, arguing for protecting Arkansas wildlife. After the Arkansas Game and Fish Commission's creation in 1915, it became the enforcement arm of the state's game and fish laws. The commission pushed for stricter protective measures, educated the public, and arrested lawbreakers.

John Muir, Teddy Roosevelt, or Aldo Leopold did not specifically campaign to conserve Arkansas wildlife. The wave of conservationism that swept across the nation at the turn of the twentieth century arrived very late to the Bear State. Hunters, both sportsmen and market, slaughtered millions of game and fish. Realizing the growing crisis, many Arkansas sportsmen, either through organizations or individually, led the way to save their game and fish from extinction. Unfortunately, their efforts came up against tremendous odds and, in the end, failed.[10]

The birth of the sportsman and a growing conservationist movement in Arkansas led to the passage of a general game bill, the establishment of the Arkansas Game and Fish Commission, game preserves, sporting clubs, and protection and propagation associations. However, a core group of Arkansawyers held fast to the idea that the wildlife belonged to the people and that no governmental body could regulate their harvest. In many cases, individuals, communities, and counties who benefited from exploiting these natural resources fought the hardest against regulation and, once regulated, against the enforcement of those regulations. For understandable reasons, these people wanted the ability to harvest and sell or eat as much game and fish as they wanted, just as a company might mine coal or cut trees. The AGFC came to represent the ideal of governmental regulation of what these people perceived as a God-given right. This core continued to battle against the commission and its work, whether through misunderstanding or outright opposition.

I analyzed Arkansas's hunting, fishing, and early conservation from 1800 to 1925. These chapters cover the evolution of Arkansas's market hunters, sportsmen, and conservationists during that period. Arkansas, once known as the Bear State because it contained so many black bears, saw their numbers drop from thousands to only thirty by 1925. This work seeks to understand how that happened and why. After the Civil War, the abundance of wildlife in Arkansas, combined with few game laws, the urban demand for meat, the invention of the pump and semiautomatic shotguns and dynamite, the increasing growth of railroads, decreased wildlife populations in other states, the proliferation of sporting publications, and the growing popularity of sport hunting and fishing made Arkansas an equal-opportunity slaughterhouse by the late nineteenth century. The burgeoning Arkansas conservation movement was too late, underdeveloped, and weak to end the extinction or near extinction of the state's wildlife.[11]

I did not set out to write an environmental, political, or economic narrative. It is important to describe the battleground and conditions leading up to the fight to save Arkansas's wildlife. Therefore, the earlier chapters are more about environmental history. The conservation fight spread into the halls of state government. Legislation became one of the primary weapons to fight the decrease in game and fish populations. Therefore, the later chapters could rightly be considered political history. I apologize in advance for the back and forth during the coverage of the political debates. Understanding where and why these politicians stood on these issues is essential. Economics, power, class, race, and masculinity play a part in this story.[12] They are bound together to provide a better overall picture of what occurred in Arkansas. It is a fight between late nineteenth-century market hunters and sportsmen, locals, nonresidents, wealthy and poor, Progressive conservationists, and traditionalists. The story of Arkansas conservation is all of those things. Some were happening openly, and others were behind the scenes.[13]

Chapter 1 describes the Arkansas wilderness during the nineteenth century and the primitive living and working conditions found there. It examines early hunters in Arkansas, providing background on early market hunting. Chapter 1 also covers European market hunting's evolution from targeting furbearing mammals to birds and fish. Furthermore, it includes the transition of early Arkansas settlers from hunters, to hunter-farmers, to farmer-hunters, and finally, to antebellum sportsmen.[14]

Chapter 2 defines sportsmanship and the characteristics of Arkansas sportsmen in the late nineteenth century. It also details the importance of hunting and fishing to the development of young southern males. During the earliest periods, the line between sportsmen and market hunters blurred, and each group impacted Arkansas wildlife negatively. There are three different types of sportsmen discussed in chapters 1 and 2. The first sportsmen, the fewest in number, lived during the antebellum period. They were generally nonresidents of Arkansas or enslavers. The second group appeared during the postbellum period, had wealth, and lived in urban centers. The final group, and the largest, was also composed of urban dwellers, but they fit into the middle class.[15]

Chapter 3 explains why sportsmen came to Arkansas to hunt and fish during the later nineteenth and early twentieth centuries. Several factors played a role in this occurrence. Primarily, new technological advancements and plenty of publicity. Sporting publications grew in popularity and compelled men to venture on hunting and fishing vacations across the South.[16]

Chapter 4 describes the earliest conflict over Arkansas wildlife, including early efforts for Arkansas game protection. During these initial days of wildlife conservation in Arkansas, the game-protective crosshairs settled on the nonresident professional hunter. It was not because citizens were concerned about the overharvesting of Arkansas game and fish; instead, they believed that nonresidents had no right to Arkansas's animals. Many Arkansawyers, politicians, and citizens did not want nonresident market hunters in the state because they thought that Arkansas game and fish were the property of the people of Arkansas. Later, the fear that Arkansas game was disappearing caused a public outcry to end all market hunting and the sportsmen to end their overkilling. Only then did all market hunters become the target of Arkansas game protection legislation. The sportsmen began an active campaign to end the practice of professional hunting for the meat markets. This effort, which stemmed from conservationists, preservationists, and sportsmen, continued until the Arkansas Game and Fish Commission was created in 1915 and the game protection laws passed with its formation.[17]

Chapter 5 explains the development of stricter game protection laws and the continued growth of the conservation movement in Arkansas. By 1900, there was enough public pressure on the Arkansas state legislature to cause them to introduce new game legislation, including placing more restrictions on nonresident hunters. At the turn of the twentieth century, market hunters, nonresident and resident sportsmen, club members, women's clubs, Arkansas sporting groups, the wealthy, conservationists, and even neighboring states sought to influence Arkansas wildlife legislation. On the table were hunting and fishing licenses, nonresident hunting, hunting and fishing seasons, nonexportation laws, market hunting bans or licensing, and other various aspects of game law.

Chapter 6 covers the Herculean efforts of Earnest V. Visart, Arkansas's first state game warden, during his first few years on the job. In September 1909, frustrated with the state government's inaction, the Arkansas State Sportsmen's Association employed Visart to enforce the state game laws. Without any authority, the association tasked Visart with stopping game violators, a difficult, if not impossible, task, as he was not a commissioned law officer of the state.

The state game warden could not arrest perpetrators, write tickets, or issue a summons. Visart could appoint deputy game wardens (already deputy sheriffs) to arrest lawbreakers, but that title did not enhance their authority. He fought a decades-long battle to save Arkansas wildlife from overhunting and fishing.

Chapter 7 continues the coverage of Earnest V. Visart, his fight against Arkansas poachers, his attempts to educate the public about game and fish conservation, and his lobbying efforts in Little Rock. It also looks at the complex relationship between Visart and avian conservationist Edward A. McIlhenny and their combined private and political efforts to influence the Arkansas legislature to pass stricter and more comprehensive game conservation laws and establish bird refuges. The chapter also explains the creation and structure of the Arkansas Game and Fish Commission in 1915. During 1915 and 1916, several wildlife conservation advances transpired, including advances against market hunters, the creation of a state wildlife enforcement commission to enforce game and fish laws, and the passage of more statewide protection laws for Arkansas game and fish. Many Arkansawyers, however, did not support these advances and believed that they should hunt and fish as they had always done, harvesting as many animals and fish as possible. These significant improvements represent the foundation of today's Arkansas Game and Fish Commission's conservation efforts and have provided the backbone of wildlife protection for over one hundred years.[18]

Chapter 8 covers the political and public backlash against the creation of the AGFC. Opponents attempted to destroy the commission for the next five years and almost succeeded. They did succeed in reducing the AGFC's authority. While Arkansas conservationists fought to preserve and expand game and fish protection, they battled crooked politicians, unscrupulous businessmen, and deeply entrenched traditional views.

Chapter 9 examines the battle between Arkansas conservationists and their opponents in the public and political spheres. Arkansas's game and fish populations continued to decrease as the AGFC tried to stem the tide while parrying ongoing attacks from those who wanted to destroy the commission. By 1925, the AGFC had existed for ten years. Although they had successfully started several conservation programs and had worked to enforce game and fish laws, they remained too weak to stop continued wildlife loss in Arkansas. By the time Arkansas experienced the effects of the Great Depression, the state's bears, deer, turkey, and prairie chickens were either extinct or existed in non-self-sustaining numbers.

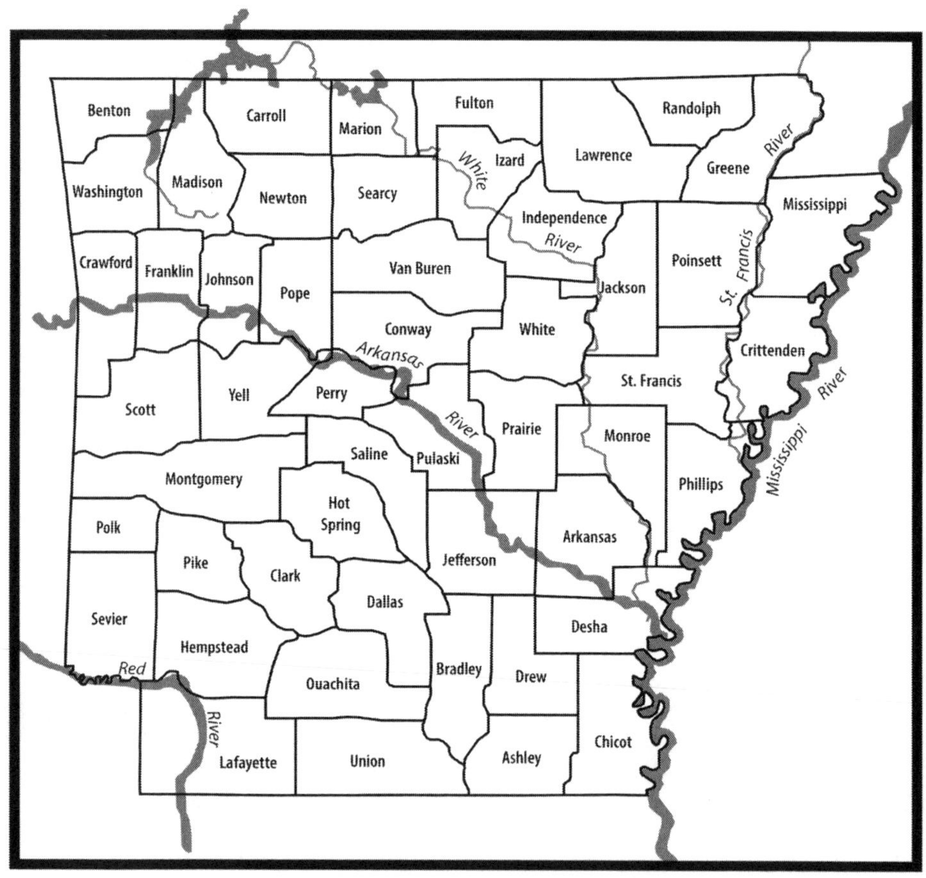

Arkansas counties and major rivers.

SO GREAT WAS THE SLAUGHTER

Ben C. Straus Meat Market in Mena, Arkansas, around 1905. Courtesy of the University of Arkansas Special Collections

CHAPTER ONE

"The Woods Furnish All the Meat You Want"

Early Arkansas Hunters, 1800–1865

> The air was excessively sultry, and the musquitoes [*sic*] troublesome to a degree, which I have not experienced before nor since.
>
> —Journal of the Rev. Timothy Flint, 1835

In early Arkansas, European hunters battled the elements, American Indians, and each other. Visitors to eighteenth- and nineteenth-century Arkansas found a land of wild animals, rough country, and even rowdier inhabitants. Many accounts depict the population as "savage" to survive and even thrive in such a brutal environment. Natives had farmed, hunted, and developed towns and cities for hundreds of years in the region. Before the Americans arrived in 1803, European colonial hunters had joined tribes like the Osage and Quapaw, chasing fur and fortune along the river valleys, swamps, and ridges. Wildlife served as Arkansas's first cash crop, unlike other southern states and territories farther west, so dependent upon rice, tobacco, and cotton. Most settlers continued this practice until some transitioned from a purely hunting lifestyle to a bit of farming, whether out of necessity or purpose. Most Europeans, and later Americans, initially settled along the river valleys, mainly the Mississippi and Arkansas. A few folks spread along other waterways like the White, Black, Red, Ouachita, and St. Francis Rivers.[1] With just a few acres of fertile bottomland, these hunter-farmers mostly ate meat from the forests and grew maize for human and animal sustenance. The hunter-farmer was less interested in fur as a primary source of revenue and more interested in obtaining household goods from the market for his few pelts, meat, or oil.

As time passed and more people arrived with agricultural pursuits in mind, another category emerged: the farmer-hunter. These farmer-hunters had the time to grow more crops and obtain and raise more domesticated animals. There was less reliance on wild game. Therefore, they moved into the farmer-hunter stage, where they obtained most of their food production from farming, supplemented with wild game. Hunting and fishing often became recreational activities that served the purpose of entertainment and proved beneficial for the family's diet and trade. Many Arkansas settlers remained hunter-farmers until the 1830s and, for some, much longer. Finally, some Arkansawyers made their living primarily selling agricultural products and only hunted and fished for pleasure. Early southern sportsmen, mainly the wealthy and landed enslaving gentry, did hunt the grounds of Arkansas before the Civil War. These categories do not include the Gilded Age sportsmen, primarily middle- and upper-class urban dwellers who came later. Indeed, from the Louisiana Purchase until the early twentieth century, Arkansawyers lived all these different lifestyles. American Indians, free and enslaved African Americans, Europeans, and Anglo-Americans all hunted early Arkansas. This chapter will provide an overview of the antebellum Arkansas hunter.[2]

The fur trade for bear, deer, bison, and beaver in Arkansas mainly had dissipated by the 1830s. That is not to say that there were no more animals left in the forests and river bottoms, but there were not enough to be economically viable for a full-time occupation for most settlers. For roughly one hundred and fifty years, Arkansas's furbearing animals served as the source of a commercial enterprise. The hills and swamps still contained a few full-time hunters who did not hoe a row and continued to feed and clothe the family with the game they caught, but they quickly became as scarce as the furbearers they chased. To fully understand the evolution of Arkansas hunting, we must begin with the environment.[3]

In 1819, the year Arkansas became a territory, Presbyterian Reverend Timothy Flint described the area around the mouth of the White River as "a dense forest of the largest trees, vocal with the song of the birds, matted with every species of tangled vegetation, and harbouring in great numbers the turkey-buzzard, and some species of eagles; and all this vegetation apparently rising from the deep bosom of dark and discoloured water. I have never seen a deeper forest except of evergreens." As they continued to move down the Mississippi River, Flint continued, "We were struck with the grandeur of the forest, the immense size of the trees, and their dark green foliage. The inundation extends itself almost indefinitely on all sides." Flint was moving through the main channel of the Mississippi but alongside the area known as the "Great Swamp" on some maps. Miles of flooded timber on either side of the river held "fever, musquitoes [*sic*], alligators, serpents, bears, and now and then parties of hunting Indian."[4]

After visiting Arkansas Post, at the confluence of the Arkansas, Mississippi, and White Rivers, Flint continued up the river, the main artery into the land of its namesake. "The Valley of the Arkansas, with very little exception, is sickly," Flint explained. He believed that most people whom he met in this area of Arkansas "had the ague," likely malaria. Eventually, Flint's family became enfevered, and two almost died. One member of the traveling party, a young African American child, perished of the fever. Years later, a Crocket's Bluff resident reminded visitors that it was not really "safe" to come and hunt in Arkansas before cooler temperatures arrived. "We are now having the last days of the sickly season along the river and overflows," he cautioned. "There is no healthier or more pleasant climate than this after frosts," as cooler temperatures and frosts drove mosquitoes into hibernation, which lessened the chance for hunters to become ill in the infested Arkansas waterways. Settlers often equated the fertility of the land with its health. They believed that health was a characteristic of the land and often judged it "sickly" or that some locations had "bad air." With this belief, early Americans extended the condition of the land to the people who lived on it and were convinced that an unhealthy environment could also make the people ill.[5]

During the same year as the Flint's party suffered, William Rector, charged with surveying the St. Francis and Arkansas Rivers in the Arkansas Territory, endured his own difficulties. He claimed that his workers "were greatly afflicted with sickness" because of working in the hottest time of the year and "bad water." At Fort Smith a few years later, at the convergence of the Arkansas and Poteau Rivers, Captain John Stuart complained about illness among his men stationed there. During the summer especially, the surrounding water "becomes Stagnant and is covered with a Green and Yellow Scum, and emits at times a very offensive Effluvia."[6]

Early sportsman Fredrich Gerstäcker cautioned his fellow visitors about the Arkansas swamps: "Under these trees the rich soil sends forth a veritable sea of vines that can drive a hunter to distraction, make the forest impenetrable in places." Settlers coming to Arkansas often experienced difficulty adapting to the rough country. Missionary Ellen Stetson, who arrived from

New England in the fall of 1821 to assist in setting up Dwight Mission near modern-day Russellville, claimed that she felt colder in an Arkansas winter than the coldest gale in the Northeast. A few years later, one newly arrived settler commented that no human could "become acclimated to this Country."[7]

The rough conditions in Arkansas could not only take a physical toll on travelers and residents but some locations, like the swamps or canebrakes, could drag on a person's mental state as well. It could also wear on their faith. Bogs, sloughs, morasses, or miles of massive cane stands could drag someone down physically, but the tedium, slow-going, endless mud, disorientation, and incessant bugs could wear even the strongest of wills down.[8]

Travelers quickly discovered the massive canebrakes that grew along the river valleys. The cane stood solidly for sixteen miles over ten feet high on the White River near Oil Trough, impenetrable unless cut with an axe or hatchet. To the northeast of Oil Trough, the Black River valley was a canebrake on both sides up to Powhatan, a distance of nearly fifty miles. According to the historian Sallie W. Stockard, "Bear, catamount, panthers, Indians, deer, and buffalo" lived in the green stalks as big as a man's leg. The canebrakes were so dense that bears living in the valleys often hibernated in their thickness instead of heading to the mountains to seek caves or crevasses.[9]

Nearly one hundred years later, little had changed about the canebrakes that the early hunters confronted. Horace Kephart, a Missouri sportsman who visited Arkansas in 1895, described what he endured after entering the blue cane on the Black River:

> Before us was a wall of fishpoles living in such exceeding close communion that they resent intrusion from all consider[ed] larger that a chigre. This fraternity was bound together by such a tangle of interwoven vines and briars, creepers and tendrils, and everything else that is twisty and thorny, that even the obstinancy of that perverse creature the razor-back hog softens before it, and piggy seeks a slash. We entered the brake boldly, proceeded bravely, [and] fought desperately. Once a man enter[s] blue cane, he is alone with his Maker. Before, behind, and all about him there is nothing but cane. He is lost.[10]

John N. Tomlinson's dogs chased a bear through the canebrakes near Oil Trough. He complained, "The cane was so thick I couldn't point the gun." One of the origin stories of the famous Bowie knife included an explanation of the real purpose of its manufacture. It was not because of duels or the killing of men. Instead, according to the 1838 account, Jim Bowie wanted the knife to fight bears in the canebrakes of Arkansas because they were so thick that his rifle was useless.[11]

European hunters and merchants had used Arkansas Post as a base of operations ever since Henri de Tonti had initially established the trading post in 1686. In 1802, just before the American purchase of Louisiana, French émigré Francois Marie Perrin du Lac described Arkansas Post as full of "hunters by profession, [but] more than half the year one finds in this village only women, children, and old people. The men go to hunt deer, the skins of which are less valued than those from the northern country; buffalo which they salt for their use; and some beaver which they still find as a little distance." When outsiders visited Arkansas soon after the Louisiana Purchase, they found hunters and hunting families who still made their living from market hunting or fed their families exclusively from hunting and fishing. These few inhabitants continued to practice in Arkansas's foremost traditional economy, hunting, firmly established ever since the arrival of the displaced Indian Nations of the East Ohio River valley during their migration to the trans-Mississippi West in the sixteenth and seventeenth centuries. In 1808, English traveler Fortescue Cuming on his "tour of the western country," observed that Arkansas Post "consists chiefly of hunters and Indian Traders, [and] of course is a poor place, as settlers of this description never look for anything beyond the mere necessities of life, except whiskey."[12]

People had moved into Arkansas at a dragging pace during the ten years following the Louisiana Purchase. In 1810, the census recorded a little over one thousand inhabitants in the Arkansas region of the Louisiana Purchase. A decade later, it registered fourteen thousand people. However, thirteen thousand in ten years was relatively paltry compared to its neighbors.[13]

Brooks Blevins explains that the "origins" of Arkansas's image "lie in the cauldron of settlement, frontier survival, and violence that was territorial Arkansas." These early Anglo inhabitants served as examples of the rough, lazy, wild, fearless, and violent images that nonresidents equated with the Bear State, Arkansas's first nickname. Many of these early accounts about Arkansas residents come to us from visitors to Arkansas. Therefore, they can offer opinions judged from the outsider's perspective based on eastern or European standards. However, regardless of whether these visitors condemned or condoned their lifestyle, the hunt was the center of their lives for most of these antebellum Arkansas residents. This was the case whether they hunted for the market or the table.[14]

What developed was a dual reputation for Arkansas. On the one hand, observers claimed that Arkansawyers were closer to the savage, stayed drunk, fought often, and were uneducated beasts. On the other, admirers called them brave, tough, free-living, fearless frontiersmen, battling the elements and winning. This dual reputation, in part, drew many nonresident sportsmen to Arkansas during the late nineteenth and early twentieth centuries. They wanted to test themselves against the Bear State, a rugged and savage land (discussed in chapter 3). A brief examination of the type of people who lived in Arkansas can give us an idea of how the reputation developed and how hunters played a significant part in its creation. This review also explains these Arkansawyers' difficulties and the evolution of hunting in Arkansas.[15]

In 1818, the Benedict family left Missouri and, after visiting their kin on the Red River, took the path from the Little Red River to the mouth of Cadron Creek on the Arkansas River. At "Frederick's Lick," the family found the four Wyley brothers, Abraham, Isaiah, William, and Obadiah, all feeding their family exclusively from the wilderness. "These men were but little better than savages, followed hunting for a living, had no farms, and never had, not even so much as a garden," later remarked one of Benedict's children.[16]

In addition to the Wyley brothers, the family discovered two other hunting families, the Flanagins and the Massengills. The Benedicts bought their first home from William "Old Bill" Flanagin near the mouth of Palarm Creek. Living the same lifestyle as the Wyley brothers, the Flanagins "plowed not, neither did they hoe or reap; made no pretensions toward farming, their sole occupation being similar to that of the Indians, hunting and trapping." The two old men, "Old Bill" and "Old Hugh" Flanagin, owned many hunting dogs and several guns. Like all the hunting families near Cadron Bluff, these families hunted all the big game and trapped beaver and otter. In 1819, Thomas Nuttall mentioned that buffalo were about a day's ride away from Cadron, and there were "panthers, bears, and wolves" in abundance in the area. He saw five deer grazing along the riverbank a few miles from "Mr. Marsongill's" place about eight miles from Cadron Bluff." The Massengill family, who "eked out a miserable existence and subsisted along by hunting and trapping," consisted of two men with two wives each and "innumerable children."[17]

When Nuttall traveled along the Arkansas River and reached the pine bluffs (twelve miles up from Little Rock) in March 1819, he stayed with "Mons Bartholome" for a day. Nuttall described Bartholome and his immediate neighbors as "entirely hunters, or in fact Indian in habits, and pay no attention to the cultivation of the soil." The botanist believed several families around the area were the "remains of the French hunters . . . and they are in all probability the descendants of those ten Frenchmen whom de Tonti left with the Arkansas, on his way up the Mississippi in the year 1685."[18]

A few free Black settlers also inhabited the rugged Arkansas Territory and scratched out a living in some of the remote countryside. During the same year that Nuttall stayed with the Massengills, a free Black man named David Hall and his wife Sallie settled on the White River in Marion County in northern Arkansas. He hacked a farm out of the Arkansas woods seven miles downstream from the mouth of the Little North Fork. Hall owned a canoe for transportation but also for hunting and fishing. Although there was an Arkansas state law concerning free Black Americans and the carrying of weapons, the Hall family owned firearms. Several accounts mention the Hall family hunting with guns and their hunting dogs.[19]

Army veteran Peter Caulder moved to the White River sometime in the 1820s and married David Hall's daughter, Eliza, in 1829. Caulder had spent time at Fort Smith as a guide, hunting wild meat for the garrison. He constructed a cabin near the Hall farm and kept a few rows of corn and a vegetable garden. He, too, had hunting hounds, and instead of relying upon farming as his main source of income, Caulder hunted. From his home along the White River, Caulder hunted not only across portions of Arkansas but also traveled into Missouri on long hunts, where he stalked larger animals like bears, elk, deer, and sometimes buffalo.[20]

While in Fort Smith, Caulder constructed a pirogue and most certainly built more after he arrived on the White River. He and his kinfolk would have used watercraft to move around the waterways, traveling, hunting, and fishing in areas where land travel was primarily accomplished by following narrow footpaths, not roads. He established a reputation as a hunter and fur trader. The year he was married, Caulder owned only one horse. Twelve years later, he owned one horse and one cow. Caulder was not attempting to grow a herd into a large commercial enterprise, but instead used his horse for hunting and traveling, his cow for milking, and his corn and garden for feeding the family. He grew no crops for the market. Caulder was a market hunter, or a hunter-farmer.[21]

One of the most interesting questions surrounding the Hall-Caulder community is how free Black residents interacted with their white neighbors and visited white towns during slavery and racial division. How could an African American market hunter move through white communities to trade? One explanation may be that in the case of Hall and Caulder and other African

American men who conducted the business of fur and meat trading, they were not required to visit a town to sell their goods. On his travels along the White River, Henry Schoolcraft remarked that keelboats carrying as much as thirty tons in goods traveled up the White River, north of Poke Bayou (Batesville), as far as the mouth of the Buffalo Fork, and brought "flour, salt, and whiskey, with some coffee, calico, and a few smaller articles. In return, beaver, deer, otter, bear, and raccoon skins, bears' bacon, fresh pork, and beef, and in the gross, venison, bees' wax, honey, and buffalo beef, are taken." He noted that many hunters lived in this area "on account of the abundance of game affords." These boatmen cut off the town markets and gathered the goods from nearer where the hunters and frontier farmers lived. Caulder could have visited these trading boats to barter or sell his season's produce, limiting his exposure to whites. In New Orleans, bear hides brought from one to two dollars, deer skins around twenty-five cents a pound, and beaver brought two dollars a pound. Therefore, we can imagine that these boatmen paid less to hunters like Caulder so the merchant could make a profit. It was easier for hunters to take less for their goods and save the time and trouble of carrying their goods into local towns or larger places like Memphis or St. Louis.[22]

Both independent traders and representatives of larger trading houses went to the source to conduct business. They loaded large flatboats and traveled into the interior of the Arkansas wilderness as far as the water height allowed. In some cases, individuals or pairs of men packed a canoe or two with goods and went even farther upriver to visit customers. Schoolcraft met a "petty trader" with a canoe full of "the remains of a barrel of whiskey, and a few other articles" that the man intended to trade for skins. This fellow played the role of a "middle-middle" man in that he had bought goods from another trader farther downriver and now carried them upriver to make a profit. It was a relay system where traders from larger settlements purchased goods either on credit or with funds, took them farther upriver, and then sold or bartered those goods with another merchant who collected skins from hunters deeper in the interior. The same goods might change hands once, twice, or even more. The profits, therefore, dwindled every time another exchange occurred. By the 1820s, there were too many middlemen, each taking a cut of the profits from the transactions, so the hunter's meat, oils, and skins brought him less, often below his costs.[23]

Likely, Hall and Caulder and the other Black men in their community got along with their white neighbors because of their rough lifestyle, frontier interdependence, and mutual regard. We can imagine that both Hall and Caulder were known for their hunting dogs, shooting ability, and hunting prowess.

That does not mean that white men considered Black men their equal in society, and the pair were more likely to hunt with fellow men of color.[24]

During the Arkansas territorial period (1819–1836), free Black men like Hall and Caulder, as heads of households, were "permitted to keep one gun, powder and shot . . . by license of the justice of the peace." Without a gun, Hall could not have defended his farm against predators, taken down larger game for the table, or collected pelts for the market. Caulder, in short, would not have been a skin hunter. The justice of the peace could issue permits to free Black men to carry firearms, and this may have happened with the Halls and Caulder. Or Izard County sheriff John Adams may have just ignored the law. However, a law hindering free Black residents from carrying firearms was something that these men contended with and was most likely one of the reasons they shied away from towns.[25]

These early Arkansas hunters, regardless of race, faced many difficulties in pursuing their chosen lifestyle. To hunt for the market was dangerous and filled with many potential problems: competition, natural dangers such as death from bears, illness, or the weather, fluctuating prices, thieves, attacks from hostile American Indians, or the depletion of game. Since the colonial period, Arkansas Native Americans had attacked other Native American and European hunters. Just because the Americans purchased the Louisiana Territory, these periodic attacks did not end. Hunting ground rivalries, Native-European alliances, and outright greed caused many attacks. If a hunter survived an assault, they often still lost all the furs and goods because the primary purpose of these attacks, especially after 1803, was to steal the pelts from the hunter. In addition, Native Americans attacked Euro-Americans who they believed were hunting without permission on their native grounds. After all, market hunters competed in the Arkansas woods.[26]

Early Anglo hunters were mainly nomads who followed their game. Usually owning nothing they could not carry on their backs or a horse or a mule, these men needed to move quickly to where the hunting was most profitable. So, once an area was "hunted out," they picked up and left for better grounds. At least once a year, they traveled to the market to offload their yearly production and obtain the manufactured wares they needed for the following season.

In 1818, Schoolcraft explained how the hunting-based economy worked on the White River: "Vast quantities of beaver, otter, raccoon, dear, and bear-skins are annually caught. These skins are carefully collected and preserved during the summer and fall, and taken down the river in canoes, to the mouth of the Great North Fork of White River, or to the mouth of Black River, where traders regularly come up with large boats to receive them. They also take down some wild honey, bear's bacon, and buffalo beef, and receive in return, salt, iron pots, access, blankets, knives, rifles, and other articles of first importance in their mode of life." The wild-game business remained essentially the same for the next hundred years. Even after cotton became the dominant crop of Arkansas, steamboats still carried plenty of wild game to market. In 1870, William E. Bevens saw "boats loaded [with cotton bales] till the guard rail was under water; dressed deer and bear meat topped the local freight, and wild turkeys hung from the upper deck."[27]

Frontier business was neither stable nor legitimate. Value fluctuations often caused one's product to lose value before they brought it to market. When goods came into the marketplace, the price fluctuations dictated the amount hunters received, for better or worse. There are many accounts of fiscal calamities in early Arkansas, and some of these misfortunes were enough to financially ruin or, at the very least, drive a hunter into deeper debt. Throughout the era, the timing was essential. When John Shaw left his family's New York state farm in the spring of 1808, he was twenty-five years old, headed to the Rocky Mountains to see the "western country." Shaw had worked, alongside his older brothers, to support the family since his father's death eleven years earlier. With his youngest brother now at the age to take John's place on the farm, Shaw welcomed the opportunity to seek freedom and adventure. After traveling for several months, Shaw reached the Cape Girardeau settlement on the Missouri River with Peter Spear and William Miller. Following a failed venture to reach the Pacific Ocean, the trio returned to the confluence of southeastern Kansas, northwestern Arkansas, and southwestern Missouri. There, they hunted beaver, otter, and bear from fall 1809 to spring 1811.

The men collected beaver and otter pelts, over three hundred bearskins, and three thousand liters of bear's oil stored in sacks made from some of the bearskins. They canoed their way down the White River to the Mississippi, eventually reaching New Orleans. Unfortunately for the hunters, President Thomas Jefferson's embargo caused Shaw and his partners to sell their bounty previously worth about $2,500 for less than forty dollars. As European markets respected Jefferson's embargo, New Orleans was flooded with such items, causing prices to plummet. Bear oil and buffalo tallow spoiled in the scorching sun while awaiting shipment.[28]

Nearly forty years later, German Charles Heinrich, a baker by trade, invested in the Arkansas fur trade, only to find financial difficulties. Heinrich and a partner went from Memphis, Tennessee, across the Mississippi, up the White River for over one hundred miles on their maiden trading expedition. The pair spent six weeks fighting river currents and the elements, trading the thirty dollars in manufactured goods they had purchased in Memphis and taking on a load of furs. Yet, as they were novices in the business, they made a grave rookie mistake. When they arrived in Memphis, Heinrich wrote, "The season had already so much advanced that we were compelled to sell our stock very cheap, and that I saw, after the division of the money, that my $30 had become 25." Beaten, Heinrich returned to his earlier vocation as a baker to work in Batesville to save money for his next venture. But he never returned to the fur-trading business.[29]

Specific European and American traders possessed a monopoly on most hunting trade. Other whites and the Indian Nations found themselves at the mercy of these men and were thus forced to take the low prices for their goods while paying higher amounts for the manufactured items that traders provided. On his way down the Mississippi River in 1819, Nuttall stopped near the Chickasaw Bluffs near present-day Memphis. He made observations of the trade conducted there. Traders conducted business at the mouth of the Wolf River for decades, dealing mainly with the Chickasaw hunters who came to trade their pelts.[30]

In Nuttall's estimation, the commerce reliant upon the fur trade was inherently corrupt. "The advance upon articles sold to the natives is very exorbitant: for example, a coarse Indian duffel blanket four dollars, whiskey, well-watered, which is sold almost without restraint, in spite of the law, two dollars a gallon, and every thing else in the same proportion," he observed. However, the Americans were not providing the native hunters' reasonable prices for the meat that they were selling. Nuttall complained that they received "no more than 25 cents for a ham of venison, a goose, or a large turkey."[31]

Time is money. Time away for travel meant time away from hunting and fishing. If dealers met the hunters out in the woods, far from town, which many did, white and native hunters were at the mercy of the trader's prices. Competition might lie far away in some distant river crossing or port town, and it was much easier for a hunter to take the lower price than to make the considerable journey. Sometimes neighbors banded together once a year to carry their goods to market, forming something of a provisional cooperative. In the 1820s, hunters who operated on and around Crowley's Ridge in Eastern Arkansas, like the War of 1812 veteran Benjamin Crowley, took their winter's produce to Cape Girardeau, Missouri, or sometimes across the Mississippi

River to Memphis. In numerous instances, hunter-farmers loaded one or two wagons with several men's pelts, meat, and bear oil, and then a few men were elected to caravan the collective goods to market. The round trip from the ridge to the Memphis market took roughly three weeks, providing much-needed manufactured goods for the permanent residents.[32]

Those men who hunted exclusively for the market also operated alongside families who moved closer toward an agrarian subsistence and used wild meat to supplement their tables. One example of a hunter-farmer family was the Billingsleys, who moved to Arkansas from Tennessee in 1814 along with two other families, first to Arkansas Post and then to Cadron for a year and eventually settled into a community of about thirty families near the Mulberry and the Arkansas Rivers' confluence. They hunted buffalo, bears, deer, elk, and raccoon, trapped beaver and otter, and gathered honey. They also fished and grew a bit of maize. They traded their pelts and meat for flour and other goods. The men and boys wore buckskin clothing, while the women wore homespun dresses. From the French traders plying the rivers, the family also bartered their pelts and bear oil and honey for foodstuffs and domestic products like teapots, dishes, and calico and checked-print material for the women's dresses. The Billingsleys tended to trade with the river men rather than carry their goods into Fort Smith because they claimed that because of the trading house's monopoly, they "solde everything vastly high."[33]

In a publication meant to inform "foreigners who wish to emigrate to this country," a Frenchman living in 1816 Arkansas provided a general description of the area. He described Arkansas in a time of transition from the hunter to the farmer-hunter, explaining, "Nearly all the inhabitants of the Arkansas post and its environs are French; many of them very amiable and sociable. All unite in wishing for us as neighbors, unless it be the few who live by hunting and trading, but the greater part have given up this mode of life for the cultivation of the land." He explained that as many as a hundred families "squat" along the Arkansas River on ground suitable for "cotton, tobacco, indigo, rice, maize, vines, fruit, and vegetables."[34]

In 1835, Fredrick Notrebe established King Cotton at Arkansas Post alongside the fur trade he had conducted since 1810. The Frenchman had advertised in the *Arkansas Gazette* as early as 1819 to purchase cotton. In 1835, Notrebe shipped a steamboat full of cotton out of Arkansas Post, bound for the New Orleans markets. Notrebe was not growing a steamboat full of cotton yearly but purchased most of it. Other Arkansawyers were thus producing it for the market too. A few years later, Notrebe was campaigning to establish a bank at the Post. At his death, the Frenchman had amassed a considerable fortune in enslaved people, livestock, and land, starting with the fur trade and evolving into more significant ventures in cotton and land.[35]

During the 1830s, America grew faster and prospered more than in the previous twenty years. The first American Industrial Revolution, starting in the late 1700s, was peaking full steam with the increasing use of the newest technology with the cotton gin and textile mills and the development of a more robust infrastructure. Thanks to new government fiscal policies, more citizens had access to capital. Cotton markets boomed and with it land and slave markets too. According to historian Joshua Rothman, America's southwestern frontier states of Georgia, Alabama, Mississippi, and eastern Louisiana prospered more than any other region in the nation because of the availability of rich fertile lands after the government forced the Native Americans out. Although the boom time had produced enthusiasm and hope for the future, this period also triggered regional hysteria concerning enslaved resistance, general disorder, and abolitionist conspiracies.[36]

As Arkansas Post was the earliest European settlement with businesses, traders, and a considerable European population, it was inevitable that the Post moved ahead of more remote areas in the territory to shift toward a more sedentary agricultural economy, transitioning from hunter to hunter-farmer to farmer-hunter to the sportsman. However, Americans outside Arkansas saw very little growth toward a more developed economy in the Bear State.

An 1835 letter from Boston to the *Arkansas Times and Advocate* noted that "the good people here absolutely shudder at the bare mention of Arkansas. Bowie knives, Judge Lynch, Captain Slick, negro insurrections, dualists, horse thieves, Indian savages, frontier—very awful!" Although the state's population grew, mainly the rich lowlands experienced it, where cotton farmers and their enslaved workers cleared ground and planted white gold.[37]

And it was the enslavers and their enslaved huntsmen who danced the intricate dance between hunting, power, and identity across the South. The landed southern gentry viewed hunting as a way to illustrate their character and station. They looked to Old World Europe and saw hunting estates and organized hunts with all the regalia and fineries of well-bred hounds, horses, and people. To show that they, too, were of finer stock, many enslavers considered themselves sportsmen and began to participate in and host specialized hunts. To the wealthy, the death of the quarry was not the prize; the chase was the proving ground. During the pursuit, these men illustrated personal grit, determination, power, and dominion over nature. The poor hunted from necessity, but these men pursued game for pleasure. The more difficult the hunt became, the more the antebellum sportsman could demonstrate his prowess. Their self-imposed limits to make the pursuit more challenging, evidence of their civility. One example of this is the concept of "fair chase," where the animal has an opportunity to get away. According to historian Nicholas Proctor, the hunt, to an antebellum enslaver sportsman, "represented control over other people, animals, nature, and even death." During an outing, they controlled their enslaved hunstmen (who participated directly or indirectly), the hounds, and themselves. The ability to exhibit self-control illustrated a more civilized person, part of the southern gentry's high moral code and culture. The power to control their huntsman in the field, among their wealthy peers, showed the white dominance over the enslaved. The success of the hunt also exhibited an individual's masculinity. A man who pursued and killed a prime buck deer gained respect from his neighbors and bondsmen.[38]

1. John E. Wadsworth calls the hunting dogs with his horn, circa 1890s. Courtesy of the Eureka Springs Historical Museum.

In January 1850, Chicot County resident Miriam Hilliard recorded in her diary that her husband and two other "gentlemen" had left for a deer hunt that early morning. The men had failed to return through the midday, and the hopeful but worried Miriam complained that the prepared supper was getting cold and that she was quite "hungry & impatient." She had almost given up dining with the men when the returning hunters appeared during the last few minutes of daylight. Her husband, "glorious to behold," Isaac rode with a "fine, fat deer slung across his horse's neck. At length, he had killed one and bears home the trophy. A shavaree [*sic*] he must have," she exclaimed. Mistress Miriam ordered all of the plantation's enslaved people into the yard to bring "horns, bells, tin pans, trumpets" to celebrate. Much to her frustration, however, one of the other men had killed the deer. "My luckless husband must still be the butt of sportsmen," she wrote disappointedly. In this case, the only event that might have caused even greater disappointment to the Hilliards was if one of their enslaved huntsmen had killed the deer.[39]

Despite many common misconceptions, enslaved people hunted and fished (sometimes with guns) to supplement their meager rations and to add to enslavers' tables. Enslaved Arkansas African Americans participated directly in the hunt as trackers, flushers, guides, drivers, or dog handlers. Other times, bondsmen dressed kills, skinned, or served as camp hands. Yet enslaved people sometimes hunted alone or in small groups with or without their captor's consent. Owners generally had few complaints if the enslaved person provided most of his catch to the big house. Most importantly, though, by hunting and fishing, huntsmen challenged the institution of slavery because they could provide additional nutrients to their diet beyond what the owner provided. Huntsmen did not have to settle for enslaver-provided rations. They could obtain more provisions, and that act was defiance of the system. Huntsmen could also, as food providers, combat the psychological "emasculating influence of the slave society" because they could illustrate their masculinity when they became the provider. Black and white relationships surrounding hunting varied. Even though they hunted together, they remained enslaved and enslavers.[40]

Historian Eugene Genovese, in his monumental study of enslaved life, claimed that most plantations contained at least one bondsman who carried a firearm. They harvested large game for the white families and used guns to drive off or kill predators. Enslavers, especially the larger ones, often tasked one or more enslaved men to hunt for their table and the slave quarters. Once enslavers realized that allowing their captives to hunt and fish could save them money on rations and provide meat for their tables, they often provided guns and equipment to their designated huntsmen. Dr. Simeon D. (Sim) Bateman of Oil Trough claimed that one person enslaved by his family, Jack, whom his father, Benniah Bateman Sr., had made overseer, hunted with hounds and a gun, and once treed a cougar. "The negro shot the panther and broke his back," explained Bateman. Those enslaved on the Bateman plantation hunted, often providing meat for the farm, according to the doctor, and the white family was unafraid to provide them with firearms. "One morning our [slaves] all went deer hunting and by 8 o'clock there were five deer lying in the yard. Every [slave] tried to kill a deer before breakfast," proclaimed Bateman. Cross County bondsman Scott Bond recalled that one of the enslaved on the plantation had the "duty . . . to hunt all night" for the plantation.[41]

These activities, providing meat or fish for the plantation, benefited enslavers. However, the huntsman used these activities to have some control over their own lives, although it was temporary. Additionally, as an exceptionally efficient and attentive hunter, a huntsman could bolster their status with the owners and possibly receive more privileges on the farm or plantation, especially if they brought meat to the main house or dispatched a troublesome predator. According to historian Nicholas Proctor, meat symbolized power in the South. Black hunters displayed their quarry after returning home to demonstrate their abilities. However, they usually had to give up the meat to the enslaver who held a "monopoly on" it. Then, the owner could consider providing the meat to the enslaved and divide it as he deemed fit, thus reinforcing who held the true power on the plantation. "A slave owner who controlled food controlled life and death," concludes Proctor.[42]

Genovese argues that if enslaved people had not secretly hunted the woods and fished the streams, they "would have suffered much more . . . than they did from malnutrition." Enslaved people often depended on hunting to survive, unlike enslavers who hunted for pleasure. In addition to potatoes and vegetables, Callie Washington of Desha County recalled, "We had . . . coon, rabbits, [and] possums." John Bates's family, who lived near Little Rock before the Civil War, added "wild turkey, deer . . . squirrel . . . possums, rabbits, and fish" to their vegetable and domestic meat stores.[43]

Tennessee-born Henry Walker, who moved to Arkansas around 1865, professed that his enslaver, Colonel Sam Williams, provided him with a gun for hunting and taught him how to shoot, the main prey being squirrels. Walker noted that Williams had twelve hunting dogs and that Walker handled the dogs on the deer hunts while Williams sat on a stand, waiting for the hounds to "jump" the deer and drive it toward him. As a young boy on the Dick Bean plantation near Lincoln, Joe Bean hunted squirrels from horseback with his owner. Joe rode ahead and served as a scout to spot the squirrels in the high canopy and alert Dick to their location.[44]

While there are numerous examples of Arkansas enslaved people with guns, they more commonly used traps to catch game and fish. There are mentions of deadfalls, fish traps, snares, and cages for capturing fish, quail, ducks, squirrels, and turkeys. Proslavery advocate Daniel Hundley, writing in 1860, mentioned that enslaved people used all these methods to capture animals for their "pot." Hundley condemned the enslaved for using such techniques for collecting meat, stating that "pot-hunting is a very sorry business, but a true sportsman will not forget, for all that, that he is, or ought to be at least, a gentleman." These "gentlemen" were not scrounging their food from the forests, tiny gardens, or the plantation owner's food ration, so they could afford to be judgmental at these effective but "unsportsmanlike" practices. Plenty of poor Arkansas white families were using the same methods to supplement their diets too. Poor whites and enslaved people hunted and fished for survival while the sporting wealthy engaged in these activities for entertainment. Because of this approach, wealthy white sportsmen condemned the poor and enslaved because of how they captured game.[45]

Through their experiences in the wild, huntsmen and their sons learned about their natural surroundings. Just as Native Americans and white settlers taught their boys to hunt, when possible, enslaved men taught their sons (and sometimes daughters) the knowledge to obtain wild game and fish. Raccoons and possums probably represented the food source enslaved people could most easily access. In some cases, an enslaved person might have a gun or a dog, but they were not necessary to catch one of these tree-dwelling mammals, and for the most part, pursuers used nothing more than their own hands or a club. Scott Bond sloshed barefoot through the snake-infested swamps in the bottomland near his home in Cross County to hunt alongside others who were enslaved. To teenage Bond, these excursions were great fun. They hunted with hounds, and once the dogs "treed," one of the boys climbed up and knocked the animal down, and the dogs then dispatched it on the ground.[46]

Like Bond, Jim Ricks of Calhoun County loved to chase possums, raccoons, and other game. "I wanted to hunt. I was a mighty huntsman," he explained. Ricks claimed that he was provided an opportunity to read and write, but instead, he wanted to be in the woods. The thrill of the pursuit provided momentary freedom, an experience that could allow Ricks to rise beyond servitude and feel free. Everything else but the chase could melt away—a predator pursuing the prey.[47]

When bondsman Doc Quinn moved to Miller County before the Civil War, he discovered a rugged country where the dense canebrakes contained bears, wolves, and wildcats. The plantation's owner required his enslaved captives to gather the farm animals nightly and place them into pens so a few could guard them. During the day, the plantation's twenty hounds chased predators away from the farmlands.[48]

Men were not the only ones who put meat on the table for their families. "My mamma could hunt as good as any man," argued Betty Brown from Greene County, whose enslaved family lived on the John and Nancy Nutt farm on the banks of the Cache River near Paragould. "Used to be a couple of peddlers who came around with their packs. My mammy would always have a pile of hides to trade with them for calico prints and trinkets," Brown explained. Mother Brown had raccoon, deer, mink, and beaver hides for trade. Sometimes, she tanned the hides and made moccasins for the family. The family, according to Betty, raised bear cubs for food in a post oak flat near the house.[49]

On a much smaller but still a considerably sized farm, Hardy Banks enslaved four people and owned two hundred acres near Magazine in Yell County in 1860. According to Bank's ancestors, the free and enslaved families "killed hogs together . . . gathered peas together, and all went hunting together on Cedar Creek." For the enslaved, who on the Banks farm had their smokehouse, the meat was vital for food supplementation. The white family, however, hunted mainly for sport on the Yell County farm. The difference in the use of wild game illustrated the contrast between those who had direct control of their food (the Banks family) and those (the enslaved) who depended upon others for the primary source of their diet (food rations). This occurrence does not mean the white families did not eat wild game meat. However, the significance of wild game meat was more critical to enslaved families than to their white captors.[50]

Just as the transition from hunter to hunter-farmer took time, the shift from hunter-farmer to farmer-hunter occurred at different rates and locations throughout Arkansas, with people living these different lifestyles next to one another. Charles McDermott observed an example of a farming-hunting community in 1834. He arrived near Edward Riley's farm on Bayou Bartholomew, about fifteen miles from the Mississippi River in present-day Chico County.

McDermott bought a few acres from the Gross family with assurances he would purchase more land later. McDermott found a settlement of six scattered cabins with maize planted and a few enslaved people working the fields. Stephen Gaston, Rease Bodwin, Easly Kurd, Billy Hones, John Smith, and Parson Gross were growing crops and "killing plenty of bears," according to McDermott, examples of the farmer-hunter lifestyle. Before leaving for Louisiana, McDermott hunted with Gaston and Reuben Smith, killing "three bears, one wolf, one turkey, and a fox" during one hunt. "This delighted me, as I had a great fondness for hunting," McDermott remembered.[51]

If the Old Southwest boomed in the 1830s, some people believed the next step to success could lie across the Mississippi River in the new state of Arkansas. The quantity of the Arkansas wild game proved especially beneficial to new arrivals. John Wilson moved from North Carolina to Pope County in the early 1840s, where he cleared the ground for a small farm. He wrote to his family that farming was productive and that he had killed deer and turkey and saw bears and cougars. He wanted the family to come to Arkansas, where good land was cheap. Like many migrants who settled in Arkansas, the Wilsons supplemented their diet with the wild game until they firmly established their farm. With the wild meat readily available, new Arkansas settlers could focus on creating their farms, clearing land, planting, and preparing for harvest. After establishing their farms, they supplemented their diet with wild game.[52]

When Scotsman Robert Brownlee moved to Arkansas in 1842, he found "much good soil and is a good poor man's country, by farming, enough to make bread and butter. The woods furnish all the meat you want, the wild turkey, deer, bear, opossum, coon, and squirrel." He observed "hundreds of backwoodsmen liv[ing] as I have now stated and raised large families." He discovered plenty of citizens living in the Arkansas wilds farming corn and wheat with a few domestic animals while hunting to provide table meat. Dallas County resident Harriet Bullock Daniel claimed that the wild animals were so abundant around her family's farm that a neighbor stood in one place and killed eight deer in one hunt.[53]

In Ashley County, the 1850 census recorded Samuel Gillian as a fisherman, Hardy Jernigan as a hunter, Josiah Davis as a bear hunter. Melvin Robinson of Newport in Independence County claimed, "The day before I got married, I went hunting and killed three bears for the wedding, the 1st of December 1853." Dallas County native James Baugh was born on his grandfather's farm in 1859. He farmed maize from a young age and hunted meat for the table.

In 1951, Baugh's son remarked that his father "did not kill for slaughter, but food. In this way, he kept fresh meat on his table when desired." Later, Baugh grew corn and fruit for the market and raised hogs. As his farm grew and he had more sources for his meat, his hunting and fishing turned to sport, as he no longer depended upon the wild game. He participated in these activities with his kinfolk, and his forays became less frequent as farm work took up more and more of his time.[54]

Farmer-hunters lived alongside their hunter-farmer neighbors throughout the Civil War. Depending upon where they were residing, how many armies had moved past their farms, how many resources they started the war with, how often guerrillas or partisans took supplies from them, and how frequently they were able to make a complete growing season with planting, tending, and harvesting, families may have looked more to game meat for sustenance. Logically, some of those who had changed to a more agrarian source for their income before the war could have moved back to wildlife for a food source during the lean war years.[55]

An excellent example of an Arkansas farmer-hunter was Yell County resident Hardy Banks. In 1860, he owned two hundred acres of land, fifty hogs, a few heads of cattle, sheep, and nine high-bred horses. He lived in the Magazine Township and held four enslaved people. To earn cash, Banks carried maize, pelts, lumber, hides, cotton, hogs, and horses to market in Dardanelle, located on the Arkansas River. Although Hardy Banks was a farmer, his descendants attested that he "perceived himself as a backwoodsman." He told the census taker in 1860 that his occupation was "Hunting and Fishing."[56]

The year after the Civil War ended, J. E. Lindsay of Dardanelle wrote to his childhood friend N. B. "Poly" Eison of Jonesville, South Carolina, to persuade Eison to join him in Arkansas. After the war, Lindsay moved to a farm in the fertile Arkansas River bottoms in Yell County. Lindsay planned to farm, but he loved to hunt too. "Poly you ought to be here to hunt with me. Theres all kind of game here. I made a terrible raid on wild cats yesterday out at the Magazine mountain," he declared.[57]

When Americans arrived in early Arkansas after the Louisiana Purchase, they found a wilderness filled with primitive European hunters and American Indian Nations. Many early American arrivals continued the market hunting lifestyle, but others transitioned from a purely hunting lifestyle to sedentary agriculture. A few of these families fed themselves with the wild game until they could clear some ground and plant a few acres of maize. These hunter-farmers ate wild game and supplemented themselves with the small number of crops they grew. Some Arkansawyers built larger farms and harvested more planted foods as time passed. Farmer-hunter families lived an agricultural subsistence life but supplemented their tables with wild meat. Enslaved and free, Arkansas African Americans hunted and fished for their captors, to supplement their tables, or for the market. From the Louisiana Purchase until after the turn of the twentieth century, Arkansawyers lived all four kinds of these ways of life: hunter (market and pot), hunter-farmer, farmer-hunter, sportsman, or some combination thereof.

CHAPTER TWO

"Covered with Glory and Blood"

Arkansas Sportsmen, New Market Hunters, and Hunting Clubs, 1800–1900

> I would rather live in a tent in the wilds of Arkansas than a palace in most places. There is more to study and more to learn for the ordinary sportsmen than anywhere I have been.
>
> —F. M. Houdlette, *St. Louis Globe-Democrat,* 1898

> Sportsman: "a man who hunts or shoots wild animals as a pastime."
>
> —*Oxford Dictionary*

On a cold morning in February 1900 in St. Louis, Missouri, over one hundred hunters filled the Union Railroad Station. Plumes of white condensation puffed from the noses and mouths of the anxious men and barking dogs as they shuffled around the platform. Encased guns hung from shoulders. Train employees loaded thousands of pounds of camping gear, tools, food, and other supplies into the vacant cars. Family members milled around and spoke in hushed tones to their fathers, brothers, and sons as the latter prepared to board. A typical hunting excursion might last a few days or weeks, and nervous wives, girlfriends, and mothers ensured that husbands, beaus, sons knew they would miss their boys. Across the trans-Mississippi western cities, this scene repeated during the opening week of hunting season each year in the late eighteenth and early nineteenth centuries, no matter where they were going or departed.[1]

Early sportsmen hunted and fished for pleasure rather than out of necessity. Market hunters killed for profit and employment. As far as their effect on Arkansas wildlife was concerned, there was very little difference between the man who hunted for sport and the man who hunted for a living. The line between the two often blurred. Before game regulations appeared, there were no limits on killing wildlife, no hunting seasons, and no restrictions for either group. However, as far as the sportsman was concerned, they were markedly different from the market hunter. From the earliest days of the sportsman in Arkansas, the two groups generally distrusted, even despised, each other. Yet, despite their mutual disdain, the nineteenth-century sportsman and the market hunter unknowingly operated simultaneously to destroy the Arkansas game and fish population.[2]

The reasons for chasing game and fish changed in Arkansas, as elsewhere. Humans have hunted for subsistence for millions of years, hunting for food and harvesting what a family could eat. The difference between life and death for a family might be a successful hunt. Later, families harvested enough game and fish to feed the family and supplement their income by bartering or selling meat or fur. Next came killing for sport or profit. During this final step, hunting and fishing experienced an evolution.[3]

Early sportsmen hunted and fished to demonstrate masculinity, obtain accolades, and develop a positive reputation among their peers. Hunting was considered critical in the transition from boyhood to manhood. Outdoor activity was good for well-being or was part of a cure for ailing health. Hunting and fishing connected them to the past. Traditions help hunters preserve their history: the land, the field, the stream, and the hunt. My father and his father hunted here. Hunting and fishing for sport showed that participants belonged to a more civilized and advanced society. Sporting activities allowed for socializing with like-minded individuals. Finally, the thrill of the chase excited the pursuer. Men who followed what they defined as sporting ethics adhered to game and fish laws, exhibited self-control in the field and pushed for stricter protection legislation, even at the expense of the subsistence hunters.[4]

After the Civil War, the significant transition from subsistence hunting to sport and meat market hunting began. Southern antebellum sportsmen were primarily wealthy landowning enslavers who equated themselves to the European gentry. In some cases, however, they were travelers, not Arkansas enslavers. Probably the earliest sportsman in Arkansas was the native Virginian,

Sir Jennings Beckwith. He was the son of Jonathan Beckwith, whose family had the reputation of being "devoted to the turf, and all kinds of hunting and fishing—laboring in search of amusement, but never known to do anything to bring a penny to the pocket."[5] Jennings Beckwith left Maryland in 1803 and arrived at Arkansas Post with his servant, where he hired a Native American guide. The trio entered the "wilderness," where they remained wandering for over four months hunting but were mainly lost while struggling through the floodwaters of the Mississippi. After a bout of sickness and near starvation, the crew limped back to Arkansas Post. Beckwith and his servant skedaddled out of Arkansas for St. Genevieve, Missouri, never to return.[6]

One of the most well-known early sportsmen who visited Arkansas, Friedrich Gerstäcker, was a German-born wanderer and hunter who came to the state in 1839. He made several trips working and hunting as he went through the state. While in Arkansas, Gerstäcker made many hunting excursions and lived sporadically with several local hunting families. He considered himself a sportsman and felt a kinship to the Arkansawyers who made their livelihood from the countryside.[7]

The Virginian Charles Fenton Mercer Noland (Fent) moved to Poke Bayou (Batesville) in 1826. Young Noland had failed at the United States Military Academy at West Point and returned to live with his father, a receiver at Batesville's United States land office. Over the next twenty years, Noland wrote letters to the New York *Spirit of the Times* filled with tall tales about hunting, racing, fighting, drinking, and general carousing in the Arkansas backwoods and waterways. Noland also wrote letters back to Virginia to family members; these accounts more accurately depict happenings around Poke Bayou.[8]

About a year after he arrived, Noland wrote to his sister Frances Berkley that he had a wonderful time killing deer, turkeys, ducks, and foxes and fishing on the White River. Most likely because many locals knew his father and because of the young man's keen wit, Noland befriended several other men who hunted around Batesville: James Caldwell, Jesse and Robert Bean, and Abraham Ruddell.[9]

But by the time Fent and these men became companions, they hunted for sport since they did not depend upon wildlife for food. Robert Bean was the territorial representative for Independence County. Jesse Bean enslaved people and owned horses, cattle, and land. James Caldwell was an appraiser and mail carrier. Abraham Ruddell enslaved people and possessed horses and several hundred acres north of Batesville.[10]

Settler Uriah Cole's family built a house in 1847 and cleared land for a farm in Lawrence County. After completing his chores, young Cole hunted for sport: "I amused myself by hunting, and never was out of sight of deer. I have seen 500 wild turkeys in a drove and killed them till I was no longer interested." Cole became a well-known bear hunter in the area, and people came from as far as Indiana to hunt with him after the Civil War.[11]

These early sportsmen and market hunters took a toll on Arkansas game. In 1845, a sportsman sent an account of his first "fire hunt" to a New York sportsmen's newspaper. Fire hunting was building a substantial fire, usually on a raised platform, to view and shoot deer after dark. The hunters built a four-by-five-foot scaffold a few feet off the ground, about ten feet from a salt lick. They took their hunting positions under the fire-topped platform, "as the scaffold casting a shadow immediately below the fire, it is impossible to be seen, the brightness of the fire dazzling the eyes to such an extent as entirely to shroud any object under the scaffold, in total obscurity." Two men hunted with this method for five nights and harvested twenty-two deer. He concluded, "We soon had our venison (hams and hides only) on our horses, returned home, and soon [were] lost in dreams all thoughts of our first fire hunt."[12]

Venison tenderloin, hams, and hides were the most valuable parts. These men were not hunting for a living or even supplementing a farm table with wild game meat. They were hunting for sport and took only the most valuable parts of the deer for their use, leaving the remainder in the wilderness to rot. These "sportsmen," in some ways, acted no differently than market hunters. Wasting or not eating what they killed are some of the worst aspects of modern sport hunting.[13]

A sportsman calling himself "Byrne" from Crockett's Bluff wrote about a hunt he and his partner went on in the "Ar-kan-saw Prairies." They went from pond to pond with double-barreled shotguns, and in morning's work, they had killed "fifty-one fat mallards, five teal, a couple of jack snipe, the same of [prairie] chickens, several quail, and a bittern, and were back to town soon after twelve noon." The fellow reported that "this is kind of 'sport' would suit a great many exactly." Byrne confessed, "I did not enjoy it with that exquisite pleasure I would have twenty-five years ago." The fire hunt and the jump shoot are examples where the new sportsman sought to kill the largest or most game and, in doing so, increased their reputation as an outdoorsman.[14]

The postbellum white sportsman connected hunting with status, masculinity, and power, just as the antebellum sportsmen did, but differently. This was not power and command over the enslaved—it was dominion over nature and himself. The countryside served as a place to challenge and prove one's abilities. A man's outdoor reputation was essential to these men. It was tied to their identity. Attributes like what club memberships a man held, the weapon he carried, the hunting partners in his group, and his shooting proficiency all characterized his self-image and identity. Race and class, too, determined how people hunted, when they hunted, what they hunted, and with whom they hunted. In more than one contemporary account, observers remarked on who had the best reputation as the best bear hunter, owned the best pack of dogs, or killed the largest deer. In 1872, a group of Louisiana sportsmen called Arkansawyer Jack Stone "the most noted bear hunter." In 1887, a sportsman remarked to another about the quality of his dogs. They are "perfect little beauties, and as true of scent and fleet of foot as ever struck a track, and all thoroughly trained."[15]

Henry R. Schoolcraft, upon observing the hunting families of Sugar Loaf Prairie on the White River, commented, "To excel in the chace [*sic*] procures fame, and a man's reputation is measured by his skill as a marksman, his agility, and strength, his boldness, and dexterity in killing game, and his patient endurance and contempt of the hardships of the hunter's life." He admired Arkansas hunters for their ability to their ruggedness and their ability to "subsist anywhere in the woods." A man's hunting reputation was important. For centuries, man has tied hunting prowess to manhood. A successful hunter meant an excellent provider on the frontier. It did not have to mean food on the table. It could mean skins or furs to sell. A boy old enough to successfully hunt meant a boy was reaching the age of manhood.[16]

Arkansas Native Americans, settlers, and enslaved people taught their boys to hunt and fish. In the South, hunting served as a right of passage into manhood. As time passed and the necessity to gather food from the wilderness became secondary or unneeded, in many cases, a boy's ability to hunt and fish remained a step toward adulthood for young southern males. Hunting introduced young men into the adult male world of smoking tobacco, drinking hard liquor, death, violence, and telling tales about everything from wild animals to wilder women. It taught boys how to interact with other males. Although some nineteenth-century women and girls did participate in outdoor activities, it was primarily a male interest. Hunting became the most popular pastime for white southern men and boys in the late nineteenth and early twentieth centuries. It often served to occupy the dull days that rural life brought. For many years, society expected the coming-of-age Arkansas male to learn to provide for his family through labor. In most cases, this included killing or catching wild game and fish. Going out into the woods also taught young males about the natural world and their place in it. Older men continued these activities during adulthood because, among other reasons, they had learned these skills during their youth, and it reminded them of their past.[17]

The May 1889 edition of *Field and Stream* featured an illustrative story about a young, inexperienced hunter and his friend who traveled to Arkansas to hunt. The narrative demonstrates a young man on the threshold of manhood, possessing the innocence and empathy of early life but wanting to be very much considered a man. After the young Illinoian had fired his black-powder shotgun and knocked a yearling deer off its feet, he "ran to him—or rather stepped, for I think that I covered the twenty-five yards in two steps," he explained excitedly. For a moment, the dying animal and the young man locked eyes, and in that instant, the boy felt "sorry that [he] had shot him." The boy continued, "But only for a moment; then the instinct of the hunter in me rejoiced in triumph." The young man threw the deer across his back and returned to camp "covered with glory and blood."[18]

Freedom and exhilaration were two experiences that hunting and fishing offered during the late nineteenth and early twentieth centuries when the nation was fast becoming more regimented. Like the old snake oil huckster says, "It is the cure for what ails you." Many men saw hunting and fishing as an escape from structure and dreariness, whether working a farm by the seasonal calendar, a factory on the manager's watch, or the office building on the economic clock—a few days to take minds away from the worry and stress of an industrialized world.

As the nation became urbanized, industrialized, and regimented, weary city folk sought relief from what troubled them outside their urban environs. Even before the Civil War, for many urban sportsmen, hunting and fishing was healthy pursuit. In 1835, German traveler George Engelmann visited a resort in Hot Springs, where the renowned mineral baths were known for their medicinal powers. A botanist and doctor of medicine, Engelmann was fascinated with the valley's healing waters and flora. The physician observed that people participated in walking and horse riding as part of their treatment activities. Engelmann observed, "Often small parties form to go hunting for weeks, specifically for buffalo (or bison, which still can be found within thirty to fifty miles of here). Many consider it a *special cure* after the end of the regular season at the baths to go in October and November with the local inhabitants on a bear hunt, and for several weeks to camp in the raw weather with no other food than bear meat nor other drink than bear fat and water." If an individual could survive the harshness of a hunter's life, then they could consider themselves well. A Memphis newspaper later agreed in the benefits of hunting excursions, arguing that "there are numerous instances of delicate and feeble Memphians, as well as others, to whom hunting has been of incalculable benefit."[19]

But Hot Springs did not have a monopoly on those customers hunting for healing waters and hides alike. In 1841, the *Arkansas Gazette* advertised an illness treatment site called Dardanelle Springs, a hot springs resort at the base of Magazine Mountain in Yell County. This health resort and club catered to the ailing sportsmen. While touting the springs' ability to cure ailments such as "Dropsey, Scufula, Rheumatism, and Dispepsia," the article also explained that the area around the resort was full of "deer, bear, [and] wild turkey." V. T. Rogers, the new owner, promised that he was obtaining "a fine pack of hounds [and] . . . a fine assortment of guns and be able to furnish gentlemen wishing to indulge in hunting with the finest ammunition."[20]

Some proponents claimed that just walking outdoors caused participants to feel better. "How refreshed he feels the next morning as he starts into the forest," explained Arkansas Judge Lee Miles, a lifelong sportsman. "He returns with new life and new energy" from his outings. "After taking an enormous amount of exercise getting all organs to working, his lungs full of November air, getting physical culture, he feels 10 years younger," explained another sportsman. Fort Smith sportsman and game warden W. S. Cochran observed that sporting "often carries with it the incidental by products of health and contentment." After a hunting trip to Arkansas, one Illinoian explained,

> I left home almost sick and though tramping from fifteen to twenty miles almost every day while there, I returned feeling like a new man; in fact, I think the trip renewed my youth from five to ten years; so that it is not all of hunting to get big bags of game. I enjoyed beyond expression wandering amid those magnificent forests which shut me out entirely from the tread-mill drudgery called business, and I reveled in a moral, mental, and physical atmosphere that dwellers in brick walls surrounded by paved streets know not of.[21]

One observer reported that lumber company owner W. O. King of Chicago "used to shoot, and then got chained to business. This week he reformed and bought a trunk full of sporting outfit." He headed to the St. Francis River in Arkansas with his newly purchased gear, including six hundred dollars worth of firearms. Most of these early sportsmen fell into the upper or middle class. Like King, they had the money to purchase outdoor equipment. Many thought it was not just money that separated them from other people but outdoor ethics.[22]

Many sportsmen believed they were of a higher class than a market or pot hunter because of their hunting and fishing methods. The lower classes butchered deer when floodwaters trapped them. They knocked pigeons from their roosts with clubs at night. The sportsman gave the animals a fair chance to escape during the chase, the idea of ethical versus unethical hunting. Sportsmen generally followed the law and did not use illegal netting or traps. Later, sportsmen became conservationists. They supported stricter game bills and placed bag limits on themselves. These outdoorsmen joined organizations like the Arkansas Sportsmen's Association, which fought against illegal activity and lobbied for more conservation laws. Sportsmen and conservationists were the bourgeois, middle- to upper-class urban professionals who did not need to hunt for a living. The conservationist usually had enough wealth to support such measures. In contrast, most working poor did not have the luxury to think in such terms as bag limits, hunting seasons, or unethical practices.[23]

In 1910, Arkansas state game warden E. V. Visart argued that game violators were harmful societal elements. He called them "a band of organized thieves" and "outlaws who are robbing the good people of Arkansas." Visart claimed that if the state legislature passed meaningful game laws, "it would drive an undesirable class of people from our borders." He referred to the illegal market hunters and fishermen, many from out of state. If the state government did not pass more and stricter game laws, then a "shiftless and roving class who have no means of support other than to subsist from hand to mouth [will] completely destroy" Arkansas's wildlife.[24]

As American cities grew more populous after the Civil War, a new hunter type appeared in Arkansas, the meat market hunter. These men were similar to the fur hunters of colonial and early Louisiana Purchase Arkansas. But, instead of hunting bison and beaver for their hides, these men hunted animals and birds (fur and feather were a bonus) and fished for the market. They helped feed the growing populations outside of Arkansas. Memphis, St. Louis, New Orleans, and Chicago's meat markets bought Arkansas game and fish from these men. Because of technological advances like trains, telegraphs, ice houses, and the pump shotgun, these hunters killed and shipped hundreds of thousands of animals and fish out of Arkansas. The meat market hunter and the sportsman later fought over Arkansas's game resources for some fifty years at the turn of the twentieth century, each for their reasons.[25]

Also, after the end of the war, American food production, processing, and distribution changed dramatically. Industrialization, immigration, and the railroad's coming created urban centers that contained thousands of workers who needed food. Millions of wild cattle grazed in the American Southwest. Cowboys drove thousands of cattle to railheads in the West and shipped them to processing plants in the Midwest. Chicago's meatpackers butchered thousands of cattle and tens of thousands of hogs every day. But the Americans still needed more.[26]

Wild game from the unsettled or semisettled areas surged in demand to supplement the agricultural products that flowed into the urban centers. Prices rose. The wild game once again became a commercial commodity. However, it was not for their hide or fur. Instead, game meat became the principal commodity of hunting. Small towns grew, and every one had a café or two. Restaurant proprietors looked for ways to provide diners with wild game. The new Arkansas market hunter provided meat for the large cities and small-town eateries.

Arkansas commercial hunters could sell locally to the town meat market, individuals, or eating establishments. Still, St. Louis, Chicago, Cincinnati, Memphis, and New Orleans served as locations to ship their harvest for the most prodigious hunters. In some cases, meat companies employed hunters, just as the Spanish and French trading companies had done during the colonial period to collect furs. These meat companies harvested duck, geese, prairie chicken, turkey, deer, and bear. They sold their harvests to the city dwellers, who bought their food instead of raising or catching it. Market hunters used hunting techniques, including running dogs, killing migratory birds as they roosted, and fire hunting. They were not concerned with the chase. Instead, they wanted the most efficient methods to harvest as many animals as possible in the shortest time. The use of the techniques placed them firmly at odds with many Gilded Age sportsmen.

From 1873 to 1886, many observers called South Water Street in Chicago "The Greatest Game Market in the World." They retailed wild turkeys at $1.00 each, $1.37 for a dozen quail, teal ducks at $2.00 per dozen, woodcock and mallard ducks at $3.00 a dozen, prairie chickens at $3.50 per dozen, geese at $4.50 a dozen, and canvasback ducks at $6.00 per dozen. Most diners considered canvasbacks the best-tasting waterfowl because they often ate wild celery, and their meat absorbed the celery flavor. A significant amount of Arkansas game and fish ended up on Water Street.[27]

By 1873, many Tennesseans viewed Arkansas as a location to exploit. As one outlet reported, "Memphis proposes to make of the state of Arkansas—a mere adjunct—a sort of fishing, hunting, trading and mining ground." Indeed, many Americans began to travel to Arkansas, searching for economic opportunities after the Civil War. The Bear State contained extensive natural resources, forests, streams, and rich river land. Arkansas coal, natural gas, oil, and minerals industries were in their infancy. Besides the coal to power steam engines and natural gas to heat homes and cook meals, timber to build dwellings, and the expansion of agriculture to feed the masses, there was a need for more meat. Both resident and nonresident market hunters were flooding into Arkansas, focusing on wild meat. Bears were basically gone, so deer was the largest meat game they sought. But birds, prairie chickens, quail, turkeys, pigeons, ducks, and geese were the primary target for these market hunters. Birds provided meat for the table, but also, certain species provided plumage for the women's fashion industry.[28]

One of Arkansas's first mentions of modern meat market hunters was only four years after the Civil War. The magazine *Sporting Times* reprinted a January 1869 article from the *Memphis Post,* which reported that every "autumn and winter, hundreds of votaries of the profession of Nimrod, the ancient, forsake civilized life [who] proceed to the wild woods and swamps of our neighboring State, vulgarly called 'Rackensack' and spend several months in hunting and trapping the abundant and valuable game." The paper claimed that these animals could provide "a handsome reward for their winter's work." Hunters could sell their "winter's work" to game markets in St. Louis, New Orleans, and Memphis at a significant gain. Memphis resident William Archer and a friend sold "a bear, four deer, four dozen ducks, four dozen squirrels, and a dozen possums" to a game dealer in the river town at $32 for the bear, $28 for the deer, $12 for the ducks, $8 for the squirrels, and $5 for the possums, a total of $85. The men only hunted for a week for their quarry. Around the same time, two other Memphians crossed the Mississippi River on a two-day squirrel hunt. Two "Nimrods" killed 946 squirrels during their forty-eight-hour Arkansas hunt. At two dollars a dozen, those squirrels fetched a market price of some $150. In 1868, the average yearly wage for an Arkansas farmhand numbered $115; in Tennessee, it measured $109.[29]

But, in 1874, the Memphis correspondent for the sporting magazine *Forest and Stream* reported that while the game on the east side of the Mississippi was "scarce, Arkansas is sending a splendid supply of venison and ducks, and I believe one of two bears, showing that gunners over in that *wild* state are having plenty of *work.*" With an emphasis on the words "wild" and "work,"

Memphis considered Arkansas an untamed place for outdoorsmen to experience the days of old, with game aplenty, ripe for the taking. The correspondent also emphasized the idea of "working" with a gun. Market hunters were again harvesting Arkansas's wild animals for the commercial markets. Yet, deer were so numerous on the "other side of the Mississippi," the Memphis reporter observed, "that our market is glutted with venison . . . which is so highly relished by epicures." Memphis was not the only market bulging with game in 1874.[30]

"The swamps and cane along both the Arkansas and White rivers in Arkansas are full of game, and the many interior lakes abound with fish and wild ducks," explained the reporter. At the same time, the "Red River is also equally supplied," and Shreveport shoppers could purchase "luscious mallards at ten to fifteen cents each" because the average market hunter killed 100 to 150 ducks per day on the river. In the fall of 1882, an Arkansas County observer remarked with each cold push from the north, "There were great masses of ducks are found here feeding on the prairie." He further claimed that the number of migrating birds was so plentiful that even "the most ardent pot-hunter or market hunter could desire until spring."[31]

Missouri brothers Charley, Austin, and Sam traveled to Arkansas in 1883, planning to trap along the White River. The brothers crossed the Arkansas-Missouri line at Mammoth Springs and saw thousands of passenger pigeons, "rivers full of ducks," possum, mink, beaver, squirrels, turkey, and raccoons along the Black, White, and Cache Rivers. Using seventy-five traps, they sold over $1,000 of furs during their first year in Arkansas, netting a profit of $448. They lived on wild meat, mainly wild hogs, squirrels, and rabbits.[32]

According to the *Memphis Daily Appeal*, most large game brought to the city's market came from across the river around Tyronza, Arkansas, with "enough to keep corps of hunters regularly employed." Yet, the *St. Louis Globe-Democrat* boasted that their city's game market was the "finest in the country." By the 1870s, many concerned Arkansas citizens raised the alarm about the damage that market hunters continued to do to Arkansas fish and game. These concerns ultimately led to a conflict between market hunters and sportsmen in the early twentieth century.[33]

The popularity of sport hunting and the growth in market hunting had taken a considerable toll on the larger game population. When economically feasible, many sportsmen formed hunting and fishing clubs to preserve acreage for their activities. Some of the first southern clubs had appeared in Virginia as early as the 1830s, and by the 1880s, several clubs formed in Arkansas.

The South played a significant role in the nonresident sportsman's vision and experienced a surge in tourism and development throughout the late nineteenth and early twentieth centuries. Northern urban and midwestern men purchased or leased millions of acres of land south of the Mason-Dixon line because they could obtain ground at low prices. Some of the earliest southern clubs bought old plantations with expansive grounds and accoutrements, including African American workers, stables, and outbuildings, in Georgia, Alabama, Mississippi, Louisiana, and South Carolina. Much of Arkansas, like certain sections of the Old Confederacy, remained largely unindustrialized and rural, full of game and fish, making them prime locations for wealthy and middle-class sporting clubs.[34]

Sportsmen sought to associate with members of their class. They joined these clubs and organizations and created relationships with like-minded hunters and fishermen. Often, the rich treated these places as social clubs, complete with lodges and chefs. Middle-class men sometimes pooled their money to purchase or lease land. At the turn of the century, American society already contained several levels of class structure—immigrants and native-born, rich and poor, country and urban, people of color and white, residents versus nonresidents, etc. Therefore, it was only fitting for these men to join other sportsmen to hunt and fish.

Historian Ted Ownby argues that nineteenth-century men hunted alone or in small groups of no more than four. That seems the case for most local or antebellum hunters, but not those Gilded Age men who traveled into Arkansas from out of the state or those residents and nonresidents who were club members. They sometimes hunted with more than four in a party. To develop an admirable sportsman's reputation, these men required as many witnesses to their deeds as possible. It was not uncommon for groups of ten or twelve sportsmen to come into Arkansas and camp, hunt, and fish together for days, weeks, or even months. A 1909 Kentucky party visiting Arkansas consisted of twenty hunters and seven African American workers. A wolf hunt near Hot Springs had eight to ten participants in 1918. A duck hunting club member might find a lodge full of fellow hunters during the season's opening week. Certainly, socialization was a vital component of sporting activities, and in many cases, the hunting club served as the center.[35] "No one can fully appreciate what a week of camp life in a secluded forest with congenial companions means," explained one gentleman. "To wind up the day's sport in camp, listening to the stories of the others and their adventures," he continued, "and then to collect around the dining table for a warm meal. It is the greatest diversion and rest for a businessman possible," he concluded.[36]

In 1884, the *Memphis Daily Appeal* ran a full-page section about Memphis-based hunting clubs in Arkansas, where to hunt if you were not a club member, and the cost of hunting camps. The article noted that "the interest in field sports . . . has been steadily growing for several years." Even if they were not experienced outdoorsmen, enough guidebooks, railroads, and articles existed to lead them to Arkansas. That did not mean that they would find success there.[37]

What compelled men to join a hunting or fishing club? According to the *St. Louis Globe-Democrat* in 1893, the type of St. Louis men who created and joined these organizations were "clubs of professional men, of business men, and of men in humble walks of life" who provided funds "to equip the grounds so as to afford visiting members every comfort and convenience and protect the game and fish for their enjoyment." Europe had contained hunting preserves for years for those of noble birth or estate-owning aristocrats. "The wealthier [American] sportsmen" now "own their own grounds which they can protect the same as similar grounds in the old country."[38]

"These urban-rural clubmen are fine, hearty fellows, the Teddy Roosevelts of the town," the newspaper later explained, and they are "believers in the strenuous life, the advocates of the theory that the survival of the fittest embraces the destruction, in limited quantities, of beast and fowl and fish." "Survival of the fittest" was more than a Darwinian concept; it stood alongside the test of manhood. They could simultaneously enforce man's control over the beasts and birds of the earth, a Judeo-Christian foundation in the Old Testament. These concepts could also link arms with the march of the Industrial Revolution with its urbanization, immigration, and civilization. The idea was that these changes would soon destroy or tame the wilderness, and if someone wanted to see the last days of some of the wildest parts of the nation, they had better go to places like Arkansas before it was too late.[39]

Memphians from across the Mississippi River established some of the first hunting clubs in Arkansas. Miles of swampy flooded areas and bottomland forests along the Mississippi, White, Arkansas, Red, Ouachita, and St. Francis Rivers attracted millions of waterfowl yearly during their migrations. Therefore, some of the first hunting clubs involved waterfowling. The Beaver Dam members paid twenty-two dollars per year and could ride the Louisville, New Orleans, and Texas Railroad to the association lease land near Clayton, Mississippi. Along the Mississippi River, the club also leased land near Austin, Arkansas. The railroad stopped one hundred yards from the twelve-bed clubhouse, with a professional cook, John Cornelius, and a fleet of five bateaux and three dugouts and "pushers" available for hire. Members paid fifty cents daily for food and fifty cents for their pusher in addition to the yearly dues. John Overton Jr., son of one of the founders of Memphis, was a member. William Arthur "Guido" Wheatley served as secretary. By 1888, the Beaver Dam Club numbered forty-seven members and ended applications. Ten years later, the club controlled six thousand acres of land.[40]

These clubs were powerful, featuring some of the wealthiest and most influential men of Memphis and St. Louis. Nationally, sporting clubs lobbied their states to pass stricter game and fish laws. Those in Arkansas could, and often did, wield a tremendous influence over Arkansas game policy. They had the money and connections to influence Arkansas politics.[41]

Another Memphian hunting club in Arkansas was the Osceola Ducking Club (established in 1882), with its "shooting lodge" roughly twenty miles west of Osceola, near Big Lake. W. Bard Edrington was president, and Dr. Robert H. Mitchell, medical director at the renowned Howard Corps, served as secretary. F. P. Poston, Judge C. W. Heiskell, and the Tate brothers were members. Club members decided to move operations and built a two-storied clubhouse in Oak Donic, Arkansas, near a rail line on the St. Francis River.[42]

Among the oldest was the Arkansas Prairie Club, another Memphis-membered hunting club established in 1872. Dr. Robert W. Mitchell was the president of the club. Wholesale groceryman W. B. Mallory served as vice president. The members took weeklong hunts that cost fifty dollars per man. They could reach their hunting grounds via the Memphis and Little Rock Railroad and wagon. Still, most of the time, they traveled by the White River packets or the Lee Line steamships, where many of the old Confederate veteran members like Mallory surely delighted in riding on boats like the *Patrick Cleburne*.[43]

The Hatchie Coon Club appeared in 1884 on the banks of the St. Francis River near Tulot, Arkansas. It, too, contained primarily men from Memphis. They had roughly 850 acres along the river and remain in operation today. The Oak Donic Club and the Hatchie Coon Club merged into one organization in 1899. They owned a clubhouse on the riverbank and fished and hunted in the Arkansas sunken lands area. They sold shares for $175 and initially limited membership to one hundred sportsmen.[44]

Another location also attracted Memphis Confederate veterans to the Bear State. Twenty miles northwest of Memphis, the Wapanocca Outing Club, where legendary author and duck hunter Nash Buckingham grew up shooting alongside his father Miles, was founded in September 1886 when a group of thirty-eight Memphis waterfowl hunters (mainly Confederate veterans) purchased Wapanocca Lake and some of the surrounding area. They limited membership to only forty. The first "large and commodious" clubhouse sat on Big Creek, about twenty-four miles from Memphis. Member James Sullivan, superintendent of the Kansas City, Springfield, and Memphis Railroad, arranged for the train to stop nearby when needed. The club paid gamekeeper and cook Phil Gwin three hundred dollars annually for his services. President of the State National Bank William D. Bethell and Memphis mayor Walker L. Clapp were members.[45]

When the first organizers for the club arrived at Wapanocca Lake, they found market hunters killing ducks and geese for the Memphis markets. Two years after its creation, members announced that "this club is strictly a 'closed club' and no one except its members and guests will be allowed under any circumstances to shoot or fish there," thus banning market hunters from the property. Gwin's job was to keep them out. A new fence assisted him. After a few years, the club also placed a fifty duck per day limit on its members.[46]

Most sportsmen, however, were not members of any club. They usually could not afford the fees. Nonmembers could find reasonable hunting grounds in several locations around Arkansas. If they liked hunting deer and bears, hunters could travel to Helena. Ducks were in the sunken lands in northeastern Arkansas. Snipe hunting was within walking distance of the Memphis and Little Rock Railroad station at Wheatley, Carlisle, Hazen, or Lonoke.[47]

Memphis sportsmen were not the only ones who rode into Arkansas, bought land, and built sporting clubs. St. Louis outdoorsmen poured into the Bear State just before the turn of the century. By the late 1890s, St. Louis sportsmen created hunting and fishing clubs like the Knobel Hunting Club, the Buffalo Island Hunting and Fishing Club at Bertig, and the Moark Boat Club on the St. Francis River. The bulk of the membership rolls in these clubs were men from St. Louis. The Buffalo Island Club, with its clubhouse located a half mile from Senecan Slough, accepted all members who could pay the annual $2.50 fee. In addition to the small annual fee, the club charged members one dollar and nonmembers two dollars as a daily use fee, including boats, lodging, and food. Experienced hotel and service manager S. Virgilio, ensured that all guests were adequately attended.[48]

Established in 1891, the Knobel Club near Bertig, Arkansas, on the Black River, was one of the most well-known of the early St. Louis hunting and fishing clubs. The sporting club allowed a maximum of 250 members, a substantial number when most clubs permitted only a handful of members to join.

The Knobel Club did not accept anyone hunting or fishing for a living, only sportsmen. Annual dues began at ten dollars but raised to twenty-five dollars, considered expensive compared to other clubs. The Iron Mountain Railroad provided a line that traveled within a few yards of the lodge, and members could board the train in St. Louis at 8 p.m. and be at the club before breakfast. According to historian Lynn Morrow, St. Louis businessman and president of the club, Alex Smith, served as the driving force behind the success of the Knobel Club. He ensured that the enormous garfish members harvested, sometimes with a rifle, were displayed in some of the Gateway City's business windows. Members returned loads of fish to their friends and pledged to provide "fish in tanks on the Iron Mountain [Railroad] for fish displays at the St. Louis World's Fair" in 1904.[49]

The Moark Boat Club sat on an island in the St. Francis River. Half of the island lay in Missouri, and the other half was in Arkansas. St. Louis architects Isaac Taylor and Will Eames, inventor Frank Mesker, packing company manager Louis Denning, Missouri Pacific Railroad time manager Fred Hugunin, and Russ Samuels started the club. They used canoes to move through the grass beds on the St. Francis and the cabin boat *Mud Turtle* to carry supplies and members to the club.[50]

A few Arkansas sportsmen established clubs before 1900. The Ouachita Rod and Gun Club out of Hot Springs had grounds outside the Spa City for its members to fish and have shooting competitions. The Pie Eaters Fishing Club's members were from Hope. German National Bank cashier E. T. Reaves, cotton buyer J. T. Haizlip, secretary of the German National Bank R. W. Wrightsell, *Forest and Stream* correspondent Paul R. Litzke, A. Chihester, and other capital city sportsmen organized the Little Rock Fox Hunters' Club. By 1890, several larger Arkansas cities had shooting clubs with many budding sportsmen as members.[51]

Banker Charles H. Drennen and Jonathan N. Brown organized the Point Gun Club in Van Buren in October 1892 "to protect game and fish and to maintain the Point Club House." Arkansas native Lewis Bryan and his cousin, William Jennings Bryan, were club members. The club built a small clubhouse and owned one acre of land. A survey of organizational records illustrates what these organizations had to pay to maintain a hunting and fishing club. Besides a small caretaker's salary, the organization's expenses included boats, wood, lumber, road building, taxes, windows, builders, a mule, a water pump, soap, lamps, and a stove. In 1899, the annual cost was $48.55; in 1900, it was $101.05. The club disorganized after former Arkansas representative and businessman G. T. Cazort bought the entire lake and started charging a so-called tax for those who wanted to hunt and fish on his lake.[52]

Many busy sportsmen tried to join clubs closer to where they lived to visit them more often. Little Rock lawyer John M. Rose, Capital Hotel manager Joseph Irwin, Dr. J. H. Lenow, and some other Little Rock men started the Prothro Hunting Club when they leased and fenced Neimeyer Marsh, a part of what they called "Hough's Inferno," four miles out of Little Rock.[53]

As more sportsmen organized these associations, they began to place stricter regulations on hunting and fishing on club grounds. The Grassy Lake Hunting Club was organized in 1894 and bought six hundred acres of duck hunting and fishing ground. Member J. M. Rose believed members could kill "1,000 ducks a day on it for a while," but they were one of the first Arkansas clubs to place a bag limit on their members. "Twenty-five ducks and two geese per day" was the limit. They also fenced part of their six hundred acres to keep "the pot-hunters . . . out of it," Rose explained in 1897.[54]

However, clubs did not have to own land or have a lodge in Arkansas to hunt there. In 1899, a Kentucky organization traveled to the sunken lands in northwestern Arkansas. According to the group, the main reason for coming to Arkansas was because "game was growing scarce and the deer was fleeing before civilization and the woodcutter's axe." Members claimed that "we have seen that sportsman's paradise fade away. Railroads, the axe, the market hunter have about finished it." The twenty members "now take their annual vacations in Arkansas and Missouri, deep down in the swamps where there is a little game left" because "flourishing towns and cultivated fields appear[ed] where once the members chased the deer."[55]

The vast abundance of Arkansas game and fish also provided the necessary resources for those newly freed from slavery after the Civil War. Furthermore, it was a way that several freedmen resisted the contract and, later, the sharecropping system. Just as hunting and fishing had allowed enslaved Arkansas African Americans to resist the slavery structure, they also allowed some freedmen to feed themselves and not be at the mercy of another white landowner-constructed labor system. Most formerly enslaved people had no monetary resources, land, or employment. Newspaper accounts reported several accounts of "negro men" hunting squirrels, turkeys, possums, and pearls.[56] Some Freedmen argued that it was easier to make a living in Arkansas through the dual combination of farming and hunting than in other states. "Squirrels, wild things, cotton, and corn, plenty of it. So, you see, the man told the truth when he said money grew on bushes," explained one observer.[57]

A white doctor provided Joe Golden of Hot Springs County with a gun and part-time job hunting, a dream occupation for Golden. "I used to like to hunt," he explained, "Hunted all over these mountains, hunted quail and hunted squirrel, and a few times I killed deers. The man what gave me the gun he promised me twenty-five cents apiece for all the quail I could bring him. Lots of times I came in with them by the dozen." Before Golden obtained the gun, he hunted what is now Hot Springs National Park with his hunting dogs: "One time I chased a deer. Got him with a knife, didn't have a gun. The dogs cornered him for me. Best dog I ever had; his name was Abraham Lincoln."[58]

Freedman Jane Hodge visited the Grand Prairie near Hazen to hunt deer. He found that the area was sparsely inhabited by people and heavily populated with deer. "It was not cleared and [it was] close to the White, Cache, St. Francis, and Mississippi Rivers," Hodge's cousin remarked, so "he [Hodge] just stayed with them [deer]." Warren McKinney also of Hazen remarked, "It was easy to live here. There were lots of game and fish." Hodge wrote to his friend George Braddox about the plentiful game, and Braddox soon moved from Texas to Arkansas and purchased a farm four miles north of Hazen. Speaking of his first experiences in Arkansas, Braddox agreed with McKinney that it was "easy to live if you liked to hunt." He hunted and trapped, and sold his skins to earn extra money. Although Braddox attended a white neighbor on his hunts for several years and cared for his dogs, their relationship remained strained. "He never paid me a cent for taking care of those dogs," Braddox complained, and "his widow never as much as give me a dog. She never give me nothing!"[59]

Some white southern landowners were not impressed with the freedmen's use of natural resources and used the opportunity to condemn them for their actions. In 1868, the *Southern Watchmen* of Athens, Georgia, reported that "many negroes in Arkansas subsist entirely by hunting and fishing and seem to be relapsing into the barbarous condition of their African ancestors." A Memphis newspaper reported during the same year that "game is abundant in Arkansas, and most of those [freedmen] who have abandoned their employers are living by hunting and fishing. All of them have either a repeater or a gun, or both."[60]

White resentment toward Black hunters abounded in the Old Confederacy. As historian Scott E. Giltner explained in his study *Hunting and Fishing in the New South*, white landowners believed that the new freedmen were using hunting and fishing for their livelihood instead of working for them in agriculture. This conception bred resentment and hatred for Black hunters who "shot and ate anything." This bitterness appeared in Arkansas as well. Enslavers, who allowed their bondsmen to hunt and fish, had not feared this activity threatened their control. After Emancipation, that attitude changed. One Arkansas legislator argued that if "all the game in this State was exterminated, it would be in the interest of the laboring man." In 1897, a Pine Bluff newspaper worried the crops would ruin because "nearly every negro is hunting pearls." The editor claimed they should go to work to provide for their families and not waste time or money hunting.[61]

Despite poor attitudes toward Black hunters, some white sportsmen sought out Black outdoorsmen. Arkansas African Americans worked as paddlers, pushers, cooks, skinners, and drivers for white sportsmen, as the enslaved had done for their owners before the war. The latter accepted African American hunting servants or guides because they were in a lesser position, not a social equal. Northern hunters wanted to experience an old-time "southern" hunt, meaning Black attendants. The social order remained the same after emancipation as before. In this example, the white man held the superior position, and the Black man was the worker. For the next seventy years plus after the Civil War ended, white people continued to use "Uncle" or "Old" before older African Americans' names, just as in the antebellum period, because they had denied enslaved people and servants the use of Mr. or Mrs. or Miss. There was an uncrossed line in the woods and the swamps just as in the cotton fields before; Jim Crow was in the wilderness.[62]

Ted Ownby contends that wealthy white hunters preferred Black guides and servants, whom they viewed as closer to animals and, therefore, possessed some sixth sense to find game. Dogs were companions and tools in the field and could find the game. White hunters viewed Black guides in the same light. Additionally, many of the white hunters were former Confederates and enslavers and expected African Americans to maintain the antebellum servile role. Scott Giltner argues that wealthy white sportsmen also used Black attendants "to publicly display their wealth and social standing."[63]

As time passed, a few Arkansas African American outdoorsmen occasionally gained limited respect from white hunters for their abilities in the field. Guides, who could place their employer in the best spot or could find the most game, held the top position in the camp worker hierarchy. A proficient guide gained respect from both his fellow workers and sportsmen. Some independent guides gained notoriety in Arkansas, and white sportsmen sought them out. The difference between a successful hunt and a poor one often relied

2. Henry Norwood (January 20, 1850–October 28, 1952) near Greenwood, circa 1870. Courtesy of Gerald Norwood.

3. Henry Norwood, second from left, and a hunting party, Sebastian County, date unknown. Courtesy of Gerald Norwood.

upon a guide's expertise. Many of these men had served as huntsmen before the war, and their previous knowledge of the woods and streams proved immense. That information, passed from one generation of enslaved to the next, served them well after the war.[64]

Henry Norwood, born into slavery in North Carolina, moved to a farm in Sevier County in the late 1850s with the Benjamin Norwood family. Young Norwood hunted the Ouachita Mountains and the prairies of southwestern Arkansas and eastern Oklahoma. After the war, he moved to south Sebastian County with his brother Bill and became widely known as a capable hunter, guide, farmer, and "horse doctor." Norwood carried a muzzleloading black-powder shotgun and harvested everything from deer to birds. One family story claimed Henry lost an eye in an altercation with a wild animal. As his reputation grew, white men called upon Henry to guide them on hunts in the surrounding area, which he continued to do sporadically for most of his life until he moved permanently to Kansas City, Missouri, in the 1940s. In one contemporary story of Norwood's life, an observer noted that "the mature men of importance of the area, from constables to our congressman, liked for Henry to be with them during their 'big game' hunting trips." The writer continued, "Being an expert woodsman, he was needed as a guide to keep the hunters, especially those lacking in experience, from getting lost."[65]

In 1912, Arkansas governor George W. Donaghey and Supreme Court associate justice William F. Kirby entered the Mississippi River bottoms after a legendary bear known as "Old Real." Hunters had seen the bear for fifteen years and had tracked the bear's grounds from Mississippi over to Arkansas. The bear was mystical, able to fight off and kill all hunting dogs that pursued him. The governor's and justice's guide was African American bear hunter Joe Gaspar, a well-known outdoorsman with a reliable reputation.[66]

In 1918, well-known Black hunter Tobe Mitchell was involved in a "big wolf drive" near Glenwood, southwest of Hot Springs, with several white hunters. When three wolves took refuge in a cave, Tobe declared he would catch them alive. In a previous attempt at a live capture, a wolf bit Mitchell, leaving a considerable wound. "Mitchell is a negro hunter from Prescott, and one of the wildest, most daredevil hunters of this section," declared one newspaper account.[67]

A few of duck hunter and author Nash Buckingham's stories featured formerly enslaved person "Uncle" Phil Gwynn as the subject. Gwynn started as a paddler and guide, and after some time, he moved to cook and caretaker at the Wapanocca Outing Club near Turrell, Arkansas, where Buckingham hunted. As time passed, Gwynn was allowed to carry a gun and hunt with the members, but mainly he worked for them. He lived at the club and remains there today in a small plot. Paddlers at Wapanocca, all African American, included Moses Holmes, Sank Davis, Osbourne Neely, Aaron Jones, Bud, and Charlie "Black" Davis, "Old Fred" Valentine, "Buster," "Sam" Cook, Crockett Winslett, "Columbus," Jim Neely, Douglas Reed, Will Smith, Kinney Reed, Rooster, and Big and Little Sammy. Five of these men remain at the Wapanocca Outing Club, buried on the grounds. It is difficult to ascertain precisely how individual white Wapanocca Club members felt about the African American workers. However, a close read of Buckingham's works, member letters, and writings gives readers an idea. Occasionally, Buckingham described the club workers as "lazy" or in some other derogatory way, as well as "strong," "vigorous," and other adjectives akin to describing hunting dogs. However, there also existed a relationship between some white hunters and Black workers that many point to as respectful or endearing. The white sportsman appreciated what the African American provided for him but did not see the latter as an equal, the common relationship for generations in the rural South.[68]

4. Pulaski County hunter and family with dogs, circa 1895. Courtesy of Historic Arkansas Museum.

CHAPTER THREE

"There Is No Game Law in Arkansas"

Why Hunters Flocked to Arkansas

> The State of Arkansas is fast becoming a resort for hunters and fishermen.
>
> —*Forest and Stream*, 1883

> The St. Louis, Iron Mountain, and Southern Railway announces that it will issue special tickets, at special rates, for hunting parties of three or more. No charge is made for dogs or guns.
>
> —St. Louis, Iron Mountain, and Southern Railway advertisement, 1883

Fascinating stories about Arkansas's wilderness intrigued nonresident sportsmen, and they wanted to participate in those adventures. Unfortunately, during this early period of sportsmen and market hunting, these men caught and killed millions of Arkansas birds, fish, and mammals. This natural bounty, with few game laws, the increasing growth of railroads, and the proliferation of sporting publications, made Arkansas an equal-opportunity slaughterhouse by the late nineteenth century. Neighboring states had experienced overhunting and fishing for decades, decreasing their wildlife supply, and greedy market hunters and eager sportsmen looked to Arkansas for new ground. Early Arkansas hunters and settlers had already established Arkansas's reputation as a wild and violent place. What better place to test yourself to prove to your fellow companions that you are a strong enough man to take what one of the most challenging known places could throw at you? Sporting publications helped publicize a mystic land overrun with wildlife. Many modern sportsmen grew up reading these sporting narratives from Arkansas country and wanted to travel there. Industrialization, starting in the 1880s, provided workers with more leisure time, and they looked for ways to entertain and challenge themselves. Urbanization caused people to feel disconnected from nature. Therefore, outdoor activities, including hunting, fishing, camping, and hiking, increased in popularity. The spread of railroads and better infrastructure brought hunting grounds closer to the sporting tourists.[1]

By then, the state's reputation as a sportsman's paradise had spread nationally. In a diary entry, Miriam Hilliard wrote that her friends from Kentucky believed that Arkansas had so many animals that a hunter could fire in any direction and kill a deer. In 1874, a Memphian claimed that "the canebrakes and swamps adjoining nearly all the bayous on the river are full of the finest game, some of them showing up handsomely in the way of bear, panther, wild cats, wolves, deer, and the occasional elk are found." Arkansas contained an "abundant and valuable game. Squirrels of all kinds, foxes, opossums, rabbits, deer, raccoons and fur-bearing animals abound in great numbers in the many vast, pristine forests, which in that sparsely inhabited State, have as yet scarcely heard the ring of the wood man's axe. In ponds, lakes, bayous, and in the great and small streams of water, great flocks of wild ducks, geese and swans are found," observed a Memphis newspaper reporter.[2]

"The swamps and cane along both the Arkansas and White rivers in Arkansas are full of game, and the many interior lakes abound with fish and wild ducks," described a *Forest and Stream* correspondent in 1874. An Arkansas County observer remarked that with each "cold spell north . . . great masses of ducks are found here feeding on the prairie." For example, a newspaper reported that Memphis sportsman W. Bard Edrington killed 114 ducks in one day, and Judge Greer shot fifty quail on Christmas day. William Arthur "Guido" Wheatley had killed 121 snipes in one hunt.[3]

Publications also noted that Arkansas contained so many animals that anyone could kill them. In 1869, William Archer and a companion "captured . . . one bear, four deer, four dozen ducks, four dozen squirrels, and two dozen opossums." A Louisianan recently returned from a week in Arkansas and offered advice to sportsmen who thought about traveling to Arkansas: "If he is used to roughing it let him take a tent, proper provisions, servants, &c and go up the St. Francis or White River and camp out. He will not only kill game

5. One-day kill at Merrisach Lake (near Tichnor) in 1917 or 1918. (*Left to right*) E. B. Minch, Bill Christine, Fred Collier, Henry Nichols, Starley White, Gilbert Franzen, Billy Reader. Courtesy of the Arkansas Post Museum State Park.

enough to satisfy any reasonable man but enjoy such health as would satisfy the most dyspeptic man in the world." Although Arkansas contained millions of birds and fish, the state legislature had passed few laws to protect them by the end of the nineteenth century. Therefore, visitors could harvest as much game and fish as they wanted.[4]

In 1876, W. S. S. of Pottsville, Pennsylvania, received an answer to his inquiry to *Forest and Stream* about quail and pinnated grouse (prairie chickens) season in Arkansas and if any laws governed the birds' harvesting. Editors of the journal replied that there were "no close[d] seasons in Arkansas and no state game laws." At the time, pinnated grouse hunting was excellent in Arkansas. Two Memphians had crossed the Mississippi the summer before and killed twenty-nine grouse in only a few minutes.[5]

The *Daily Arkansas Gazette* answered another letter about prairie chicken hunting season the following year. The newspaper responded, "There is no game law in Arkansas, though it certainly would be a good idea to have one." This instance was one of the first times that the paper publicly supported the passage of any game law to protect the wild animals of Arkansas. Likely, one of the reasons that so many nonresident hunters, both sportsmen, and market hunters, came to Arkansas was that the state had no seasons or bag limits. Arkansas had passed a license law in 1875, but local authorities rarely enforced it, and the license pertained only to market hunters. Arkansas did not have a game warden system. Therefore, anyone could harvest any number of animals at any time in any manner, and they did.[6]

A group of well-to-do St. Louis sportsmen, including Colonel J. G. Prather, Major Grimes, Mr. William Hyde of the *Republican*, Captain George W. Ford, Daniel Able, and Captain Joseph "Joe" K. Rickey, traveled to Arkansas in October 1875 for a two-week fishing trip on the Black River. When they returned home, they had "2,000 pounds of bass, besides other fish, and bagged a few deer, wild turkey and squirrels . . . and an otter."[7]

Many historians have stressed the importance of sporting publications in developing the sporting ethic and the popularity of hunting and fishing as a leisure activity. By the 1880s, the flood gates opened to both sportsmen and market hunters, and sometimes, they were the same people. Sporting newspapers published account after account of sporting in Arkansas. In the first month of 1882, Walter Childress of Greene County, Ohio, wrote from his camp about his trip with three friends on a "season-long" hunt in Arkansas. They intended to stay for months, camping, trapping, and hunting along the White River. "We enjoy fine sport," he explained, but only when he was referring to killing deer. They killed deer, but only for camp meat. His party was trapping for the market. During this transitional period, it was difficult to determine who was a sportsman and who was a market hunter. The vernacular of the time was sometimes used interchangeably, at least for a while. Childress mentioned that J. Lee Smedley was coming to hunt with them from Pennsylvania but that "he comes down solely for sport." We must assume that Smedley was only there for the chase, not the months-long trapping expedition. Therefore, when the Ohioans deer hunted for fun, it was sporting. When they trapped for fur, it was market hunting.[8]

The same year Childress and his friends trapped and hunted, a sporting newspaper listed state game laws in the United States. These publications published state game laws at least once yearly because states constantly added, amended, and abolished them. Three states had no game laws in 1882: Arizona, Arkansas, and Florida. The other thirty-five states had some game law. Although hundreds of Missourians traveled into Arkansas, killed the game, caught the fish, and shipped it home, their state had made it unlawful for nonresidents to kill wildlife for exportation from Missouri. Nonresidents continued to arrive in Arkansas via rail, wagon, horseback, keelboat, and even private steamers. Because of the growing popularity of recreational hunting and fishing, many sporting publications sprang up, and many newspapers started a hunting and fishing column. Novice sports from across the country could not get enough of these Arkansas stories, and sporting publications had provided those stories since the 1830s.[9]

William T. Porter created the *Spirit of the Times: A Chronicle of the Turf, Agriculture, Field Sports, Literature and the Stage* in 1831, which covered everything from horse racing and the theater to fox hunting and cricket. It later absorbed the *American Turf Register and Sporting Newspaper* and continued to grow. After the first decade of publication, the paper's citizen correspondents submitted more fictional stories than true ones. Readers found some of the earliest American literature published in its pages.[10]

Sprouting sportsmen could find information about hunting dogs and the latest fishing techniques from many of the new "outdoor newspapers." *Spirit of the Times* was a favorite of many Arkansas sportsmen, and several, like Fenton Noland, wrote letters to the publication. Unfortunately, many budding authors sent letters mixed with truth and fiction, rumors, tall tales, folklore, and the like. These publications transitioned from tall tales to more serious information and valuable material as time passed. Some publications took on a more informative tone with less fiction, trying to weed through the more outlandish stories and publish something more practical. *American Turf Register and Sporting Newspaper*, the nation's first sports newspaper, represented one of these publications.[11]

In 1831, the *Arkansas Gazette* contained an advertisement for *American Turf* that explained that it was "for the use of those who are fond of good horses and dogs—who would study the natural history and habits of American game—whether animals, or bird, fish or fowl." The newspaper promised coverage of sporting topics such as game animals and fish, hunting and fishing, but also to cover horse racing on the "turf." J. S. Skinner, postmaster of the City of Baltimore, edited the Maryland publication.[12]

In 1841, the *Spirit of the Times* printed one of the most famous and influential Arkansas bear stories. Thomas Bangs Thorpe's "The Big Bear of Arkansas" is considered southwestern humor along with the works of Mark Twain and certainly was popular during the 1840s. Several periodicals in the United States and abroad republished Bang's tale. The fictitious story begins on a Mississippi River steamboat with protagonist Jim Doggett talking about an Arkansas bear hunt. His piece included the relationship between animal and man, the ruggedness of the state, and the exaggerations of a country full of "twenty pound turkeys . . . big mosquitoes" containing soil so rich that corn grows to full height overnight. Thorpe's stories, he also wrote about Arkansas bee hunters, provided readers a view of a wild country where the animals were plentiful and enormous, both in size and in legend, a near-mythical place. Doggett concludes, "Natur [*sic*] intended Arkansaw for a hunting ground, and I go according to natur." Fent Noland's and Thomas Bangs Thorpe's stories influenced many new sportsmen to visit Arkansas to find a taste of what Pete Whetstone and Jim Doggett experienced. Fifty years later, a Texan hunting in Arkansas commented on this legendary place, "we were as much in the wilds as though the past two hundred years of Arkansas history had been only a dream."[13]

Almost a year after the publication of "The Big Bear of Arkansas," the *Spirit of the Times* republished an article from the *American Eagle*, a sportsmen's journal published in Fort Pickering, Tennessee. The piece was about a group of starry-eyed sportsmen, possibly under the influence of Jim Doggett or Pete Whetstone, who came across the Mississippi River "for a grand Arkansas Hunt." The author explained the group's high hopes for seven days of productive hunting: "Never were men much better prepared or more determined to have sport and game" and that the men expected to harvest "half-a-dozen deer, a bear, a wolf or two and a panther, with other game proportionally thrown in, was the least quantity some of us expected to have back into camp each night." However, they saw only a few turkeys, two deer, and a dead cougar on the first day, but they killed nothing. One of the crestfallen members had brought along a few ducks, and they had those for supper. After six days of hunting and ten miles of exploring through flooded lands, they killed five deer but little else, even after befriending a local "bear hunter" with dogs. "After wading about in quest of better luck, for six mortal days . . . we finally gave up the chase and turned our horses' heads towards the Mississippi," the disappointed hunter wrote. The inundated waters along the Mississippi River, observed by the Tennessee sportsmen, contained thousands of ducks and geese, but the "general surface of the country was low and sickly-looking." The flooded lowlands, lack of game, and harsh terrain caused the disillusioned hunters to conclude that "Arkansas is unrivaled for canebrakes, lakes, and possums." This experience was not what the Tennesseans expected to encounter on their "Grand Arkansas Hunt."[14]

Nevertheless, possibly because of the sporting magazines, more accessible avenues of travel, word of mouth from other hunters, or the glorious stories from men like Noland and Thorpe, by the 1870s, sportsmen began to travel to Arkansas in more significant numbers. Although some hunters came from much farther distances to Arkansas, most sportsmen who traveled to the Bear State were from Tennessee, Mississippi, Louisiana, and Missouri. Some old Arkansas resident hunters sometimes made a few dollars guiding and housing these men.

In 1872, a Louisiana expedition of four men came to hunt with Jack Stone, a legendary bear hunter. He hunted in Bolivar and Washington counties in Mississippi and Chicot and Desha counties in Arkansas. Stone "was the most noted bear hunter, not even excepting Gen. Hampton, in the Mississippi [River] bottoms," explained one participant. According to contemporary accounts, the old hunter was an excellent marksman, an owner of many fine hunting dogs, and an honest man. According to one of his clients, he owned a thriving Arkansas plantation and was "well to do." In a letter to the sporting newspaper, *Turf, Field, and Farm,* the Louisiana sportsman explained that if anyone wanted to hunt in Arkansas or the Southwest, they would have to hire men like Stone and compensate them adequately. "A stranger desirous to see the State of Arkansas, and have the best hunting in America," he expounded, "can do no better than to get a letter to Jack Stone."[15]

In November 1874, the Memphis correspondent for the New York–based sporting newspaper *Forest and Stream* published an article proclaiming the excellent hunting across the Mississippi River.

> Sportsmen are just now reaping rich harvests with the dog and gun, as well as the rod . . . in Arkansas, where the prairies abound with game, such as chickens and quail, and plenty of deer can be found in the cane, while the lakes abound with game fish, and the bayous and river bottoms afford ample amusement for those who are fond of burning powder behind huge flocks of ducks, mallards, teal, and almost every other known species, as well as wild geese, which have just begun their Southern flight," he announced.

With reports like these, readers of the sporting newspaper wanted to know more about the wilds of Arkansas.[16]

"John and Mort" from Poughkeepsie, New York, wrote to *Forest and Stream* inquiring about traveling to Arkansas to hunt. "Another person and myself talk of going to Arkansas to rough it for six months or more," John explained. "We propose to strike out in the woods and live on whatever we shoot," he spiritedly wrote. The New Yorker wanted to know the best place in Arkansas to visit for his adventure and, in typical northeastern fashion, inquired about the winters there. "What kind of game shall we find, and is it enough to keep us from starving," John questioned. In response, the editor advised the pair to write to Edward A. Linzel of Little Rock for information. Linzel was a Prussian gunsmith who owned a sporting goods store on East Markham Street in the capital city.[17]

In terms of wildlife conservation, the most influential publication was *Forest and Stream*. Charles Halleck started the journal in 1873, and conservationist George Bird Grinnell later edited it for thirty-five years (1876–1911). Besides promoting sportsmanship and fair chase through backing the passage of stricter game laws, the newspaper supported the expansion of the Audubon Society and fought to create a migratory bird conservation law. Although national publishers printed plenty of pulp for the sporting mind who wanted to imagine a trip to Arkansas, it was another thing to get there. However, changes in transportation had opened the door more widely for sports wanting to visit the Bear State.[18]

In 1885, Nicholas Longworth, a Cincinnati financier, built a steam yacht named the *CO* (pronounced Sue) and used it for traveling on various waterways for fishing and hunting excursions. Later that year, Captain Wes Doss piloted Longworth's eighty-one-foot stern paddle wheeler up the St. Francis River from Helena seventeen miles with a load of sportsmen on a bear hunt. Unfamiliar with the river, Doss had enlisted the assistance of Captain Weeks for his knowledge of the Arkansas River. Fourteen huntsmen and their double-barreled shotguns and many hunting dogs entered the canebrakes and sloughs along the St. Francis River bottoms. They were photographed, not with a bear, but alongside the "knees" of some massive cypress trees. These men had entered some of the richest hunting grounds in Arkansas and, possibly, the United States.[19]

A few years earlier, chief engineer with the Pacific and Great Eastern Railway Company M. La Rue Harrison had produced a survey report describing northeastern Arkansas, including the region where the *CO* traversed. He assessed the feasibility of constructing a railroad connecting Arkansas and Missouri. Alongside his comments about the placement of bridges and markets for freight with mineral, crops, and timber lists, Harrison was stunned to see the quantity of wildlife during his excursion. "I have read romantic stories about game and fish but the hunting and fishing grounds of St. Francis and Little River in the realities outromanced any fiction I ever read—bears, panthers, deer, opossum, coons, foxes, otter, mink, muskrats, turkey—myriads of geese, ducks, swan, pelicans, brants, and in the rivers enormous catfish,

6. Nicholas Longworth's steam yacht *CO* on the St. Francis River near Helena, 1885. Courtesy of the University of Arkansas Special Collections.

pike, trout, etc, etc find a home—trappers and hunters spend their lives here, and keep off immigration by telling bug-a-boo stories about the swamps," explained Harrison.[20]

Sometimes the wealthier sportsmen rented or owned unique train cars to follow the waterfowl migration. An affluent hunting party from Chicago, Illinois, including executives from a coal company, construction company, Montgomery Ward, and the Illinois Steel Company, followed migrating ducks in their special car. The group hunted several places in Arkansas but remained the longest at the St. Francis River hunting grounds.[21]

In 1883, eleven sportsmen traveled on the Iron Mountain Railroad 220 miles from St. Louis to fish and hunt along the Black River for two weeks, but floods caused them to leave after only eight days. According to a group member, the Missourians caught over two thousand pounds of fish, most of it to rot because they could not obtain adequate ice. On their excursion, the party also killed deer, raccoons, minks, geese, squirrels, and ducks. These men were not hunting for the market. Instead, they considered themselves "sportsmen" but thought nothing about killing as many fish, birds, and animals as possible. Even though they knew they could never eat all the meat or give it away, they did not stop killing.[22]

The railroad industry's growth significantly impacted the rapidity of wild game and fish destruction in Arkansas. With more efficient transportation, more game killers, whether market hunters or sportsmen, came into the Bear State. In addition, those men who harvested wild game could now ship more weight in less time than before. Sportsmen could send additional meat home before it spoiled, thus making it more likely that they would kill or catch more. Market hunters now had broader markets to reach in a timelier manner and the ability to kill as many ducks, deer, or fish as possible without much fear of spoiling before making it to market. Railroads did not openly cater to the market hunter, but they shipped their game to distant meat markets. They did not post advertisements for cheap shipping to the slaughterhouses. However, men who killed for a living knew that freight companies initially asked few questions and were happy to have their business. The railroad companies also saw an opportunity to make greater profits if they appealed to the traveling sportsman. Icehouses, increasing demand, and faster shipping opened the wild game and fish exploitation in Arkansas to their highest levels.

Serious railroad construction started in Arkansas after the Civil War ended. By 1895, thirty-one different railroad companies operated in Arkansas. There was no longer a need for rough travel to the Arkansas wilds. Railroads now traversed the forests and fields of Arkansas. Outdoor novices need only purchase a ticket and climb aboard—no need for horses or mules. Pick up a railroad guide, choose your game or fish species, and get off at the station where indicated. The railroad pamphlets sometimes even told travelers how far to go and in what direction from the station for best results. According to the railroad pamphlets, there was no need for a local guide. Outdoorsmen only had to feed themselves once they arrived, and if they were a decent shot, they could complete that with meat from the land.[23]

F. Chandler, general passenger agent for the Missouri Pacific and Wabash, issued an order for Missouri Pacific and Iron Mountain agents at various stations, including Arkansas, to fill out questionnaires containing information about "express and post offices, local game laws, dealers in sporting goods, markets for same, and hotel and other accommodations for hunters." In this manner, the railroad planned to gather the information to hand out to outdoorsmen "as inducements for sportsmen to travel via the Missouri Pacific and Iron Mountains lines in search of game."[24]

By 1888, the Missouri Pacific Railroad advertised the Iron Mountain Route in several sportsman's newspapers and magazines: "From St. Louis to Texarkana with a network of branches through the State of Arkansas, touching all the famous Hunting grounds in Arkansas, The Sportsman's Paradise." Another advertisement said, "This region abounds in game and game fish" and "reduced round trip hunters' rates." It listed all the rivers and streams sportsmen could reach via the Iron Mountain Route. The M. P. R. had "special hunting cars" with places to sleep and cook. St. Louis power players like Budweiser heir Adolphus Busch Jr. and the heir of one of the founders of St. Louis, John B. C. Lucas III, used the Iron Mountain Route for some of their hunting expeditions. Son of early St. Louis pioneer Alvord Easton, world-famous St. Louis cyclist George Easton, came to the Cache River to fish and hunt.[25]

After the turn of the century, railroads continued to purchase advertising in magazines and newspapers. But, instead of just sporting publications, they bought space in daily and weekly local newspapers. "You can find better duck shooting on the St. Francis, than anywhere else in America," the *St. Louis Republican* proclaimed boldly. With an image of a hunter shooting at a flock of waterfowl and a dog retrieving a duck from the water, the ad explained that a St. Louis hunter could leave at night from the city and at daybreak the following day, be on the river hunting.[26]

William C. Harris, the *American Angler* sporting newspaper editor, published *The Sportsman's Guide to the Hunting and Shooting Grounds of the United States and Canada* in 1888. Harris's guidebook listed hunting locations with a brief description of the terrain, how to get there, nearest station of transportation, game varieties and numbers found there, the best time to hunt, livery and guide charges, lodging, whether the guides had hunting dogs available, if landowners charged a fee, and "character of the sport." The guide listed fifty-four locations for Arkansas, starting with Altheimer and ending with Wheatley. All a visiting hunter had to do was find the kind of animal he wanted to hunt. The guidebook provided all the information the hunter needed to achieve his goal, even estimating a trip budget.[27]

Arista Shewey, a cartographer and publisher from St. Louis, seized on the rising popularity of sport hunting and fishing. Shewey had created and published illustrated guides and pictorials of St. Louis. In 1892, he published *Shewey's Guide and Map to the Happy Hunting Grounds of Missouri and*

Arkansas, an annotated guide explaining everything travelers needed to hunt or fish in Arkansas. His guide included game laws, available transportation, and information on what wildlife to pursue from each railroad depot. Shewey also published present sportsmen railroad rates and, depending upon the railroad company outdoorsmen chose, their trips' starting and ending points. For example, if you picked the Iron Mountain Route, left St. Louis with a hunting party of three or more, and went to Hoxie, Arkansas, the cost was $9.05 per person. Each party member was allowed up to 150 pounds of supplies, including guns and dogs, at no charge. There was room, if needed, to charter a particular car for up to thirty hunters, with sleeping berths and everything needed to cook on board, including pots and dishes. If hunters traveled on the St. Louis, Arkansas, and Texas Railway from St. Louis to Clarendon, Arkansas, (330 miles) to hunt duck or deer, the cost was $13.20, but each passenger was allowed 200 pounds of gear.[28]

A few years later, the Paragould and Buffalo Island Railroad Company published a guide to "The Sportsman's Paradise" in Arkansas. The handbook was a railroad advertisement, information piece, and recruitment booklet. It contained pictures and information about locations along the Buffalo Island Route, including hunting and fishing clubs like the Knobel Hunting Club, the Buffalo Island Hunting and Fishing Club, and the Moark Boat Club on the St. Francis River. This route also stopped at Hornersville, Missouri, located on the northern edge of the vast body of water known as Big Lake, one of the South's most famous waterfowl and fishing lakes.[29]

Although they did branch out with their advertising, railroad companies continued to self-publish hunting and fishing guides. Charles S. Nichols wrote *With Rod and Gun in Arkansas* in 1900 for the Chicago and Eastern Illinois Railroad, advertising their Cotton Belt Line. The CBL ran across the state from the northeast to the southwest, shadowing the old Southwest Trail. It connected St. Louis and Chicago with Dallas and Waco, Texas, and made stops in several places across Arkansas. "You now have a new and better way to get to the St. Francis River. The Cotton Belt service is explained in our book," one newspaper advertisement advised. According to the ad, this complimentary guide "describes and illustrates the finest sporting grounds in America." Chicagoans arrived in Little Rock in twenty-four hours, "but quite clear of the snow and ice."[30]

The St. Louis & San Francisco Railroad published their guide the same year entitled *Feathers and Fins on the Frisco*. The SL&SFRR operated in the northwestern part of Arkansas near the King's, Big Piney, White Rivers, and the Boston Mountains, the area the book called "High Arkansaw." Following

SHEWEY'S

GUIDE AND MAP

TO THE

HAPPY HUNTING GROUNDS

OF

MISSOURI AND ARKANSAS.

The Paradise of the Hunter and Fisherman.

ARISTA C. SHEWEY,

MAP PUBLISHER,

AND DEALER IN ALL KINDS OF MAPS,

714 PINE STREET. ST. LOUIS, MO.

7. Arista C. Shewey, *Shewey's Guide and Map to the Happy Hunting Grounds of Missouri and Arkansas: The Paradise of the Hunter and Fisherman*, 1892.

the established architecture, this guidebook listed game laws, hunting and fishing locations, and town and farm descriptions. Railroads could only get someone so far into Arkansas. Sportsmen had to find another mode of transportation for the remaining distance to certain hunting or fishing locations. By the turn of the century, easier-accessed hunting and fishing spots near the depots were gone, forcing a growing number of outdoorsmen to seek better grounds. The guidebook explained that travelers could hire a livery team and driver for three dollars per day and just two dollars a day in some places. Near War Eagle, "Will Kiser offers his services as guide, with a team, for $2 per day." He could also furnish hunting dogs for an extra fee. Lodging at War Eagle was just one dollar a night. Therefore, by the early twentieth century, sportsmen could acquire any number of guidebooks for their Arkansas excursions depending on where they wanted to go.[31]

Railroads and entrepreneurs were not the only ones publishing information about hunting or fishing in Arkansas. The United States Cartridge Company, operating out of Lowell, Massachusetts, printed a hardcovered book entitled *Where to Hunt American Game* in 1898. This guide had hunting information from all forty-eight states, the Indian Territory, and the District of Columbia, including black-and-white plates of examples of birds and mammals from each state. The guide explained the different types of animals and game birds for various locations in Arkansas. A special section on hunting in the cane informed: "The sportsman will often have to get within ten feet of his game before he can shoot successfully."[32]

Arkansas attitudes toward nonresidents, in some cases, began to change more toward the suspicious and unwelcoming by the end of the 1800s. By the time the United States Cartridge Company published *Where to Hunt American Game*, the sour attitude was well enough known that the author made a note of it. Many Arkansawyers had long blamed nonresident market hunters for destroying their wildlife. Resident market hunters did not want the competition. Locals became tired of out-of-staters buying up some of the best grounds and waters and posting them. This attitude caused nonresidents to rely upon one another for information and spurred several wealthy to buy hunting and fishing land and hire friendly locals to protect it.

Where to Hunt American Game explained that "there are a number of duck-shooting clubs throughout the state, the membership being made up of sportsmen from all parts of the United States." But the most significant problem for nonresident sportsmen in Arkansas was "securing a good guide" because "as a rule, they are shiftless, indolent fellows, exerting themselves as little as possible. Most of the sportsmen of the state, when they hunt, select a place and rely upon themselves to find their way about." Another guidebook mentioned that "there was considerable prejudice entertained by the natives along the St. Francis River towards sportsmen who came here to hunt and fish, as the taking of fish and the killing of game for the market was their chief means of subsistence. They looked upon the coming of strangers as the destroyers of their opportunity for making a living."[33]

Two Missouri friends, Tom and Richard, decided to try their luck on a hunting excursion to Arkansas since it was the "proper thing to do." When the pair arrived in Cairo, Illinois, the two greenhorns purchased provisions (including a large stove) and went to the meat market for some research. "Several deer hanging there . . . were the most attractive thing we saw, and we carefully inquired where they came from, studying carefully where the fatal shots pierced them, for we wanted to know where to place our bullets when we found deer in the woods," Richard explained. After arriving at Meredith, a few miles from Little Rock, the small party hurriedly entered the woods before setting camp. "We did good stalking," Richard described until they saw a deer. "Perhaps I had buck paralysis . . . but I have a faint idea that when the loggers get one of those trees to the mill the big saw will snag itself on a chunk of lead intended for the doe that got away," he concluded. For the remainder of the weeklong excursion, the pair never fired another shot at a deer. They did, however, shoot several skunks and pleaded "cannot some reader of *Forest and Stream* tell me some way to skin these truly beautiful animals?" Richard wanted to make a rug. Veteran hunters who traveled to Arkansas usually had better luck.[34]

Neophytes continued to find instructions in their local papers on what to purchase for their hunts. The popularity of duck hunting grew so quickly that the *St. Louis Globe-Democrat* published instructions on preparing for a duck hunting trip. The article "A Ducking Outfit: Important Advice as to What Guns and Other Articles to Buy" encompassed a list of what to buy. It included basic advice: "The first thing a duck hunter needs is a gun." The article covered everything from guns, waders, decoys, vests, jackets, trousers, calls, paddles, poles, boats, and even how to budget. After a lengthy discussion about weaponry brands, lead ball size, powder, and wadding, the article closed the ducking-gun conversation with a tongue-in-cheek summation: "A gun is a good deal like a woman. The more you steady it, the more you don't know about it, and the more you have to humor it. It may be quite a while before you strike the proper adjustment in your cartridges for your particular weapon, but it will be a great pleasure for you to experiment until you have found it."[35]

Railroads, private steamboats, and new wagon roads allowed the penetration of Arkansas's desolate swamps, backwaters, and vast prairies where previously only the heartiest of travelers could go. The rise of publications specifically catered to the budding sportsman offered incredible tales about hunting and fishing in Arkansas's wilds, enhancing the state's reputation as a sportsman's paradise. Commercial publications encouraged and guided novices and nonresidents to experience their Arkansas adventures. The growth in outdoor activities and the demand for meat to feed the urban markets during the late nineteenth and early twentieth centuries sent thousands of seasoned market hunters and greenhorn sportsmen into Arkansas. Unfortunately for Arkansas wildlife, these men devastated Arkansas's game and fish populations.

CHAPTER FOUR

"So Great Was the Slaughter"

Early Arkansas State Game Laws and the Birth of the Arkansas State Sportsmen's Association

> Now, brother knights of the trigger, you who love the whispering fields when autumn's blush is on leaf and blade, what can be done to silence this greedy occupation?
>
> —J. M. Richardson, *DeVall's Bluff*, 1895

Market hunters, and sometimes sportsmen, gauged success on the number of kills. The more deer, ducks, and quail killed, the better the hunt. The main difference was that the professional hunter sent his game to market, and the sportsmen sent it home; even then, some of their meat ended up in the town butcher shops. Only the fear that the game was disappearing caused a public outcry to end market hunting and the sportsmen to end their overkilling. Later, Arkansas sportsmen began an active campaign to end the practice of professional hunting for the meat markets. This effort, which stemmed from conservationists, preservationists, and sportsmen, continued with creating the Arkansas Game and Fish Commission in 1915 and the game protection laws that went with its formation.[1]

But before this general movement started, the first game-protective crosshairs settled on the nonresident professional hunters and sportsmen. It was not because the masses were concerned about the overharvesting of Arkansas game and fish; instead, they believed that nonresidents had no right to Arkansas's animals. Many Arkansawyers, both politicians and citizens, did not want nonresident hunters in the state because they thought that Arkansas's game and fish were the property of the people of Arkansas. Citizens wanted to hunt and fish at will. If any nonresident wished to travel into Arkansas to hunt or fish, Arkansawyers thought they needed to pay for the privilege.

Most Arkansawyers did not like market hunters, whether locals or nonresidents. One Arkansawyer who followed the Arkansas wildlife population closely for over twenty years claimed that market hunters killed "90 percent of the game [in Arkansas]." Most southern states experienced the same situation. In Arkansas, market hunters became the first target of game conservation legislation. Historian Daniel Herman argues that southern locals were the ones who engaged in the most pot and market hunting when forced during difficult economic times. The public tolerated these men (at least for a while) because they provided a service to citizens and restaurants, but residents despised outside professionals. Market hunters were in the business of removing the resources from the Bear State and shipping those resources to out-of-state markets.[2]

8. F. M. Smith & Company Game, Chicago, Illinois. Library of Congress, Prints and Photographs Division.

Arkansas passed the nation's first statewide nonresident hunting law, a direct shot aimed at market hunting. Approved on March 6, 1875, the Nonresident License Law stated that "a tax of ten ($10.00) dollars is hereby levied upon all non-resident trappers, hunters, seiners, or netters of fish . . . in this state." The state charged ten dollars, payable to the county clerk, to any nonresident who wanted to trap, hunt, or fish for a living in Arkansas. If the state found any non-Arkansawyer without a license, it charged them with a misdemeanor and fined them twenty-five dollars. Local authorities often overlooked nonresident sportsmen and did not require them to purchase a license, believing that the law was meant for professionals. Furthermore, the act contained no language about how the law was enforced or who would enforce it. The legislation fell under the criminal code, so it was up to the county sheriffs, clerks, or the justices of the peace to enforce the law.[3]

Physician and early amateur ornithologist William H. Deaderick claimed the law, "at least at first, [gave] greater protection to the native pot hunter than the birds." Now, the local market hunter experienced less competition from outside the state, but not much. Deaderick blamed the law for the cause of so many market hunters becoming Arkansas residents in the following years. Theodore S. Palmer, assistant in charge of game preservation for the United States Department of Agriculture, wrote in 1900 that with this law, Arkansas was a "pioneer in the restrictions on market hunters."[4]

In 1877, Arkansas's neighbor to the north, Missouri, outlawed nonresident market hunters entirely. During the same year, Tennessee banned nonresident deer-hunting market hunters in certain counties and, ten years later, prohibited all market hunting statewide. Furthermore, by 1879, the Show Me State ended all nonresident hunting outright. Interestingly, most visitors to Arkansas came from those two states and complained the loudest when Arkansas started charging a fee to hunt or fish. Arkansas remained behind in almost all areas of game protection legislation nationally and regionally, especially in enforcement.[5]

Nonresident sportsmen and railroad companies continued to grumble over the Arkansas license law for some time afterward. One visitor argued that "the act . . . was dead at birth, owing to a variety of causes." He claimed that the attorney general of Arkansas had informed Arkansas county officials not to enforce the law. Instead, he charged, local law enforcement was using the flimsy regulation to line their pockets from unsuspecting nonresident sportsmen who "were pounced upon by these deputies and forced to pay." Law enforcement gave visitors a nondescript receipt, not a license.

He concluded that the sportsmen just wanted to shoot a few quail and return home.[6]

A January 1887 letter from a Carlisle resident to the *Arkansas Democrat* editor scolded the paper for reporting that many Arkansawyers wanted the license law repealed: "It should by no means be repealed," the letter claimed, but instead be made more stringent. He argued that game had multiplied since the passage of the license law, and he credited the legislation directly for the results. For many years, observers viewed game in abundance throughout the countryside, but "not a few years of invasion of the 'pot hunters,' who think not, nor cared, for the final results of their destructive methods, have well nigh destroyed it all."[7]

Unfortunately, in 1887, circuit court judges in both Craighead and Poinsett counties ruled the 1875 license law unconstitutional. Indianan M. Adaml sued the local sheriff in Craighead County and won. In the Poinsett County Circuit Court, eighteen nonresident cases, including those of A. L. Fisher and W. E. Bingham, went against Sheriff Addison "Add" Harris. The Poinsett County ruling caused Sheriff Harris to remit the ten-dollar fee to all eighteen individuals who had paid him for the licenses. He had "enforced it vigorously and collected hundreds of dollars from sportsmen." Few county officers were enforcing the law anyway (with one exception), and few nonresidents knew the law.[8]

The *Southern Standard* in Arkadelphia called the court ruling "common sense, at least." It continued, "Now let the law be amended to sharpen and ensure the punishment of pot-hunters." One nonresident hunter declared, "Thanks to the officials of the St. L, Ark. And Texas R'y sportsmen can now visit this splendid game region without fear of molestation." This instance was not the last time nonresidents and powerful outside lobbyists swayed Arkansas game legislation. When the ban on all nonresident hunting occurred in 1903, wealthy outsiders influenced lawmakers.[9]

These events represented a pattern in Arkansas legislation during the early days of game and fish law. In many cases, the state assembly passed legislation, and the laws were either not enforced, eventually gutted by exemption legislation or court rulings, or both. Pressure against game legislation came from nonresident sportsmen, meat processors and handlers who bought and sold the game and fish, railroad companies that shipped the animals and carried the sportsmen into Arkansas, and residents who hunted for a living or catered to nonresidents. The failed 1875 law was the first attempt at game protection in Arkansas, and the discussion surrounding a modicum of game protection at least brought the subject to the halls of the Little Rock legislature.

A few years later, in an article concerning the proposed passage of a nonresident license in Maine, the writer cautioned that the Pine Tree State should not follow Arkansas as an example. He claimed that the Arkansas law was "in effect employed chiefly for the purpose of blackmail by various county authorities." The Arkansas law was designed for market hunters only and not to draw revenue for game protection into the state, as Maine license proponents wanted in their state.[10]

In 1879, Arkansas legislators argued about game laws concerning the birds and deer. A glimpse into the proceedings surrounding the bill provides an understanding of the split between those who supported early game laws and those who did not. Lee County representative William Hines Furbush introduced House Bill No. 22 to "aid in the preservation of birds, bird's eggs and deer." Primarily aimed at protecting quail and grouse, the bill also outlawed the harvesting of bird eggs and placed a season on deer hunting. Independence County representative Thomas J. Stubbs said that the bill did not "suit him," and he moved to remove any protection of grouse or partridges from the language. After Stubbs, a Lyon College teacher, pondered a moment more, he decided that his first motion was not strong enough and moved, instead, to indefinitely postpone the bill.[11]

Furbush, an African American Republican, used a three-pronged argument to stump for this bill, in parts an argument that game-law proponents used repeatedly. First, the state government needed game laws to counter their backward image. Arkansas possessed more progressive thinking, he argued, and "if Arkansas ever expects to get out of the toothpick sphere which has been hanging to her for twenty years, she must adopt some of the laws in effect in older States." Second, Furbush wanted to attract money to the state, stating that "sportsmen are turning their attention west, and we could induce those men of capital to come [here]." And finally, he emphasized the benefits that the animals provided Arkansas, explaining that "the birds save us from grasshopper plagues. The birds were killed off on the prairies, hence the affliction of the 'hoppers. When the birds are killed out, the worms and germs take possession."[12]

Sebastian County representative R. H. McConnell claimed some merits to the bill and did not want it postponed. He probably wanted an opportunity to bring it to the floor, so those opposed to it could defeat the bill before Furbush's supporters could convince others that it was worthy of passage. William Fishback, the other representative of Sebastian County, had been at odds with Furbush over other matters, so McConnell probably worked on a tit-for-tat basis.[13]

Washington County representative W. C. Braley moved to refer the bill to the Committee on Agriculture. Many Americans believed that wild game and fish were the state's property. They considered them food, which fell under the purview of the Agriculture Department. This was the case on the local, state, and national levels at the time. This attitude caused great trouble during the battle for state and local game laws in the future.

James H. Frazer, a representative from Van Buren County, then rose and said that he "was from a mountain county, and his horny-handed constituents demanded the passage of such a bill. At present, the loafers killed the game." Stone County representative J. H. Morris then bellowed, "This bill is worthy of the respect of the house." Stubbs withdrew his motion to postpone, and instead, the representatives sent the bill to the agricultural committee.[14]

The evening edition of the *Daily Arkansas Gazette* read, "We are glad to see that Representative Furbush has introduced in the house a bill for the protection and preservation of our wild game." The editor challenged naysayers, arguing, "Preservation of our wild game is a measure of public policy [and] does not infringe upon nor curtail any reasonable liberty of the citizen, but guards against wanton destruction. No man is prohibited from killing as much game as he can find in the proper season, but the idle and vicious are prohibited from useless slaughter and disturbance during the breeding season." He closed with an impassioned plea: "We hope the bill will pass without opposition, and not be murdered by foolish or ridiculous amendments intended to kill it."[15]

As it became clear that his bill would fail, Furbush rose and moved to indefinitely postpone the bill and amendment: "I don't believe Arkansas is *quite* ready for the game law," he said defeatedly. Instead of the postponement, they voted Furbush's game bill down 38 to 24.[16] The *Arkansas Democrat* editor James Mitchell asserted that "Representative Furbush was right: 'The house is not *quite* ready for the game law.'" "Nearly every other State in the Union has rigid law on this subject," the editor lamented, "[and] it is a great pity that it did not [pass] because it was a first-rate bill." Ten days later, Furbush wanted a revote on his game bill. The motion did not carry.[17]

During the House debate, the Senate considered their own game bill: "An act for the preservation of deer, turkey, prairie chickens, commonly called grouse, and other wild game." When this bill was defeated, some Arkansawyers expressed a different take on exactly who the state should prevent from butchering Arkansas game. However, this time, the target was not market hunters but nonresident sportsmen.[18]

Arkadelphia's the *Southern Standard* argued "that some action should be taken to prevent the sportsmen of St. Louis, Memphis, and other places from coming to this State in large camping parties at any and all seasons and slaughtering the game by wholesale." "Legitimate hunting of game is all right," the paper conceded, "but these city gentlemen come down with their double-back action repeating guns, nets, traps, and devices of all kind (and whilst the farming community are busy with their crops), work destruction to thousands." It seems that many citizens were for some sort of game law to bar the slaughtering of Arkansas game and fish. While a few lawmakers introduced game legislation, many other Arkansas politicians voted against progressive game protection.[19]

Perhaps a quote from one legislator who opposed game laws exemplifies the opponent's mindset. "I shall vote against any 'game laws,'" stated one anonymous representative, "my reason is this, that I believe that if all the game in the state was exterminated, it would be to the interests of the laboring man; many who have labor to perform in order to make a support for their families, leave their legitimate occupations to go hunting." He continued, "Every pound of game secured by them costs, them, in loss of labor and ammunition (including 'snake medicine'), at least two dollars." This politician likely represented a delta county where laborers were needed, and white landowners complained that poor Black and white laborers supported their families through hunting and not working in their fields.[20]

After William Furbush's 1879 game bill flopped, a few citizens pointed out that its defeat was a failure to modernize. Arkansas citizens were ready to progress and shed their backward image. The Warren brothers, editors of the *Prescott Dispatch* in 1879, were furious about the failure of the game bill and its implications. Much like Furbush's argument concerning the backward "toothpick" image that Arkansas maintained nationally, they lumped the game bill with many other progressive ideas that Arkansas lawmakers had failed to act upon. "The propriety of a game law—or a law protecting game in our State; a law taxing dogs for the protection of sheep, and for the increase of our free school fund; a law for devising a more efficient and practical system of free schools, public roads, and to encourage agriculture, immigration, and the building and equipping [of] an asylum for the insane," they argued, "would all meet with due consideration, and public mind, would thoroughly settle upon the propriety of making laws to accomplish these objects." But "progress" was not forthcoming in the area of wildlife legislation and the killing continued unabated.[21]

Interestingly, Furbush, an African American, pushed for game protection because his actions contradict the accepted narrative concerning race and the passage of conservation laws in the South. Several historians, from Tara Kelly to Scott Giltner, have illustrated that southern whites pushed for wildlife protection laws due to their racist attitudes toward southern Black hunters and fishermen. White sportsmen blamed indiscriminate Black wildlife killing for reducing the game and fish population. One Louisiana sportsman, hunting in south Arkansas in 1872, commented that his party had an excellent four-day hunt. He believed that the reason for this exceptional hunt was that the animals they harvested were "the game negroes cannot hunt."[22]

"The colored brother with his smooth-bore musket or pot-metal breech-loader is the gentleman we want to stop," argued Clarendon native William E. Spencer. The South needed a gun tax, which "would afford much relief . . . so much as the South is concerned." He argued that a tax on weapons would lessen the likelihood that southern African Americans could afford a gun and, therefore, less likely to hunt or trap anything more than possums and coons. Just as poll taxes served to disenfranchise Black hunters, the gun tax could serve to disarm them.[23]

Even the African American–owned dog was not safe from criticism. In an article about quail, including the woodland and prairie quail of Arkansas, the author wrote that "in the south, the cur dogs of the negroes—every family owning one to as many as they can possibly maintain, and all kept in a kind of half-famished condition—prowl through the fields seeking food, and they are the worst egg destroyers." He concluded that "with such rapacious enemies to contend against, the destruction of the quail must be great." It seems, according to this article, that only African Americans owned egg-eating canines, and everyone else owned dogs that would not eat eggs.[24]

Around the same time that Arkansas white sportsmen and lawmakers began to pass game protection laws, more wealthy and middle-class sportsmen joined private hunting and fishing clubs. Stronger public game protection laws and growing white control of the best hunting and fishing grounds tightened restrictions on Black hunting. These actions suggest that the conservation movement in the South went hand and glove with Jim Crow in some parts of the state. White control of African American hunting and fishing meant controlling Black labor, and according to one historian, white "dominion over Southern fields, forests, and streams."[25]

Whether he was an anomaly or not, Furbush was right about Arkansas not being ready for a game law in January 1879, but the state was prepared for a fish protection law. This development may have occurred because men, women, and children fished while generally only men and boys hunted. Fish served as easy food to obtain unless unscrupulous men destroyed them. Farmers, who might resist game protection laws, were more apt to support fish conservation for these reasons. Arkansas legislators also passed Act 51, "for the Preservation of Fish in the Rivers and Other Waters Within This State," making it illegal to use dynamite or poisons on fish. Also, landowners could not have a dam across a stream, even on their property, without a way for fish to pass through from March 1 to June 1.[26] Six years later, Arkansas policymakers passed a law compelling the building of "chutes or aprons" to allow fish to pass over or around dams built for powering waterwheels. This rule was a year-round requirement with no restrictions. Within two years, they repealed it, another example of pass and repeal.[27]

Some Arkansas citizens remained angry over the state government's slow action. "The total want of protection has done its deadly work," declared a Jacksonport citizen in July 1883. "Not even the surley bear can hold his own. Deer are butchered during the overflows (flooding) on the ridges and high places to which they resort for safety," he explained. Even during the hot summer, bucks and does were killed, regardless of size. "The result is that venison is very scarce in Arkansas," the man concluded. Writing from the banks of the White River, the Arkansawyer commented on the influence that sporting newspapers had on public sentiment toward game laws in the state, maintaining that "through the salutary influence of *Forest and Stream,* a strong favor of game protection has grown up, and in many places the lawless acts above complained of are restrained solely by the force of public opinion."[28]

Arkansas continued to hemorrhage game and fish, either through waste or out-of-state shipments. Eleven sportsmen traveled on the Iron Mountain Railroad 220 miles from St. Louis to fish and hunt along the Black River for several days. The group caught over two thousand pounds of fish, leaving most of it to spoil. The Missourians also killed deer, raccoons, minks, geese, squirrels, and ducks on their excursion. In another Arkansas location, "Two hundred pounds of game fish have been taken by a single person in a day, with the fly rod, in the neighborhood of Lone Gum Island" in the St. Francis River, bragged one St. Louis news article. While some continued the slaughter, other outdoorsmen had had enough. As news about the removal of Arkansas wildlife increased, so did some sportsmen's resentment toward the action.[29]

In 1888, a correspondent named Earnest wrote to *Forest and Stream* about his recent trip through Arkansas and what he observed at a train station on the edge of the "sunk lands" of Arkansas. "On the platform," he explained, "were twelve barrels and two sacks of mallards for the produce house in St. Louis. On the night before, nineteen barrels were shipped from the same place to the same house. The smallest barrel was marked 'sixty-four mallards.' Must the tens of thousands of true sportsmen all over this land give up their sort in order that a few market hunters and a few produce houses may make money?" he complained. "Why in the name of heaven can't we get some laws to stop the sale of game? Year by year, we go further and get less; and it will be only a question of time till all American game will go just as the buffaloes have gone," he concluded.[30] In response to Earnest's letter, "Garganey" from Cleveland noted that "club men . . . are the [market hunters] worthy coadjutors in the work of game extermination." Most clubs initially set no limits on the number of waterfowl killed.[31]

During the 1889 session, whether from the constant complaints, the growing animosity toward market hunters, the idea that the Bear State needed modernization, or the pressure from Arkansas conservationists, Arkansas legislators finally passed a whole series of game and fish laws. They approved a bill that stopped the use of dams, fish traps, seines, or nets in any Arkansas navigable stream unless the property owner was using them. They set net mesh size and length limits. Anyone caught breaking these laws received a fine of no less than five dollars and no more than two hundred dollars.[32]

Furthermore, policymakers finally passed an act that set deer, turkey, quail, and prairie chicken seasons for their harvest and the taking of the birds' eggs. Other states had hunting seasons for many years. With pressure from the Arkansas State Sportsmen's Association and other state conservation organizations, Arkansas legislators believed that seasons might protect Arkansas wildlife, following examples from other states. Hunting seasons protected animals in their most vulnerable time, primarily when they raised their young or were on the nest. Several residents, however, thought that the legislators had acted too late and that what they had done was not enough.[33]

The most controversial law passed during the 1889 groundbreaking session was "An Act to Prohibit the Exportation of Fish and Game From This State." The law supposedly challenged both resident and nonresident market hunters who were devastating Arkansas wildlife. It also was a blow to nonresident sportsmen, who had a habit of shipping their spoils back to friends and coworkers (and sometimes markets too). Finally, a law prohibiting the shipping of Arkansas living resources to out-of-state markets instead of a weakly enforced license law that allowed nonresidents to purchase permission to ship the game out of the state.[34]

This new law stated that all game and fish, except those privately held in landowners' ponds and lakes, were the state's property. Taking or shipping said animals were considered a "privilege" for individuals or shipping companies. The act made it illegal to export game or fish for the following six years. If authorities caught an individual or company shipping game, they charged the violators with a misdemeanor and a fine of no less than fifty dollars and no more than two hundred dollars, with half of the fine going to the county treasurer and the other half to the informant. Jurisdiction for any of these cases fell under the justice of the peace, which was the case for most minor law violations. The *Southern Standard* proclaimed defiantly that the "Chicago and St. Louis pot-hunters will now have to seek other preying grounds." But this was not the case because the law was not well enforced, and the shipping continued, even though declared illegal.[35]

The following year, an observer passing through the sunk lands of Arkansas observed more barrels of ducks headed to Chicago. He discovered from the railroad employees that the Cotton Belt shipped, on average, ten barrels a day out of that one station. The traveler claimed the market hunters used swivel guns to butcher ducks at night. "Was there any sport about this?" he questioned. Many sportsmen believed that to ethically hunt, the game needed a chance to get away, making the chase or pursuit of the game more challenging. For waterfowl, it was killing birds while they flew, not sitting on the water. One 1894 sportsman noted, "Roost-shooting is the worst thing in the world to do, and that night-shooting is usually branded as unsportsmanlike." Interestingly, early sportsmen did not consider killing an unlimited number of birds or animals unethical or unsporting. However, after the federal and state governments introduced bag limits, changing attitudes and the sporting ethic evolution caused modern shooters to evaluate their sportsmanship with other hunting aspects (some had already done this). The way a man killed, not the number of animals slaughtered, was the yardstick by which they measured themselves as sportsmen.[36]

Many changes occurred during the two legislative sessions following the groundbreaking 1889 session, including exemptions to the 1889 game laws. It was common for counties to exempt themselves, and they wasted no time. For example, it was still unlawful to ship some game out of the state, but now legislators had *excluded* "beaver, possum, rabbits, ground hogs or wood chucks, raccoons, squirrels, snipes or plovers, ducks and geese, when shipped openly." It was still unlawful to ship fish out of the state, *except* catfish and buffalo from "Mississippi, Crittenden, St. Francis, Poinsett, Jackson, Phillips, Desha, Chicot, Lincoln, Prairie, Craighead, Greene, Clay, Ashley, Cross, Marion, and Jefferson." These exemptions are only two examples of how lawmakers gutted laws of their power.[37]

All market hunters had to do was throw a couple of pounds of catfish on top of a whole barrel of bass, crappie, or other game fish, and no one was the wiser. Shippers were in business to profit and could claim plausible deniability to what was in the barrel to the bottom. Besides, who would come to check an entire barrel of stinky fish? An 1896 guide to state games laws published in *Forest and Stream* listed the Arkansas exportation laws as only pertaining to "professional hunters, who follow hunting as an occupation, and not to apply to the sportsmen who hunt for pleasure." Within the decade, a Craighead County circuit court ruled that none of the exportation laws could stop anyone from taking their catch out of the state. Therefore, legislative exemptions and court rulings made current Arkansas state game laws useless for protecting wildlife.[38]

Although numbering in the tens or hundreds of thousands before 1850, prairie chickens or pinnated grouse had disappeared from much of Arkansas during the nineteenth century. They did not migrate, lived in most of the state, and were popular among diners. Because of their docile character, mating habits, and how they traveled together in large groups, the birds made easy targets. Market hunters were killing the birds at a staggering rate. Sportsmen enjoyed hunting them too.[39]

In 1876, W. S. S. of Pottsville, Pennsylvania, had received an answer to his inquiry to *Forest and Stream* about quail and pinnated grouse season in Arkansas and if any laws governed the harvesting of the birds. Editors of the journal replied that there were "no close[d] seasons in Arkansas and no State game laws."[40] A sportsman from Little Rock reported that, a few years earlier, he observed "a party of Memphis shooters [had] had killed 1,500 [grouse] in a few days. [and] piled them up and left them to rot." The "shooters" claimed they were sportsmen. However, the Arkansas legislature blamed only the market hunters for the destruction.[41]

"The pot hunters from Memphis came in July before the chicks were half grown, and one party killed twelve hundred, every one of which spoiled," cried one Arkansawyer. "Thousands were trapped that fall, and the result is that they are now very nearly extinct, and they are still being further trapped, and unless they are soon protected, there will be none very soon," an alarmed

Arkansas Democrat declared. People gathered eggs and brought them into the market by the barrel loads, and many spoiled before they were purchased and eaten. Market hunters also killed hundreds of quail during each trip. In March of the previous year, three Memphians crossed the river for a seven-day hunt and killed over one hundred quail during the hunt.[42]

Although the legislature, in several bills, had authorized "sheriffs, constables, coroners, marshals, market masters, and police officers" to enforce the few game laws, very few prosecutions occurred. Finally, in 1893, as numbers continued to plummet, Arkansas lawmakers banned the hunting of prairie chickens for five years. The lack of prairie chickens must have indeed been grim to get a slow-moving legislature to pass such a rigid law. However, these men failed to see the growing crisis for Arkansas game and fish for the most part. The lack of the birds, not the law, ended the killing of prairie chickens.[43]

Many sportsmen expressed concern at the end of the five-year Arkansas ban on prairie-chicken hunting in 1898. *Forest and Stream* correspondent Emerson Hough claimed that the birds "were very much scarcer and wilder" and he certainly hoped that the Arkansas legislature would extend the ban. A hunter from Carlisle agreed, wanting another five years at least.[44]

A 1901 federal report concerning game laws noted that although twenty-one states required nonresident hunters to purchase licenses, only Arkansas and Oregon made market hunters buy licenses while other nonresident hunters did not. Out of the southern states, only Florida and South Carolina charged licensing fees to nonresidents. The remaining southern states either did not allow nonresident hunting or had no license law. Arkansas game laws, or lack thereof, failed to protect. Fish and game continued to experience danger from sportsmen and professional hunters alike.[45]

"Byrne" in Crockett's Bluff observed, "First-class legitimate sport on ducks, to wit, wing-shooting over decoys, has been hard to find—our visiting sportsmen killing 129 one afternoon, nearly all single and double wing shots," the correspondent complained. These were not market hunters. They were sportsmen, firing with muzzleloading shotguns. Later, after pump shotguns appeared, the shooters slaughtered wholesale. At an Arkansas lake near Memphis in 1894, six "sports" killed 430 ducks in one day. In 1895, chief clerk of the United States Sub-Treasury Charles G. Ricker spent two weeks in Arkansas hunting waterfowl. As a gift to his employees at the US Treasury offices in St. Louis, Ricker shipped 120 ducks for distribution and then continued to hunt.[46]

A noted hunting guidebook refused to include any listing of Arkansas state game laws in its publication. "The frequent and often absurd and conflicting changes made in these laws render any compilation published in an annual work of this character misleading rather than instructive," the editor explained. Instead, hunters should consult with the railroads for a list of current game and fish regulations for the location where the sportsman was traveling.[47]

Little Rock sportsman J. M. Rose agreed with the guidebook about the absurdity of the few game and fish laws that Arkansas had on the books in 1894: "Some of our streams have been badly trapped and netted. It is impossible to secure any convictions under our fish law, for it is too loose that about all a court can do with it is to decide under which exception an offender shall be discharged."[48]

A blistering indictment appeared in the pages of an Arkansas newspaper on Valentine's Day in 1895. "I desire through the columns of *The Gazette* to call the attention of the sportsmen of Arkansas to an evil," DeVall's Bluff resident J. M. Richardson wrote, "which if not abated with remedial legislation, will result in the total decimation of every game interest in the State." Richardson was referring to market hunters, or "every person who kills, traps, or snares game for the purpose of selling the same." To men like Richardson, it was of no consequence from where the market hunter hailed, resident and nonresident killers were alike. They slaughtered Arkansas game, and "nothing that wears hair or feathers has escaped these human vultures," he exclaimed defiantly. He pointed out that the prairie chicken was gone. "Now, brother knights of the trigger, you who love the whispering fields when autumn's blush is on leaf and blade, what can be done to silence this greedy occupation?" Richardson implored. The only way to end market hunting was to pass a law prohibiting game sales in Arkansas and any wildlife shipment out of the state. He guessed that the market hunters would slaughter the game past the point of return in two more years unless something changed.[49]

This was the case for the prairie chicken. When Richardson said they were "gone," he meant that only a few remained of a vast population. When the five-year ban on grouse hunting ended in 1898, "hundreds of shooters were afield on opening day." Five years was not enough to bring back the numbers even partially, and the only pinnated grouse left in Arkansas in any numbers lived in only three counties (in the Grand Prairie region). With so much pressure from a shooting public that had waited five years for a chance to kill the birds, the meager prairie chickens had little opportunity to escape extermination.

The Arkansas legislature did not meet again for two years, and according to one sportsman, "By that time, the work of destruction will be complete." There was no execution pardon for the prairie chicken.[50]

After hearing news of the reopened hunt and the large numbers of shooters killing the remaining prairie chickens in Arkansas, Edmund Hough wrote angrily that "no State is more prodigally reckless in squandering its game wealth than is Arkansas, and no State leave its game interests more unprotected." He explained that no one enforced the paltry list of Arkansas game laws that did exist because the regulations never had any legal statute attached for their administration. Since Arkansas had a provision barring any creation of new offices, he complained, "it would be impossible to appoint a game warden without the necessary amendment to the constitution." When lawmakers had passed the five-year ban, the bill contained no enforcement section. Instead of improving it, they let the law lapse, and hunters legally killed the few birds they could find during a five-month season.[51]

The Arkansas government was not going to protect Arkansas game unless there was a grassroots movement pushing them to do it, and even then, it took several years. One organization, founded in 1891, had worked toward stricter game laws since its founding but had found little success. For four days in August 1891, two hundred Arkansas sportsmen met at the Little Rock fairgrounds for a shooting tournament. Most of the attendees were considered members of the upper or upper-middle class and were often leaders in their communities. They could afford to purchase $250 trap guns and the ammunition necessary for competitive shooting. However, this shooting competition was not their only purpose for traveling to the capital city on those hot August days.

The men who attended that meeting created an organization that laid the foundation for the legislation and protection of Arkansas wildlife for the next century, but the process took time. During the fight for wildlife protection, Arkansas lawmakers defined the Bear State's place in the new national conservation movement that appeared in the late nineteenth and early twentieth centuries. The argument over who owned the wildlife and could harvest it also shaped Arkansas game legislation and provided the basis for the present system.[52]

The Arkansas State Sportsmen's Association contained members who recognized the need for a conservation effort to protect Arkansas wildlife and stop the wholesale slaughter. These men wanted to save the animals for their and future generations' use, and they made conservation the ASSA's primary mission.

In 1925, Dr. George Bird Grinnell, an outdoorsman and early conservationist, wrote that the movement to protect wildlife went through three stages during the previous fifty years. The first stage came from the sportsmen who, for their selfish notions, wanted to end the slaughtering of game because they wanted enough animals to continue their sport hunting and fishing into the future. The second stage came from an emotional or romantic belief that the animals were beautiful and that all future generations should have the same opportunity to view these majestic creatures. The third stage developed from the idea that wildlife had an economic value to the people, and spending money on their protection was justified because it protected a more significant investment. Theodore Roosevelt relatedly believed the movement was an exercise in democracy— the protection of all-natural resources for the people and by the people.[53]

If grassroots support for game conservation was to materialize in Arkansas, organizations like the ASSA had to convince farmers that it was in their financial interest. Tara Kelly argues that sporting publications, which spread the concept of the sporting ethic and conservation, influenced "middle-class Americans . . . to support game laws in the late nineteenth century when previously they had (generally) opposed them." Sportsmen, who published their stories in these magazines and journals, "had a massive impact on the speed and direction that it took."[54]

The ASSA recognized that Arkansas wildlife represented an incredible value and state resource. They also realized that the wholesale slaughter of these resources, an action that nonresident market hunters primarily carried out, had to end. The ASSA pledged its members to push for stricter game and fish laws to protect Arkansas wildlife. They continued this fight for the next thirty years, and their efforts continue to influence game legislation today. The ASSA released a public statement concerning their stance that read:

> Whereas, The state of Arkansas is supplied by an all-wise Providence with a variety of game and fish, worth many thousands of dollars to her citizens and Whereas, Wanton destruction of the same is being indulged in by a class of people known as 'pot' hunters, who pay no taxes in the state, who hunt and fish in season and out of season for the dollars and cents they can make out of it, and in gross violation of our game law; now, therefore, be it resolved that we, the subscribers hereto do hereby form ourselves as the Arkansas State Sportsmen's Association for the better protection of our game and fish by adopting from time to time such measures as may seem best for the rigid enforcement of our game laws.[55]

In addition to the resolution explaining the group's overall purpose, the members also voted on officers and their duties. The most interesting caveat was that the president would appoint a vice president in every Arkansas county. Those men were to observe and report any illegal game abuses to the local law officials and demand prosecution and conviction. This was a way that individual members could directly assist in wildlife protection. Furthermore, all members vowed to press their state representatives to enforce the game laws and pass new laws to protect Arkansas game and fish. The members agreed to meet annually in the last week in August in different locations across the state.[56]

The ASSA, more than any other group, was responsible for prowildlife legislation in Arkansas from 1891 to 1915. The organization they formed that summer was responsible for creating the first state game warden position and the Arkansas Game and Fish Commission. The efforts of this organization, coupled with those of other organizations like the Arkansas Audubon Society and the Arkansas Federation of Women's Clubs, pressed Arkansas lawmakers into passing laws that helped save many species of animals from extinction in Arkansas.

When the ASSA formed, Arkansas had few game laws on the books, and those that did exist were primarily worthless. Although the passage of several game laws during the 1889 session had provided hope for protection advocates, enforcement of those laws proved inadequate. Starting with the 1875 Nonresident License Law, Arkansas conservationists and game protection supporters experienced continued disappointment in Arkansas's efforts to safeguard game and fish. Several of these citizens decided that the best way to compel the legislature to act effectively was to create an organization that could lobby as a group instead of individually.

The *Arkansas Democrat* endorsed the ASSA's 1891 platform, particularly the call for the preservation of game and the enforcement of the game laws, noting, "We believe the association will do good work on this line." The *Democrat* also pointed out that while the meeting was occurring, word came to the organization from Camden about four men who were dynamiting fish in a local river, and citizens of the town had demanded that these law violators receive prosecution. Local authorities fined the men when they returned to town. "Camden [is] the first to put themselves on record as prosecutors of violators of the fish and game laws of the state," the paper declared.[57]

The following summer, a Fort Smith man wrote a letter to the editor of the *Daily Arkansas Gazette* asking for information about the 1891 meeting in

Little Rock. He could not attend and had not received any information about whether the call for the state sportsmen association was fruitful or not. "Every sportsman in the State is interested in this organization. We must have laws enacted for the protection of our game. For if ever game needed protection, it is in this State," he pronounced. Early examples of men from Little Rock, Camden, Hot Springs, Dardanelle, Fort Smith, and other towns pushing for a state organization for game and fish protection illustrated concern from across Arkansas, not just larger cities or the wealthy. Although these men most likely would not have identified as "progressives," the push for game and fish protection in Arkansas fit with the national crusade occurring during the Progressive Era (1890s–1920s) with more Americans becoming concerned with environmental issues like clean water and air, the creation of state and national parks, and wildlife protection.[58]

The progress came slowly, and the ASSA employed several measures to achieve its goals. In July 1892, the second annual meeting in Hot Springs reaffirmed the original aim. As the program announced, "The object of this association is to secure better protection for the game . . . and to promote a friendly intercourse among the sportsmen. It costs only $1 to become a member of the association." It seems, however, that for the first five years of its existence, the ASSA had no real direction about how to achieve these aims. They wanted to promote game protection; but other than lobbying their local politicians, there was no overall plan.[59]

For example, in 1897, ASSA president James T. Lloyd offered a five-dollar reward to anyone in the state who provided information for the arrest and conviction of game violators. The *Pine Bluff Daily Graphic* argued that Arkansas sportsmen should be "extremely gratif[ied]" and that the offer was a "move in the right direction." It is unclear how much, if any, moneys were paid out in rewards. Instead of paying for information, the organization decided to use their men to gather evidence on lawbreakers.[60]

The following year, the ASSA annual meeting in Little Rock contained a great deal of discussion on a new topic and one that might provide a more specific cause that the organization could rally behind—a state game warden. Little Rock Mayor James A. Woodson, a charter member of the ASSA, suggested that the position of state game warden belonged under the commissioner of mines, manufacturers, and agriculture.

Additionally, the president had previously appointed a committee of attorney John M. Rose and association secretary Paul R. Litzke to outline recommended both changes to Arkansas's current game laws and additional laws (including establishing a state game warden), and present them at the 1898 meeting. They were read to the group, discussed, and voted on. While not overly progressive compared to some other states, particularly those on the eastern coast, Rose and Litzke's plan was remarkably different from Arkansas's paltry game protection laws before 1898. The foremost matter was to create a law enforcement position for the prosecution of game violators. Previously, once lawmakers had passed a game protection law, it was the duty of the county sheriff to enforce these laws, but few did. Many citizens believed that was ridiculous because the local authorities were breaking the regulations themselves in many cases. In other cases, what most believed to be more serious crimes, like property theft, took precedent over game violations. Most people did not understand that violating the game laws was theft. It was stealing the game from the people of Arkansas. According to Rose, only Pulaski County authorities enforced game laws. Pulaski County probably enforced the regulations more strenuously because of its proximity to the state capital.[61]

Rose provided a copy of the association's recommendations for game and fish protection to the *Arkansas Democrat*: "It creates a game warden and imposes the duties of the office on the Commissioner of Agriculture since the moss-grown constitution forbids the creation of a new office. The warden may appoint as many deputies as he chooses, whose compensation shall come by receiving one-half of the fines in case of convictions." It also prohibited the sale of game and fish or feathers and bird skins, barred dynamite, poisons, nets, or traps to collect fish and set seasons for hunting deer, turkey, pheasant, quail, and ducks. The ASSA believed a warden was as important as passing new game protection laws. If no authority existed to arrest and prosecute violators, any new law was worthless. Each measure complemented the other.[62]

Besides supporting the passage of new game laws and the enforcement of those laws, some members of the ASSA were interested in introducing new animals into the Bear State. They supported the introduction of game and fish into Arkansas, but lawmakers had to pass legislation to protect the new wildlife. In October 1898, newly elected president W. R. Duly suggested that the organization sponsor the introduction of the ring-necked pheasant into Arkansas. These white-meat birds were considered extraordinarily beautiful fowl that challenged the hunter with their celerity and cunning. The pheasants raised chicks two or three times per year with as many as twenty eggs per session, making them quick to propagate an area and sustain a population. As an experiment, Duly placed twenty-four birds in fields near Big Lake in northeast Arkansas and believed they could thrive in their new home with legislative protection. The association tied propagation and protection together.[63]

On December 21, 1898, the *Helena Weekly World* republished a story from the League of American Sportsmen's magazine, *Recreation*, reporting exactly what the ASSA was fighting against. Back in July, nine Illinoisans had come to the Macon County Hunting Club in Arkansas for their ninth annual hunting and fishing trip. The author reported that these men caught over two hundred pounds of fish and had killed at least four deer, in addition to numerous squirrels, geese, and ducks. Instead of giving the meat away, they buried the whole lot, wasting it. Many in the association thought that maybe instead of a nonexportation law, Arkansas needed laws dealing directly with all nonresidents.[64]

Mrs. Louise M. Stephenson, writing in the *Helena Weekly World* in response to the *Recreation* article, asked, "Can nothing be done to prevent it?" She claimed that a national organization like the one-year-old League of American Sportsmen had only a couple of chapters in the west and no branches in any southern state and therefore had little impact. She argued that state organizations might serve as the best chance to accomplish something. "Arkansas Sportsmen made a step in the right direction at their annual meeting . . . where they formulated a new Fish and Game Law which will be presented at the coming session of the legislature," Stephenson observed. She believed that the most crucial part of that new plan was the establishment of a state game warden to enforce those new laws.[65]

Despite the ASSA efforts in 1898 and 1899, the five-year ban on prairie-chicken hunting had lapsed. Paul R. Litzke reported that in 1899 "so great was the slaughter of these birds that they could be purchased for the same price as the domestic bird in our market." Because of the high temperatures, some of the chickens spoiled before making it to the city. A year later, Lizke remarked that there were a few birds left, but not many, and unless there was a special session of the legislature called to act, "there is but little chance of averting their utter extermination." The ASSA had made all efforts to persuade the Arkansas legislature to do something to no avail, and the whole affair "was treated with indifference, and of course, remained unenacted," he concluded.[66]

CHAPTER FIVE

"Humans Cannot Live without Birds. The World Will Perish"

Arkansas State Game Laws, 1900–1909

> The movement for the conservation of wildlife and the larger movement for the conservation of all our natural resources are essentially democratic in spirit, purpose, and method.
> —Theodore Roosevelt

By 1900, there was enough pressure, mainly through the growing conservation organizations, on the Arkansas state legislature to cause them to introduce new game legislation, including placing more restrictions on nonresident hunters. Local officials had not earnestly enforced the 1875 or 1897 nonresident licensing laws, and many Arkansas citizens were tired of the confusing, weak, and poorly enforced wildlife laws. A tug-of-war ensued between several groups, each fighting for their interests and influencing lawmakers to act in their favor. At the turn of the twentieth century, market hunters, nonresident and resident sportsmen, club members, women's clubs, Arkansas sporting groups, the wealthy, conservationists, and even neighboring states sought to influence Arkansas wildlife legislation. On the table were hunting and fishing licenses, nonresident hunting, hunting and fishing seasons, nonexportation laws, market hunting bans or licensing, and other various aspects of game law.[1]

From 1875 to 1900, several factors caused the paltry few game laws that the Arkansas government passed to remain useless for wildlife protection. They included a lack of enforcement, general indifference toward wildlife legislation, a belief that animals and fish existed for the use of humans without consequence, overhunting, and overfishing. In the first ten years of the new century, state representatives introduced, debated, and passed a few new game laws with similar results. Most of the time, the bills never made it to a vote. And although most of the passed legislation found itself in the dustbin because of court challenges or political gutting, each time one of these wildlife bills made it onto the floor, the discussion influenced the general thinking toward conservation laws across the state. These debates and increasing public pressure laid the groundwork for more robust, longer-lasting, and more significant wildlife legislation.

By 1900, twenty-two states required hunting licenses, and the remaining twenty-three had no license law. Arkansas was one of two states that distinguished its hunting license conditions. Not all citizens were happy with the Arkansas licensing law, particularly those who saw it as government overreach. While challengers argued against government intervention, two Arkansas social societies advocated for the government to do more for wildlife protection.[2]

In April 1902, the Arkansas Federation of Women's Clubs gathered in Fort Smith for their annual meeting, where Mrs. Louise Stephenson provided the keynote address entitled "The Economic Value of Birds." One has only to view a field in winter to observe hundreds of birds pecking and gathering noxious weeds that would taint future crops, she explained. Little birds, working year-round, snatch insect eggs and larvae out of the rotted wood patches and grounds that "if left undisturbed would hatch enough creeping, crawling, and flying things to make waste of every garden, orchard, and farm in the neighborhood." Stephenson provided the Department of Agriculture statistics that determined that insects destroyed $200 million worth of American crops per year. She contended that the small amount that birds consume of some of the harvests was a paltry sum compared to the enormous good they did with ridding the countryside of crop-devouring bugs.[3]

Stephenson informed the group that several organizations in the state had pressed for stronger game laws and the creation of the position of state game warden. She believed that the Federation "should (as powers behind the throne) decree such an office be created [and] that the farmers and fruit growers will be more grateful than the sportsmen." Previously, the Federation had created two thousand "bird pamphlets," which contained information about the importance of the birds to Arkansawyers.[4]

Stephenson further called upon the women to create an Audubon Society in Arkansas that would educate the masses, particularly children, about the importance of birds. Twenty-five states had Audubon Societies, so, she charged, why couldn't Arkansas be number twenty-six? Stephenson asked two critical questions, answering one but leaving the other for contemplation: "What is their value?" "It cannot be estimated," she replied. "Humans cannot live without birds. The world will perish," she continued. Stephenson questioned the club's women about their character: "May I be permitted to ask to which class the women who compose the Arkansas Federation of Women's Clubs belong?" she concluded.[5]

Stephenson's last question is both interesting and telling. She believed saving Arkansas's wildlife was up to the middle and upper classes. In her view, the cultured and civilized protected and conserved animals and fish, while the uneducated, uncivilized lower classes wasted and killed indiscriminately. She believed the state government and organizations like the Arkansas Federation of Women's Clubs must work together to make a difference.[6]

During the same year that Mrs. Stephenson pressed the Federation ladies toward a more active role in protecting the animals, the Arkansas State Sportsmen's Association men decided to make another lobbying effort toward the Arkansas legislature. Since the organization had failed to influence the Arkansas government to enact a broader, more sweeping set of game and fish laws during the previous ten years, they focused on a more piecemeal approach to obtaining stronger game laws.[7]

In 1902, the ASSA decided to push heavily for three items. The first, and they deemed most important, was ending the exportation of game and fish from the state. This plan was a direct action to prevent market hunters from selling and shipping game and fish to out-of-state markets like Memphis, St. Louis, Chicago, and Kansas City. The *American Sportsman* reported that over $500,000 of game had sold in Chicago in one year alone. Most Arkansawyers considered the market hunter the most significant threat to Arkansas wildlife. The second area of their focus centered on nonresident hunters. For several years, nonresident hunters could hunt on their land in Arkansas. A new measure would stop all nonresidents from hunting and fishing in Arkansas legally, whether they were on their own ground or not.[8] Third, the ASSA pushed for a change in the open and closed hunting seasons for deer, turkey, and quail. They wanted animals harvested only during certain months, protecting them while they raised their young. Hunting seasons would allow for the recovery of the population and reduce overhunting to the point that the populations could remain self-sustaining. The Arkansas legislature acted a year after the ASSA agreed to pursue these three areas.[9]

In 1903, Arkansas politicians introduced a series of wildlife protection bills, including legislation that dealt with hunting and fishing licenses, game exportation laws, and seasons. Two licensing bills passed the House, and both infuriated nonresident club members. The first raised the licensing fee to three hundred dollars for nonresidents, and the other banned them from hunting entirely in Arkansas. Governor Davis, himself a sportsman, vetoed the licensing fee increase but accepted the ban. Therefore, laws prohibited all nonresident hunting in Arkansas except in Mississippi County. The influential Big Lake Shooting Club contained many wealthy Tennesseeans. It controlled several thousand acres in the Mississippi County lowlands. The club had enough political clout to obtain an exemption for the county.[10]

On February 6, 1903, the *Arkansas Democrat* accused the legislature of supporting the wealthy nonresidents who purchased land in Arkansas for preserves. "It is like locking the stable door after the horse has been stolen," the authors argued. The *Democrat* concluded that what Arkansas needed most was a law to stop the game shipment out of the state.[11]

While the *Democrat* argued for new legislation, several new laws were already in the works, both in the Arkansas House of Representatives and the Senate. With the word leaking out of the state, many people outside of Arkansas, especially those who owned land in Arkansas for hunting and fishing, began to clamor about the potential regulations. Members of several Memphis clubs were "alarmed at the proposed radical changes in the game laws . . . and the proposed radical legislation was causing general uneasiness." Several of the new proposals included provisions to ban nonresident hunters from Arkansas and end the transportation of game and fish outside of the state. The *Memphis Commercial Appeal* argued that these measures were aimed directly at the hunters of Memphis and that, if passed, would "cause unnecessary hard feelings and endless litigation." The newspaper claimed that many Memphis citizens owned land in Arkansas and that these laws would prevent them from using their property as they intended. "Every sportsman in Memphis is opposed to the bill," the author explained. Several prominent Memphis men were meeting to determine a course of action if the bills passed. "It is not supposed that a condition of affairs will be tolerated in which a man will be deprived of the use of land which he has lawfully purchased," they promised. The Memphis clubs intended to challenge the laws in court.[12]

The Memphians had grounds for worry. Nonresidents owned most of

Arkansas's hunting and fishing clubs, controlling thousands of acres of land they barred locals from using. To many Arkansawyers, stopping nonresidents from hunting and fishing or transporting game outside the state was the right decision.

Adding to the distrust between locals and nonresidents, in St. Francis County near the Missouri state line the year before, a Memphis-owned outdoor club on Mud Lake had stopped local county citizens from using the lake by claiming they owned it. Because it was part of the St. Francis Levee District, locals contacted the levee board and obtained permission to fish the lake. Based on the lake's position in the levee district, Forrest City citizens attempted to use litigation to drive out the Memphians. Because of this conflict, Arkansas state representative Frank W. DeRossitt of St. Francis County became convinced that the state needed more stringent game and fish legislation. He wanted to end nonresident hunting in Arkansas altogether and any exportation of game or fish from the state.[13]

In what was probably an ill-advised move, on February 12, 1903, the president of the Mud Lake Hunting and Fishing Club, Robert Galloway, visited Little Rock to appear before the agriculture committee in the Arkansas House of Representatives (the committee oversaw game and fish matters at the time). Galloway denied that he was there to argue against the game laws then in consideration. He claimed that his club and the citizens of St. Francis County were "in perfect harmony." Moreover, he was only in the capital to learn more about the proposed game laws. Galloway contended that he had the "greatest confidence" in the Arkansas lawmakers and that if any laws were passed, he believed firmly that they would be fair to everyone. Despite Galloway's claims, most observers, like the members of the ASSA, believed this wealthy outsider was attempting to sway the Arkansas legislature.[14]

Two weeks later, the *Arkansas Democrat* printed what many Arkansas outdoorsmen believed—if the Arkansas government did not stop outsiders from coming into the state and "slaughter[ing] our game and fish, it would only be a question of a very short time until there would be neither game nor fish left for the home people." The newspaper demanded a "game law that will be final, that will fix the responsibility for the violation, and make it the well-defined duty of some officer to the enforcement of the law."[15]

Representative William W. Whitley from Polk County introduced the first game protection bill to make it to a vote in the House of Representatives. House Bill No. 39 made it illegal for a "person, corporation or company to purchase or have in possession for barter, exchange or sale, or to sell any buck, doe, fawn . . . wild turkey or prairie chicken or any quail . . . or any other kind of game, wild fowls or birds whatsoever within the state, except bear, rabbits, and squirrels." It also made it illegal for shipping companies of any kind to "export, ship, removed, or carry" game out of the state, setting the penalties for no less than ten dollars and no more than one hundred dollars for each offense. In March 1903, the bill passed the House and moved on to the Senate.[16]

In the same month, Senator Thomas E. Mears of the Fifteenth District introduced the first significant bill to originate in the Senate in 1903. The Mears bill called for all nonresident landowners who wished to hunt or fish in Arkansas to present their paid land tax receipts and take an oath that the game and fish they were harvesting were for their own use. They could bring a friend to hunt on their land without the friend owning land in Arkansas. The bill also made the local sheriff the county game warden. It prohibited the shipment or sale of any game other than bear, rabbit, or squirrel.[17]

It is essential to note the nonresident influence on the debate and the role of the Arkansas State Sportsmen's Association. Most of the legislators recognized the danger of market hunting. In many cases, they were attempting to end nonresident market hunting but keep nonresident sport hunting. Over several legislative sessions, lawmakers tried to walk this tightrope and failed until the 1920s. Without any experience in conservationist legislation, Arkansas lawmakers did not know how to accomplish what they wanted. In addition, some congressmen wished to maintain the traditional way of hunting and fishing and stood against protective legislation altogether.

Senator James W. Wood of the Fourteenth District challenged the Mears bill, stating that it "seems that an attempt was being made to railroad this bill through." He wanted the debate postponed until the Senate conducted more research.[18] Senator Joseph L. Short of the Twenty-Third District agreed with Wood, arguing that the Senate, ten days before, had unanimously decided to end nonresident hunting. Now, why were they changing their minds so quickly? Short claimed that he had private information about the matter and charged that Judge Joseph H. Acklen of Tennessee had brought this bill to Senator Mears. It was, in his view, essentially a nonresident sporting club–written document.[19]

Senator Creed Caldwell from Jefferson County stepped forward and claimed that Short was out of line with his statements, and if someone had told him about this information in private, he should keep it as such. Short replied that he had nothing to retract and stood by his statement. Senator Frank Smith of the Twenty-Second District stood and confessed that he was the senator that Short had mentioned about speaking in private. He did not mind the information being public or private because he, too, stood by what he said. He stated that Acklen had contacted him and urged him to support an amendment to another bill (House Bill No. 35, Whitley bill), which allowed nonresidential hunters the same rights as written in this new Mears bill. The amendment did not carry, and since the Mears bill carried the same conditions for nonresident hunters as the failed amendment, Smith believed it came from the same source: Acklen and his hunting club friends.[20]

Senator George Sengel of Sebastian County said that he did not like the bill and did not want the state to become a big game preserve. A man had told him that he had recently bought twenty thousand acres for a game preserve, and Sengel stated that those twenty thousand acres should be farmland. He would fight against this bill and any other legislation that resembled it.[21]

Senator Oliver N. Killough of the Seventh District declared that he preferred the Whitley bill, which was better than anything before 1903. He agreed with Wood that more research was warranted.[22]

The Whitley bill came to the floor in proper order during the following week. Senator Hal N. Norwood of the Thirty-Third District rose immediately after the bill's introduction and moved to strike everything in the Whitley bill, except the enactment portion, and replace it with the Mears bill. Senator Killough opposed Norwood's motion. Norwood asked Killough if the senator "claimed the geese and ducks belonged to the state when they only [come] here in the winter." Killough answered, "Yes, sir, they do. We have a lien on them for their winter board." The Senate erupted in laughter.[23]

Killough continued after the laughter died down and explained that he had read in that week's *Arkansas Democrat* that C. F. Lane, a beekeeper from the St. Francis and Little River area, and J. W. Short had recently returned home from a hunt in Arkansas County with over 1,400 ducks and sold them at market for three hundred dollars. The *Paragould Soliphone* reported that the two "made a great slaughter on the feathered tribe while in the swamps." Killough had also made a trip to Memphis and had seen the markets filled with Arkansas game. Hunters wanted this bill, he argued. They do not want the pot hunters to slaughter the game for the big city markets. He even had heard stories, he contended, about market hunters stuffing hog carcasses with quail and shipping them secretly to market.[24]

Senator Killough said that he supported the Whitley bill and thought that if it failed, then the legislature might not pass any game laws in that session. "Arkansas need[s] a game law," and they needed to pass this bill, he argued.

Senator Smith, too, supported the bill's passage and stated that he favored especially the section of the legislation concerning the banning of selling or shipping wildlife out of the state. Smith said he found nothing wrong with nonresident hunters killing game on their land for their own use, and he wanted to protect Arkansas game from all pot hunters whether they were residents or not. He believed the Whitley bill was the best choice to achieve this aim.

Senator Caldwell stated that he did not like the law and would not vote for it. "You can never walk into a restaurant in the state and get even a quail on toast" if this bill passes, Caldwell blasted. "If you found a man in Arkansas with a bag of game he was liable to arrest, no matter if the game came from Kalamazoo," he said sarcastically. The senators laughed again at this statement. "This bill is nothing but a freak," he continued, "There was no more comparison between this bill and others . . . than there was between the Declaration of Independence and the Republican platform of Arkansas." He would support a "good game bill," but this was not it. "This bill is not worth the paper it is written on," he concluded.[25]

Senator Calvin T. Cotham of the Seventeenth District, like Frank Smith, stated that his constituents wanted game laws to end the slaughter of Arkansas game and that he was in support of the Whitley bill. Senator William P. Fletcher of the Twelfth District followed Cotham with a declaration of support. Fletcher was not a hunter, he said, but the citizens of his district wanted a game bill too. Fletcher claimed he knew a man who opened a restaurant with money earned from killing for the markets. He challenged Caldwell, saying he "thought the bill would not prohibit the hotels and restaurants from shipping game into the state, [but] it would prevent the wholesale slaughter in the state for market." The Whitley bill passed twenty votes to eleven.[26]

Within a few days after the bill's passage, Senator Caldwell introduced legislation that ended the transportation of game and fish out of state or between counties. Caldwell argued that this bill was better than Whitley's because it rewarded informants for reporting violators to authorities. The arresting officer and the person who reported the violation would split the fine. Senator Wood contended that it was a repeat of the Whitley bill, and he did not like it. Senator Smith favored this bill, except it exempted private ponds, which meant the private hunting clubs could ship their fish and game out of the state, and he opposed it.[27]

Caldwell replied that the legislature could not outlaw catching fish on private property. He also confessed that he was not an outdoorsman, so he had no direct interest in the law. Caldwell admitted that representatives of the

Arkansas State Sportsmen's Association had provided him with the bill. A vote was called, and the proposal failed five votes to eighteen.[28]

Finally, after over a decade of lobbying, the ASSA found a champion in St. Francis County representative Frank W. DeRossitt. In early March 1903, farmer-politician DeRossitt introduced House Bill No. 367, referred to as the DeRossitt game bill. It contained a stricter nonexportation law (and more considerable fines for the shipping companies) and the nonresident prohibition law that the ASSA thought most critical; it also listed new seasons for some animal species and stipulated that hunters could not shoot ducks and geese at night. To clarify who would enforce these updated game laws, the bill officially announced that county sheriffs and constables were also game wardens. The legislation set penalties at no less than fifty dollars and no more than one hundred dollars for each offense for individuals. Courts could fine railroad companies or express agents up to five hundred dollars for each act of shipping game out of the state. Any levied fines went to the arresting officers, and the courts could charge costs. This seemed precisely the sort of law that the *Arkansas Democrat* had clamored for in February. The bill passed sixty-eight to fourteen.[29]

Governor Davis signed the bill into law a few days later. There was probably little discussion in the Senate because the legislation was sent to the Arkansas Committee on Agriculture in February and had emerged two weeks later with only a slight change. One interesting item that made no news at the time was something that happened only minutes before the passage of the DeRossitt bill. Senator John J. Mardis of Harrisburg amended the bill, making Mississippi County exempt.[30]

Wealthy Tennessean Joseph H. Acklen, the same man accused earlier of influencing Arkansas game legislation, was most likely behind Mardis's action. Acklen's Big Lake Shooting Club was headquartered in Mississippi County. Or Mississippi County market hunters may have pressured Mardis. Even though the two groups despised each other, they both wanted the ability to export their game. Unfortunately, outlawing the shipment of game and fish alone would not solve the issue of game depletion. Lawmakers needed to create a daily limit for game and fish to control the exploiting hunters, which did not happen for another twenty years.

This continuous stream of game laws represented an effort for lawmakers, market hunters, sportsmen, and club members to legislate the interaction between humans and the environment. Some early conservationists, including Teddy Roosevelt and Aldo Leopold, believed abusing natural resources was undemocratic and immoral. The spirituality between man and his environment, a connection, was foreign to the market hunter. However, many newly created sportsmen related their experiences in the woods and on the water on another level. Animals were not just resources for man to exploit, like coal or oil; they were there for reverence, admiration, and connection. They were God's gifts of wonderment and mystification. Wildlife needed protection for future generations to have this same connection with the wilderness. Market hunters represented the opposite of this way of thinking. They made hunting and fishing dirty, impersonal, and a sin against nature. Grover Cleveland called the true sportsman "serene." Aldo Leopold claimed, "Perhaps no one but a hunter can understand how intense an affection a boy can feel for a piece of marsh." Sportsmen, argued that they, more than farmers, were closer to the land. Sportsmen had a deep connection with their environment that only they could comprehend. Many of the sportsmen of the ASSA believed that the wildlife needed protection, and they might have considered the battle over wildlife protection a spiritual war. While the debate was an economic one for the market hunter, in many instances, for sportsmen, it was divine.[31]

On the morning that the DeRossitt bill passed, Little Rock attorney and sportsman J. M Rose had mailed a letter to the editor of *Forest and Stream* magazine that included copies of the Whitley game laws. He had explained that the new regulations fit perfectly into the magazine's mission against market hunting. Rose believed that the nonresident hunters in Arkansas were a recurring issue, and pot hunters were killing all the game. Non-Arkansawyers had "come in the hundreds, stay[ed] all winter, and in many instances have bought up the large lakes and kept the natives out." This, of course, angered the locals.[32]

Forest and Stream had earlier taken "the sale of game should be prohibited in all seasons" as their "platform." For years, the periodical published news in a section they called "Nails Driven," about states who had passed laws "driving nails" into the magazine's plank on the platform of ending game sales. On May 2, the journal commended Arkansas for its nail with the reprinting of the section of the Whitley bill that banned the sales and shipping of wildlife. Together with Massachusetts and New York ending the sale of woodcock and grouse, *Forest and Stream* argued that the prohibition of game sales in these three geographically divided states meant their platform was catching on nationally.[33]

The Whitley laws received widespread accolades. The DeRossitt laws, however, drew mixed results. The Arkansas State Sportsmen's Association praised the Arkansas legislature for the passage of both sets of game regulations. When the ASSA formed in 1891, members had pledged themselves to lobby the legislature to improve the game laws and arrange for their enforcement, and now, they had accomplished the first step. Hunters from other states, many of whom owned land in Arkansas, were not so celebratory. In a few cases, defiance came even from within the Bear State.[34]

"The Whitley game law which is aimed solely at the pot-hunter is a good law and will greatly benefit Arkansas," one Arkansas newspaper article read, "but the DeRossitt bill . . . is selfish and hide-bound." In the author's opinion, we should not stop "our neighbors and friends in other states" from coming to Arkansas to hunt and fish. These laws will "breed ill feelings between the adjoining states, where friendship should exist," it concluded.[35]

In May, Representative Harry E. Cook from Lake Village in Chicot County explained in an interview that he thought the new game laws were good for the entire state but that many of his constituents had asked him why Chicot County was not exempt. He answered that he believed that "90 percent of the nonresidents who hunt and fish in this state spend nothing here. They come prepared with food and beverages, kill our choicest game and fish and are gone until another periodic vacation." Cook argued they were most often wealthy and looked to Arkansas as their private hunting grounds. He knew that Lake Chicot fish and game appeared on restaurant menus in places like Greenville, Mississippi, and Memphis. Cook represented many Arkansawyers, although he was not wholly correct. He did not ultimately see the difference between a nonresident club member and a nonresident market hunter. They were outsiders, all of them, and they stole Arkansas game from Arkansas people. It was the market hunters, however, who put the Lake Chicot game (and game and fish from other parts of the state) on the Greenville and Memphis restaurant tables, not the nonresident sportsman or club member.[36]

Lake Chicot was famous, both in Arkansas and in surrounding states, for some of the best sportfishing in the South. The Lake Village mayor Allen Beadel agreed with Representative Cook that the new game laws were beneficial. Lake Village accepted all fishermen and hunters who came for sport, but not if they were market hunters. Beadel rebuked the law barring nonresidents, announcing, "We will continue to welcome the true sportsmen, and public sentiment will prevent the prosecution of any of our visitors. If anyone is prosecuted by our over-zealous officials, the citizens of the town will take pleasure in paying any fines assessed." He pledged city tax dollars for penalties.[37]

A *Forest and Stream* article with the headline "Inhospitable Arkansas" appeared on May 9, 1903. It noted that Arkansawyers and nonresidents had sparred for many years. Many hunting clubs from Tennessee, Illinois, and Missouri owned thousands of acres of Arkansas land, and in most cases, barred locals from the property. Some club members believed the DeRossitt game laws were retribution for the treatment. The article pointed out that many renowned men, like J. M. Rose of Little Rock, had attempted to have all the hunting clubs exempted from the nonresident law. Unfortunately, the magazine claimed, "just at the critical moment a certain nonresident appeared on the scene and unwisely tried to bully the Legislature, and thus undid the work of the mediators." Around the same time that this outsider appeared in the capital, the Mud Lake Hunting Club members were driving off the locals near Forrest City, striking a two-blow combination to almost guarantee passage of the DeRossitt bill.[38]

George L. Babcock of Warren, Arkansas, wrote to *Forest and Stream* about joining the Saline Fishing and Hunting Club, a club of fifty Arkansawyers. He reported that game laws did exist in Arkansas but were not enforced, and not "one person in ten" knew the hunting seasons. The Warren resident declared that his club promised to prosecute all lawbreakers and that they had already posted reward signs for information leading to the conviction of violators.[39]

The headline of an article in the *St. Louis Sportsman* the same month decried "Disgraceful Arkansas Legislation." The piece read, "The mossback legislators of Missouri's southern neighbor have seen fit today that non-residents of their state shall be prohibited from hunting and fishing in the future, and already one St. Louisian has been fined for daring to cast a line into one of the lakes in the swamp region." The author argued that the recent prohibition on nonresident hunters hurt Missouri sportsmen, especially since men from the Show Me State had purchased thousands of acres of land for a place to hunt and fish and, for generations, had come into Arkansas. These men built clubhouses and cabins and spent money in Arkansas, and now, this law "worthy of the dark ages" kept them from hunting and fishing on their property. The Missouri sportsmen, the magazine argued, stood with Arkansawyers for game protection laws. This exclusion, however, was not the solution: "In plutocratic Europe any gentleman had the right to shoot and hunt anywhere provided he secure the permission of the owners of the property. Even this provision is denied in democratic Arkansas!" The *Sportsman* concluded that any court would not hold this law constitutional, and it was simply a matter of time before they threw it out. If readers followed his argument, outdoorsmen from Missouri could come into Arkansas, take whatever game and fish they wanted using whatever means they wished, and if any Arkansawyer attempted to stop this activity, the Missourians damned them.[40]

The Arkadelphia *Southern Standard* reported that the nonresident clause in the new law had "aroused bitter antagonism among certain parties who live outside the state but who make trips within the state for the purpose of hunting and fishing." All American men have the right to hunt and fish on their own land, and any law that says otherwise is unconstitutional, the out-of-staters argued. This was not a new argument. The Bible, in Genesis 1:26, states, "Then God said, 'Let us make human beings in our image and likeness. And let them rule over the fish in the sea and the birds in the sky, over the tame animals, over all the earth, and all the small crawling animals on the earth.'" Animals existed for man's use, and any man-made law that said otherwise was either unconstitutional (secular) or against the word of God (religious). It was simply a matter of time before someone challenged it in court.[41]

In the June 27 edition of *Forest and Stream*, one reader argued that most states deal with issues within their boundary. But Arkansas had just passed a law that all Americans must follow. This law "is not limited to the state of Arkansas; it is comprehensive and all-embracing [and] in the name of several millions, we protest, flout the authorities of Arkansas to their face, and declare that we shall still hunt and fish," he avowed. The only laws he would follow were from his home state. He intended to continue hunting and fishing in Arkansas, whether laws existed against it or not.[42]

Within weeks, several Memphis hunting clubs threatened to challenge the law's legality by sending one of their members into Arkansas, inviting arrest, and then contesting it in court. Rumors swirled that the Memphis men had retained United States senator James P. Clarke from Helena, who practiced in Little Rock and the Little Rock law firm of Rose, Hemingway, and Rose as representation. The Arkansas State Sportsmen's Association members announced it would provide attorneys to battle against the Memphis hunting clubs. Association members believed the DeRossitt laws were the "salvation for the game of the state." Even if challengers won, the law would at least stand for a while, they argued. Any reprieve from the overkilling was beneficial.[43]

The Memphians decided that Captain William B. Mallory, Confederate veteran and owner of W. B. Mallory and Sons Wholesale Grocery, was their man to challenge the Arkansas nonresident game law. Mallory was a member of the Wapanocca Outing Club, which owned several thousand acres, including the six-hundred-acre Wapanocca Lake, in Crittenden County across the river from Memphis. Before his trip, Mallory had contacted the sheriff's office and provided information about where and when he would be fishing and hunting in Arkansas and invited the sheriff to meet with him. Mallory crossed the Mississippi River from Memphis and entered Crittenden County, Arkansas, on club grounds where Sheriff Williamson arrested him. Williamson took Mallory before the justice of the peace, who fined him fifty dollars for each of the two charges, hunting and fishing, and then released him on bond. Mallory appealed his conviction to the circuit court. The previous rumors proved true, and the Memphis club announced that Senator Clarke and the Rose Law Firm would indeed represent Mallory. The ASSA announced that their attorneys would assist the prosecution.[44]

The case went to trial in Marion in the summer of 1903, with circuit court judge Allen Hughes presiding. Both sides agreed that Mallory had fished and hunted on his property. Mallory's attorneys argued that Section 4 of the DeRossitt game laws prohibiting nonresident owners from hunting and fishing on their own land was invalid, calling it a "violation of both the state and the United States constitutions." Prosecutors contended that the state owned all the wild game and fish within its boundaries and, therefore, could stop all nonresidents from taking them because they were not citizens of the state and were not entitled to the state-owned property, even if they owned the land. Judge Hughes agreed with Mallory, saying that the United States Constitution guaranteed a citizen's right to own property in Arkansas. Hunting and fishing on their own property was an extension of that clause because the landholder owned the birds, game, and fish contained on that land. Therefore, Hughes ruled the DeRossitt game laws unconstitutional. Arkansas could, Allen pointed out, outlaw nonresidents from hunting on land that they did not own.[45]

State prosecutors appealed the case to the Arkansas Supreme Court. The ASSA hired attorneys Henry M. Armistead and J. Fairfax "Fax" Loughborough to assist. Not only would the pair be involved in the higher court fight, but they would also help prosecute any game law violators on the circuit-court level. "The sportsmen's association . . . is settled in its determination that the law shall be enforced," said association representatives. When the case came before the Arkansas Supreme Court in November, the justices declared the DeRossitt game laws unconstitutional with a split decision three to two, sustaining Hughes's ruling.[46]

Associate justice Edgar A. McCullough provided the majority opinion, concluding that the central question in the case was who owned the game. In ancient Rome, he said, animals were common property, but the owner of the land where the animals lived could prohibit others from killing them. After the Magna Carta in England, wild animals were held for the public good, and the authorities only oversaw their protection. English common law allowed landowners to take game on their land. But, in his opinion, no definitive modern answer existed about game ownership. McCullough concluded that the DeRossitt statute "prevents the same enjoyment by the appellee [Mallory] of the property rights afforded the more fortunate resident land owner, it is a denial of 'equal protection of the law' within the meaning of the constitutional guaranty and cannot be enforced, and the taking away of this right because of his non-residence is 'without due process of law.'"[47]

When the Arkansas Supreme Court upheld Judge Hughes's decision, the only line needed in the DeRossitt game laws to make them constitutional was in Section 4, which read: "It shall be unlawful for any person who is a non-resident of the state of Arkansas, to shoot, hunt, fish, or trap at any season of the year, *except on his own premises*." They were more concerned with the shipping of game out of the state, and for now, the prohibition stood on that matter. The *Osceola Times* in Mississippi County reported the news from the Supreme Court, declaring, "We are of the opinion that a majority of the people will endorse the idea that any man acquiring land in the State should be permitted to hunt and fish without molestation." When the dust settled at the end of 1904, nonresidential hunters and anglers could come into Mississippi County whether they owned land or not, and they could hunt or fish anywhere in the state if they did own land.[48]

However, it remained illegal for nonresidents to hunt or fish in areas they did not own. The *Memphis Scimitar* reported that a local sheriff bagged over five hundred ducks without firing a shot in Forrest City. Almost every family in the town ate duck that night because the St. Francis County sheriff J. D. McKnight dumped his confiscated cargo near the town square, offering a share to anyone who wanted it. Three Chicago sportsmen had killed a wagon full of ducks on land they did not own. The sheriff discovered them at the depot, along with their impressive kill, waiting for the train out of town. McKnight arrested the Chicagoans but allowed them to leave on a Chicago-bound train when they agreed to relinquish the ducks in exchange for nonprosecution.[49]

Just before quail season in November 1904, the *Daily Arkansas Gazette* announced that although the bird forecast was promising, the only way folks could get them was to get out and harvest them for themselves because the Whitley laws did not allow the sale of quail. Even the hotels and the restaurants could not serve Arkansas quail legally. If most people knew the actual consequences of the Whitley bill, it would have never passed, but "[it] was passed at the instigation of the State Sportsmen's Association, and that body is so strong that it is believed it would be able to defeat any attempt to repeal or amend it," the *Gazette* explained. With the passage of both the Whitley bill and the DeRossitt bill, the paper declared most people believed that Arkansas had some of the strictest game laws in the United States. The newspaper complained that citizens could not get a quail dinner without shooting it themselves because of game laws and then reported that wild turkeys were essentially extinct in the state (because of the lack of game laws earlier) in the same article and did not recognize any correlation or irony.[50]

In the southwest part of the state, T. L. Pennington of Washington, Arkansas, complained that the rabbits "have played havoc on the young fruit trees." Arkansas contained nearly seven million acres of improved farmland by 1900. The few larger game animals like deer, bear, and turkey (even predators) had difficulty finding adequate habitats to survive. However, row crop farming and the overgrown field edges provided excellent cover for smaller game like rabbits and quail. Therefore, many sportsmen shifted from hunting the few larger animals to pursuing the more populous small game. Migrating squirrels and resident rabbits were abundant, and they had damaged many crops. Pennington claimed that hunters had bagged hundreds so quickly that shooting them was no longer fun. He complained there were not many bears, but they were "long" on squirrels and rabbits. "We are thinking of asking the legislature to amend the Whitley law so as to permit the shipping of rabbits," he concluded. However, even small game began to disappear within a few years.[51]

Perhaps a 1903 predator law might account for the increased numbers of small animals like rabbits, mice, and squirrels. An April 1903 law provided a bounty for killing hawks and foxes. Farmers and, by extension, lawmakers considered these predators a nuisance species. They are, however, necessary for maintaining rodent populations. With this action, policymakers decided between those animals that deserved protection and those that did not. "Game" fish and animals deserved protection. Predators and "trash" fish were unimportant and damaging to domestic and game animals. Raiders that might find their way into a chicken coop or pig pen did not warrant protection. On the contrary, they deserved extermination. In doing so, they opened the door for the increased killing of species essential to maintaining the balance of an ecosystem.[52]

9. Judsonia wolfhunters. The hunt took place north of 14 Mile Creek toward the Hickory Flat community near the Little Red River. J. C. Gibson, photographer. (*Left to right*) George Donnell, Louie Hilger, Noah Hilger, Garland Robbins, Gerbert Robbins, Asa Robbins, Wiley Caruthers, Will Hern, and Mr. Rutherford. Two adult wolves with five or six pups. Courtesy of the Old Independence County Museum.

In December 1904, Arkansas State Sportsmen's Association members in Pine Bluff contacted prosecuting attorney William D. Jones's office. They reported that restauranter W. C. Culley, owner of Culley's Cafe, was selling mallard, prairie chicken, and quail dinners and had a display of birds in his front window. His birds, he claimed, came from St. Louis, and therefore, the federal interstate commerce law protected him. The prosecuting attorney countered that the interstate commerce law did not nullify the section in the Whitley law outlawing the sale or exhibition of game. Arkansas newspapers reported this as the first test case against the Whitley law.[53]

The prosecution argued to Pine Bluff justice of the peace A. J. Stewart that the birds' origin was of no consequence. Culley sold the birds in Arkansas, and that was against Arkansas law. Culley decided not to appeal the case, ensuring that this litigation would end in a much less dramatic fashion than the challenge to the DeRossitt law earlier. However, other attempts to change Arkansas game laws appeared on the heels of the Culley and Mallory cases.[54]

During the 1905 session, the Arkansas legislature introduced more than ten bills that attempted to alter or remove Arkansas game laws. The most significant change in the Arkansas game laws in 1905 was repealing Mississippi County's nonresident exemption from the DeRossitt game laws. However, Arkansas conservationists had little time for celebration because, within a few years, Mississippi County added an exemption to the Whitley shipping laws.[55]

One of the most progressive measures attempted in the 1905 senate session was Senator Rison's bill, which barred the use of semiautomatic shotguns for hunting. Many national sporting magazines had filled with angry letters damning the use of "automatic" shotguns. Nationally, the League of American Sportsmen had taken up the cause to outlaw these types of shotguns. They believed that semiauto shotguns allowed market hunters a more efficient way to kill. Editor George Shields called automatic shotguns "a disgrace to the nation" and mailed form bills nationally to League of American Sportsman members to push in their states. Theodore S. Palmer of the United States Biological Survey and many other national conservation advocates supported these laws. "A true sportsman cannot use such a gun," argued one Arkansawyer. The Rison bill passed the Arkansas Senate with a vote of fourteen to twelve, with nine senators not voting. Unfortunately for its advocates, the legislation died in the House.[56]

That same year, Theodore S. Palmer, assistant in charge of game protection for the United States Biological Survey, wrote a five-year retrospective about state and federal game protection. In this publication, Arkansas and her sister states in the south faired rather poorly. Only four states in the nation did not have a game warden or commissioner in 1905: Texas, Arkansas, Mississippi, and Alabama. Arkansas, Mississippi, Alabama, North and South Carolina, Kentucky, and the Indian Territory had no bag limits on game or fish. Both Arkansas and Missouri had recently halted the spring shooting of ducks, a particularly devastating practice on the waterfowl population, but it still went on. The rest of the southern states had not ended the practice. Both Missouri and Louisiana forced nonresidents to purchase a license to hunt and fish, and Arkansas had at least partially banned nonresident hunting.[57]

Chicago and St. Louis, Palmer claimed, were the largest game markets in the west and were filled with wildlife from the southern states. As game disappeared in those southern states, some market hunters had moved farther west and plied their trade in places like Nebraska and Montana. When Missouri enacted a law barring the sale of most of the game there, many avenues available to Arkansas market hunters dried up. They were forced to send them to other locations or go underground completely with their sales. Palmer thought that federal law passage had assisted in pushing some states into passing their game laws or strengthening and enforcing those already on the books. For Arkansas, howevever, enforcement continued to plague the proper protection of wildlife.[58]

The 1907 legislative session brought some additional challenges and changes to Arkansas game and fish laws, and not for the better. All were tremendously confusing. Some counties opted out of one or more laws. Often, the very rivers that market hunters abused the most received the least protection. County officials passed laws contradictory to state law. Two different lakes in the same county had different rules in some instances. No printed law booklets existed outside the state government minutes for many of these regulations. Sometimes, the Arkansas legislature passed laws that conflicted with rules already in place. It was a mess.[59]

For example, two laws passed banning the shipping of squirrels out of White, Dallas, Lafayette, and Craighead Counties, regulations that conflicted with the Whitley game exportation law that allowed for squirrel shipments out of the state. The representatives approved legislation that allowed fishing with hook and line or hoop net only in Chicot, Ouachita, Washington, Madison, and Woodruff Counties, hook and line only at Old Town and Long Lake in Phillips County, and the banning of seines, nets, or traps for fish in public waters in Lee County and any seine larger than three inches square in Miller County. The only exception to most of these laws against using nets or seines was when the users were catching fish for picnics, family use, or stocking purposes; then the law allowed it. However, no matter the reason, fishermen remained barred from using poisons or dynamite.[60]

In a devastating blow to Arkansas conservationists, Mississippi County representative Anthony George Little introduced a petition from the inhabitants (market hunters) of his district, asking for exemptions from the Whitley laws for three of the county's townships to allow shipping game out of the state. The residents argued that the Big Lake Shooting Club members, who owned a large clubhouse on Big Lake and controlled thousands of acres of Big Lake, frequently harvested game and fish on their land and shipped it out of the state. Why should residents not have the option?[61]

Bill No. 397 added the exemption to the Whitley 1905 illegal shipping law of "Big Lake, Neale, Hector, and Bowen Townships in the Chickasawba District of Mississippi County," one of the state's more notorious market hunting regions. This passage was significant because it essentially made shipping game out of the entire state legal, nullifying that section of the Whitley bill. If Mississippi County could ship wildlife out of the state, anyone from another part of the state could first send their game and fish to Mississippi County, and then the parcel could go outside of the state legally. The news horrified members of the ASSA.[62]

On September 14, 1908, Governor Donaghey provided his inaugural address to the Senate specifying the need for a "more comprehensive law concerning its game and fish." He hoped that the legislature would pass a bill that could protect the wildlife and not provide any "individual or corporations" any sort of exclusivity for taking game and fish in the state. However, the laws should protect the landowner's exclusive rights to the game and fish on their lands.[63]

Among the continued stream of game and fish bills in 1909, Senator DeRossitt, bolstered with the success of his first significant game bill six years earlier, tried again. Senate Bill No. 369 was "an act to protect the fish, game, and birds, regulating the hunting and taking of same, and to providing for the carrying out for the provisions of this bill." This time, the bill included provisions for daily bag limits, a major ASSA initiative, the first time a daily limit appeared in an Arkansas game or fish bill. Other states had daily bag limits, becoming more accepted as part of wildlife protection. This legislation also created a new nonresident licensing system, charging them fifteen dollars per year. Even a "non-naturalized foreigner" could get a license for twenty-five dollars. It did not require residents to obtain a permit. DeRossitt stipulated that the governor appoint a state game warden with a salary of $1,800 plus travel expenses. Therefore, this bill appeared as the complete package. It created a licensing system, a game warden system, and a daily bag limit system. Like several other fish and game bills, the Senate referred it to committee, this time the Committee on Agriculture, and it never emerged. Although this bill did not pass, it contained language that was the most aggressive in Arkansas wildlife protection up to that point.[64]

All of the 1909 general game laws failed to pass. Instead, the Arkansas legislature continued to assail the DeRossitt and Whitley laws, passing local county laws, often under the auspices of game protection, to exempt their counties from the two major game laws. First and foremost was Mississippi County's exemption from Whitley shipping laws. Legislators created a county licensing program for residents and nonresidents in Phillips and Chicot Counties and nonresidents only in Stone and Clay Counties, thus undercutting the clause in the DeRossitt law barring nonresidents from hunting and fishing unless they owned land in Arkansas. For nonresidents in Clay County, a provision in the law allowed them to take their harvest out of the state for their use, another exemption of the Whitley law.[65]

On the other hand, Arkansas lawmakers passed a few bills that proved positive for game and fish protection, but only on the local level. They outlawed the sale of fish from any waters in Hot Springs, Union, Calhoun, Ouachita, Chicot, and Bradley Counties (in-state sales only because the Whitley law barred out-of-state sales), limited fishing to a hook and line in Phillips, Washington, and Madison Counties (a strike against market hunters), affirmed fishing rights for all citizens in Faulkner County, restricted hunting in Calhoun County to residents only, and created a county game warden system in Chicot County. Despite the limited success on the local level, these laws caused confusion.[66]

State organizations had the most effect on game legislation change. Through its lobbying efforts, the ASSA was primarily responsible for creating the foundation of many modern Arkansas game laws. Their tireless efforts spurred Arkansas lawmakers to pass meaningful legislation to protect Arkansas wildlife. Sometimes, the ASSA wrote legislation and handed it to friendly senators and representatives to introduce in their respective houses. Although the ASSA had labored since 1891, lawmakers were not fast in passing protective measures. The prairie chicken and the passenger pigeon were gone. Quail were almost wiped out. Bears, once numbering in the thousands, were essentially gone. Turkeys were nearly extinct. Arkansas eventually had to restock much of the wildlife during the modern era. It took another few years to see the fruition of the ASSA's work.

By the beginning of the twentieth century, organizations like the Arkansas State Sportsmen's Association and the Arkansas Federation of Women's Clubs had pushed for more vital game legislation to protect Arkansas wildlife against difficult odds. Politicians, club members, nonresidents, market hunters, sportsmen, conservationists, preservationists, and farmers attempted to influence these regulations for their benefit and interests. During these earliest years of the conservationist movement, organizations like the ASSA and the Women's League battled against difficult odds. Individuals and businesses that profited from a state with few game and fish restrictions combined with Arkansas citizens who either did not trust the government or believed that the wildlife belonged to the people and that state officials had no right to regulate them put up a strong enough defense to stem this growing push toward conservation. Both sides courted legislators, some more successfully than others.

Arkansas passed the most convoluted and confusing game legislation in state history in the first few years of the twentieth century. Although the Arkansas state government had spent hours debating hunting and fishing licenses, nonresident hunting, hunting and fishing seasons, nonexportation laws, market hunting bans or licensing, and other various aspects of game law, no enforceable general game law existed after the gutting of multiple regulations. No state enforcement arm existed. The only protection came at the local level in just a few counties.

Although no enforceable state-level regulations existed, the discussion influenced the general thinking toward conservation laws across the state. Enough of the state representatives realized that county exemptions to game regulation damaged the legislative process and served as the main reason for so much confusion surrounding wildlife rules. These debates, along with increasing public pressure, laid the groundwork for more robust, longer-lasting, and more significant wildlife legislation.

CHAPTER SIX

Earnest Vivian Visart

Arkansas's First State Game Warden, 1906–1910

> I will guarantee that if the good farmers of the country will assist in securing the passage of a bill that will protect the game and fish of the state, as they have assisted in the enforcement of our present laws for the past six months that within five years our fields will swarm with pheasants and other game birds that will aid them in the destruction of insect life, and our streams will be filled with the choicest fish, and their children may enjoy a few hours sport with hook and line as like their forefathers did a few years ago.
>
> —Earnest V. Visart, 1910

In 1909, Arkansas did not have a game warden system. Arkansas state legislators came close to implementing an enforcement system a few times over the previous three years but failed. In September 1909, the Arkansas State Sportsmen's Association, frustrated with the state government's inaction, hired their secretary, Earnest Vivian Visart, as the first Arkansas state game warden. They had no legal authority to employ Visart as law enforcement. The association tasked Visart with stopping game violators, a difficult, if not impossible, task, as he was not a commissioned law officer of the state. The state game warden could neither arrest perpetrators, write tickets, nor issue a summons. Visart could appoint deputy game wardens (already deputy sheriffs) to arrest lawbreakers, but that ability did not enhance their authority.

Despite these limitations, Visart's efforts bore fruit. During the first year, he reduced the number of Arkansas game violations, produced programs with long-lasting effects on wildlife protection, and provided a foundation for developing a professional game law enforcement and management system. Visart formed several county sportsmen's groups, started a public education program, and lobbied the state legislature for sensible game and fish legislation. He vigorously pursued evidence against poachers and gathered information about express rail companies that illegally shipped game. Visart stocked many Arkansas waterways with fish and farmlands with birds. Indeed, his management programs served as the foundation for modern-day Arkansas's game and fish wildlife management plans.

In January 1906, the Arkansas State Sportsmen's Association created a subgroup to directly assist officials in enforcing the game laws and arresting and prosecuting violators. After some success, they wanted an "auxiliary" group within the ASSA to assist in prosecuting and enforcing the new regulations. These men were not lobbyists but took to the field, watched, recorded, and reported violations.[1]

Weekly reports started to flood into the ASSA offices. Violators continued to poach all over Arkansas, particularly for shooting game out of season, as the hunting season dates changed nearly every legislative session. Previously, local law enforcement viewed game and fish violations as a part of the more extensive criminal code. The Whitley bill contained no language about authorized enforcement. The DeRossitt bill had clarified that all Arkansas sheriffs served as game wardens. However, the ASSA believed that turning local law enforcement into game wardens was insufficient. By July 1906, the ASSA initiated a campaign to create the office of state game commissioner.[2]

Arkansas sputter-started its development of a sustainable, adequate game law enforcement system. Local officials either did not possess the manpower and resources to police the forests and streams or demonstrated a lackadaisical attitude toward game law enforcement. While some citizens pressured Little Rock to pass enforcement legislation, the Arkansas State Sportsmen Association took it upon themselves to prosecute violators with their limited resources, hiring and paying a state game warden and prosecuting attorneys. The ASSA had wanted a state game warden since 1891. Fifteen years after first lobbying for an official state game warden position, ASSA members were fed up with the government's lethargy.

Because of the continuing reports coming into the ASSA during the fall months of 1906, the association decided to offer a ten-dollar reward to anyone providing information leading to the conviction of game-law violators, specifically the shooting of quail out of season. The *Daily Arkansas Gazette* reported that some hunters did not fear any consequences for breaking the law. Others thought the regulations did not go into effect until the following year. As in other states, Arkansas newspapers sometimes printed the names of poachers along with their court verdicts to expose or shame violators.[3]

Finally, in 1909, it appeared that the ASSA's continuing efforts might finally bear fruit. During that year's Arkansas legislative session, Representative James T. M. Holt of Hempstead County introduced House Bill No. 73, calling for a game commissioner (governor-appointed), a state game warden, and deputy game wardens to enforce the game laws. The bill specified one deputy game warden per county with full power to act as a law officer when enforcing game and fish statutes. The bill set the state game warden's salary at $1,500 annually, paid from a so-called game warden fund. District wardens received three dollars per day while on duty and half of any fines collected for violation convictions. Holt's bill also created a licensing system for all participants except those hunting on their own land. Those not hunting on their ground needed written permission from the landowner before doing so. License fees were collected and added to the warden fund.[4]

Soon after Holt introduced his legislation, letters poured into the capital from interested parties asking for the appointment as state game warden. Official responses to the correspondence reported the man who had the Arkansas State Sportsmen's Association's endorsement would most likely win the position, reinforcing the ASSA's influence over Arkansas game legislation.[5]

A few days after Holt introduced his legislation, ASSA secretary Visart penned an impassioned letter to the *Daily Arkansas Gazette* encouraging the bill's passage. He claimed that the reason for the continued violations was because no enforcement arm of the game and fish laws existed. Visart argued that other states recently reported that game wardens successfully slowed or halted game violations, and Arkansas should use a sensible hunting license system to pay for wildlife protection. According to Visart, outdoorsmen would welcome such a system.[6]

Visart also appealed to nonhunters, mainly farmers, stating, "The strongest tendency of game laws toward conserving the public is the effect they have of taking guns out of the hands of the shiftless and roving class, that loiter most of the while, and under the pretenses of hunting, congregate for the purpose of committing a crime." If left unchecked, these men would commit other crimes, like arson, vandalism, and livestock and crop theft. Visart believed that the written-permission requirement of Holt's bill was one of the best parts of the act. This could help curtail some of the additional crimes, Visart contended, and the farmer should have the right to "choose his guests." The farmer was a game warden, he argued, because the farmer protected the animals on his land, and with Holt's bill, those men had the backing of the law to keep others from killing the wildlife on his property. He called upon agriculturalists who "constitute[d] a major portion of the reserve patriotism of the State" to report game violations. It was their civic duty. Farmers and sportsmen should work together, he declared, to "put a stop to the unlawful hunting and fishing."[7] Authorities learned that lawbreakers involved in moonshining, theft, or other illegal activities often broke game and fishing laws. In Warren, during a raid, law enforcement officials burned a poker table, nearly twenty gallons of homemade whiskey, and six illegal fishing nets. When Arkansas outlawed game shipments from the state, runners used the same avenues to transport illegal fish and waterfowl that they did to smuggle white lightning.[8]

Visart also claimed, "Many hundreds of undesirable citizens have been making their livelihood by capturing the fish from the streams, the game birds from our prairies and fields, and shipping them to the largest commercial center, to the extent that the once abundant supply is now almost extinct." He argued that safeguarding the birds of the air and fish in the water was the duty of all good citizens. Since the state is supposed to protect the people's welfare, it is the government official's job to pass laws to safeguard the game and fish because wildlife is for the benefit of the state's people. Visart then suggested his overall plan to the public, and he hoped the Arkansas legislature:

1) Create an office of state game warden and provide for its maintenance.

2) Provide for a state hatchery and its maintenance.

3) All of this could be had without cost to the state by [instituting] a non-resident hunter's license fee of $15 and a resident license fee of $1 per year, at a low estimate would probably be not less than $40,000 per year, and fines not less than $10,000 per year, making a total of $50,000 that would be paid annually to the state game protection fund.

4) The salary of the state game warden and expenses of the office not to exceed $5,000, one hundred acres of land suitable for hatchery, $3,500: improvements of the grounds, $15,000, distributing cares and other incidental

> expenses, $15,000, making a total of $28,500, leaving a balance of $11,500 from the first year to credit of the state game protection fund, to be used in stocking our fields and streams.[9]

Two weeks later, the *Daily Arkansas Gazette* published a letter from "Arkansas Traveler." Traveler provided an overview of licensing and shipping laws in the United States as compared to those passed in Arkansas. Traveler noted that Arkansas ranked as the only state in the nation to outlaw nonresidents from hunting and fishing and one of only five states that did not have a licensing system. Visitors could hunt without a license in those other four states (Georgia, Oklahoma, New Mexico, and Nevada). Other states with a licensing system charged between ten and one hundred dollars for nonresident licenses and seventy-five cents to five dollars for resident licenses. The article contended that the money gained from licenses could fund game and fish protection. Traveler argued that more than twenty-five states had nonexportation laws but allowed limited shipments for nonresident license holders. In several states, landowners, residents or not, could ship game out without restrictions. Arkansas Traveler argued that the Bear State remained behind many others regarding progressive and effective wildlife laws. Arkansas needed to protect its game and fish but had to look to other states for examples of accomplishing this mission properly.[10]

Because the Arkansas government had not created a statewide game warden system, the ASSA decided to take matters into its own hands in the late summer of 1909. The association intended to hire its own game warden for the next two years. This action would give them time to build support in the state legislature to create a game warden system. Arkansas needed enforcement as much or more than additional game protection laws.[11]

The ASSA decided on a two-part plan: The first part called for raising funds for the salary and expenses of the temporary warden. The second part continued a focused lobbying effort to obtain support for the warden position and more game protection laws. The lobbying efforts also included a grassroots campaign to rally farmers to their cause.[12]

The ASSA believed they needed five thousand dollars for the warden's two-year employment. As part of the warden's responsibilities, besides enforcing the game laws, the ASSA expected him to visit all the state's counties and recruit deputy game wardens. The county sheriff could deputize these men if they were not already law enforcement. If the ASSA warden could not continue his duties after two years and if the state legislature did not create a state warden system during the following session, then county game wardens could continue their work with funding from fines for local violations. Some congressmen claimed they had not created a warden system already because of the "rush of other legislative matters and the avalanche of local bills."[13]

On September 14, 1909, ASSA president Tip T. Omohundro called together his association officers and club representatives from across Arkansas to solidify plans for the game warden operation. The group met in the office of the ASSA attorney, Baldy Vinson. They agreed that ASSA secretary Earnest V. Visart was the man for the state warden position. The association also formed a committee to meet with Governor George Donaghey to request that he appoint Visart as a state game warden with the authority to carry out his duties.[14]

Omohundro, Visart, Paul Litzke, R. D. Crenshaw, Dr. James L. Dibrell, and Vinson met with the governor the next day. Donaghey, a contractor from Conway, was an avid hunter. He sympathized with the ASSA and supported game legislation. Donaghey charged the ASSA committee to create a summary of game and fish laws that the game warden intended to enforce and deliver this document to Arkansas attorney general Hal L. Norwood for his opinion.[15]

By 1909, the ASSA boasted a membership of more than one thousand. They claimed that members had pledged over $25,000 to employ a state game warden in less than thirty days. Because of these association funds, the committee assured the governor that the organization would provide financial support for the new position and even provide ASSA attorneys to assist in prosecutions. The governor only had to give the game warden the lawful authority to perform duties as an "agent of the state."[16]

The *Arkansas Democrat* interviewed Baldy Vinson about the new state game warden. According to Vinson, "The rivers of Arkansas are literally crowded with shanty boats occupied by non-residents who float down the streams continuously and kill, by sein and dynamite, all kinds of fish; trap and shoot game in and out of season, ship to market out of the State, and by the time attention of the authorities is called to their actions, they have moved down to other counties and are seldom brought to punishment for their crimes." The attorney complained that many of these men were inexperienced and wasted the animals with their careless actions. "It is a well-known fact," he explained, "those wanderers have great difficulty in distinguishing between a bear and a fat hog or a buck deer and a bull yearling." Although Arkansas banned most nonresidents from hunting and fishing in Arkansas, these men, Vinson claimed, "pay no attention to it whatever, pay no taxes, [and] acknowledge allegiance to no sovereign."[17]

10. Earnest Vivan Visart, Arkansas's first state game warden. Author's collection.

Besides these "wanderers" who haunted the riverways, Vinson continued, thousands of nonresident sportsmen entered Arkansas in the fall and winter "with guns and ammunition enough to [suppress] the Filipinos and slaughter ducks and geese by the carload, ship them out as baggage or by other clandestine methods, and then go back home and write long alleged funny articles for such magazines as *Outdoor Life, Sports Afield, Field and Stream, The Outers Book* and other, sarcastically referring to the good people of Arkansas who have tolerated them as the 'natives' until it makes every decent law abiding sportsman sick and mad."[18] Local authorities could not react quickly enough to catch these men. The state needed someone with authority to continue to chase poachers across county lines with statewide jurisdiction.

A few days after their meeting, Vinson submitted the digest of Arkansas game and fish laws to the governor's office for inspection. Arkansas secretary of state Oswald C. Ludwig and Attorney General Norwood reviewed the document to determine the lawfulness of the ASSA request that the governor appoint a state game warden and how the warden's authority was defined. By the end of September 1909, Governor Donaghey named Ernest V. Visart state game warden with full authority to appoint deputy game wardens but no power to arrest anyone. Visart relied on local law enforcement for the apprehension of game violators. This predicament hampered efforts to arrest and prosecute lawbreakers because some sheriffs would not allow deputies to go after game violators. The state did not pay Visart a salary or expense money.[19]

In retrospect, Visart did not possess any particular skillset to serve as a game warden. He possessed no law enforcement training, did not hold political office, nor did he have a college degree or any specialized training. He was interested in sport hunting and fishing; many people knew and liked him. His family had lived in Arkansas since before statehood, and he was well-known in some areas around the state. He was tall and slender, with light skin, dark eyes and hair. Visart always seemed on the go, energetic, and outgoing. Although determined to end the slaughter of Arkansas wildlife and disciple wrongdoers, Visart was empathetic to his fellow citizens. Rather than take the children's money, he once paid the fine for five orphans he caught illegally fishing. While warden, he provided confiscated meat to several houses of charity.[20]

In 1836, Earnest's grandfather, Julian J. E. Visart, left Belgium with his father, Maria Phillipe Julian Joseph Visart, count of Bocarme, and moved to Arkansas. After staying at Arkansas Post, the family eventually put down roots in Arkansas County. Using family money, Julian J. E. Visart acquired land

and raised cattle. His son, Edward Visart, served in the Confederate army and became a physician after the war. Dr. Visart opened an office in DeWitt and built a reputation as an exceptional general practitioner and a malaria specialist. His son, Earnest Vivian Visart, born March 28, 1872, lived in DeWitt through his teenage years. During that time, Colonel Robert H. Crockett, grandson of Davy, took young Earnest on several hunts before the Visart family moved to Eureka Springs in the late 1880s.[21]

Reverend H. A. Tucker joined Visart and Flaurance Lola "Flora" Tucker, a twenty-year-old from Texas, in marriage in Eureka Springs on December 11, 1895. The couple resided in Eureka Springs, where Earnest worked as a traveling salesman, wholesaling cigars for the Shibley-Wood Wholesale Grocery Company in Van Buren. In 1901, his employer closed. Now out of a job, Visart decided to go into business in Van Buren. He partnered with Mr. John of Johns Photo Company to open a cigar factory in the river city in January 1902.[22] Unfortunately, Flora died in October 1902, leaving no children. Two years later, Earnest married Frances Ellis from Jacksonville, Illinois, and the couple moved into the Meyer addition in Van Buren. Earnest's mother, India, also lived with the newlyweds but died the following October.[23]

During a 1905 murder investigation of law enforcement officer Will Jones, Van Buren sheriff D. W. Moore appointed Visart a "special deputy." During this investigation, Visart and two other deputies entered the African American section of Van Buren known as "The Colony." They arrested two men on suspicion of complicity in the killing. This is the only instance where Visart was involved in law enforcement before serving as a game warden.[24]

With the cigar factory failing, Visart entered into a string of business ventures in Van Buren and Little Rock during the next few years. He partnered with Frank Woolum, J. W. Hodges, and Al Belt in the Fort Smith and Van Buren Realty Company, which purchased one thousand acres between the two cities for $100,000. They created four hundred residential building lots in an addition called Mountain View. This venture too failed, and the company finally surrendered most of the lots to the county for taxes.[25]

Visart and his wife sold out and moved to North Little Rock in 1907. He established the Argenta Wholesale Cigar and Tobacco Company in Argenta and became well-known around Little Rock and the surrounding area through his connections with the company. Another venture brought Visart even more notoriety, mainly through newspaper publicity.[26]

Newspapers trumpeted the incorporation of the Argenta Baseball Association in March 1908, and Visart served as vice president. However, Visart sold his stock in the organization within a few months and ended his affiliation. He also resigned from his management position at the tobacco company. Why Visart made these decisions is not clear. He informed friends that he was considering becoming a traveling salesman again. By October 1908, a rumor claimed that Visart planned to open a wholesale grocery business in Conway.[27]

Around the same time, Visart became involved in a sporting club. A growing number of Little Rock sportsmen began searching for fishing or hunting locations closer to the city they could lease or buy. Several men complained about the lack of game and fish near the capital because of the "number and method employed by [other] hunters and fishermen." Owning and controlling the fishing hole was the only way to guarantee fish.[28]

In August 1908, Visart joined George Clementa, Lee Omohundro, and W. A. Baxley to create a fishing club on leased land. They set their sights on Baxley's Mill Pond and the surrounding 250 acres. The men planned to recruit enough members to raise money to purchase the leased land eventually. This venture was Visart's first official outdoor club membership, although he had hunted with family and friends as a young man. His associates evidently took a shine to Visart, for they elected him ASSA secretary.[29]

When Visart left the September 1908 meeting with the governor as state game warden, the ASSA leadership gave him two objectives. First, Visart was to solicit donations to fund game law enforcement. Second, he was to appoint deputy game wardens across the state. Visart believed he also needed to change the public mindset.[30]

Visart targeted illegal seines first. They were easier to find and remained in a static location. Illegal fishing was also one of the first few state regulations. The warden intended to "break up the use of illegal seines in the rivers of Arkansas" and wanted the public's help. Going after the nets was the first move in his two-part approach to combating fish removal from the state.[31]

Game warden Visart's first assist would prove whether the new system would meet immediate challenges. Visart and newly sworn Crawford County deputy game warden J. W. Smith arrested Ralph Cazort, son of former Arkansas senator G. T. Cazort, for illegal netting near Van Buren. The two officers observed hundreds of dead fish and found several hundred pounds of fish in Cazort's wagon. Visart and Smith piled up the illegal net and burned it. Cazort, according to the two officers, intended to take the fish to market.[32]

Baldy Vinson arranged for the employment of a Fort Smith attorney to prosecute Cazort in front of a local justice of the peace. After a determined argument from both sides, the official ruled that Cazort was guilty and fined him fifty dollars for his actions. Members of the ASSA were elated over the triumph of their first arrest and successful prosecution. With continued success like this, the ASSA pledged to continue its operation.[33]

The *Arkansas Democrat* commended the ASSA's work to enforce Arkansas game laws. If organizations like the ASSA did not continue to fight to protect the fish and game, the wildlife would soon be "wholly exhausted," one article contended. Arkansawyers must continue this fight, it continued, against the "slaughter of game out of season, either by local or non-resident hunters, the depredations of the pot hunters, and the unlawful catch of fish by dynamite, small-mesh nets or other means not sportsmanlike."[34]

A few days later, Visart "caused the arrest" of Fort Smith saloon keeper Charles Echer for displaying wild quail and doves in his storefront window. At trial, the prosecutor and game warden reached a deal to drop the charges if Echer freed the birds. The bartender opened the door, and the birds "were turned loose on the street of the Border City." Because the bar kept the birds in cages for so long, they probably did not last long between the alley cats and stray dogs. Nevertheless, the state game warden succeeded in his second case.[35]

The state game warden traveled to Pine Bluff a few days later and paired with Deputy A. L. Cason to accost Andy Crossett for killing quail out of season. Crossett admitted to the crime. Armed with a confession, Visart trusted that Cason would follow through with the prosecution. Accordingly, he left Pine Bluff and headed toward Lake Village. "Warden Visart Continues Crusade Against Game Law Violators," read one headline in the *Daily Arkansas Gazette*. Another paper called his work a "vigorous campaign for the enforcement of the game laws." Visart did have his critics too.[36]

Some Arkansawyers, like ex-senator Cazort, did not like Visart and his activities. Word went out across the state that a challenge toward the legality of an organization appointing a game warden without a state-legislature-passed enforcement law might be in the making. The ASSA returned to Attorney General Norwood to ask that he review the situation. The association representatives also inquired whether Visart possessed the authority to arrest violators independently.[37]

A few months later, Jackson Walker of Little Rock sued Visart for destroying his fishing net. Acting upon a tip, Visart and some deputies had gone to Old River, a cutoff of the Arkansas River a few miles from the capital. They discovered Walker's illegal nine-hundred-foot-long net, which they pulled out and burned. They arrested Walker, but a jury acquitted him. Walker made his living with the net; with its destruction, he was unemployed. Walker sued the state game warden, justice of the peace L. W. Roberts, and the Arkansas State Sportsmen's Association lawyer Baldy Vinson for one hundred dollars in damages. Walker lost the case. Yet, this was not the last time a disgruntled fisherman sued Visart for burning nets. But they never won a lawsuit against the warden.[38]

Other adversaries sent anonymous complaint letters. Visart publicly stated that he would investigate legitimate reports but not answer correspondence without "bona fide signatures." He was not one for idle talk.[39]

Visart also received supportive letters. County sportsmen's associations and clubs invited Visart to speak at their meetings. Visart's first address to a local club was in Brinkley in October 1909. The meeting's attendees were quite vocal in supporting Visart's efforts. They pledged to report any game violations they observed. These "deputies have commenced a war," claimed Visart, and in a "short time, [the] great part of the illegal fishing and hunting will be stopped." The following spring, local sportsmen in Warren and Morrilton showed their support at meetings with the warden. He even organized the Garland County Sportsmen's Association in 1910, with over two hundred members. Local organizations like these provided the backbone of the Arkansas conservation movement.[40]

Visart also experienced support on the national level. In 1910, the United States game wardens invited him to their fifth annual meeting in New Orleans. He also arranged for a meeting of the game wardens from the surrounding states in the spirit of cooperation. His friend Tennessee game warden Joseph Acklen and Mississippi, Missouri, and Oklahoma wardens attended.[41]

The new game warden sometimes found that local officials backed his efforts. In Morrilton, Sheriff T. C. Hervey "offered all possible aid." While Visart was in Lonoke County, Sheriff Tom Fletcher declared his support for enforcing game laws, and Visart appointed Deputy George H. Sellers as Lonoke County's game warden.[42]

In 1909, Arkansas lawmakers passed a law barring nonresidents from hunting, fishing, or trapping in the state unless they owned the land. In a newspaper article, a party of nonresident friends of the general manager of the Meto Valley Railroad S. M. Savage, announced they planned to visit Lonoke County for a quail hunt. Visart went to catch them. He and Deputy Sellers found the group had returned from their hunt. After a few questions, Savage admitted to hunting and offered to show the two officials the game they had bagged. When entering the barn where they had stored their kill, the officers saw one rice bird hanging from a rafter, the sum of the hunt. No law was against nonresidents harvesting rice birds, so Sellers made no arrests.[43]

Three points are evident in this occurrence: These men held Arkansas game laws in such contempt that they gladly published their intentions without caring about the consequences. Second, Visart was willing to go anywhere and challenge anyone, even the wealthy or well-connected, over violations. Finally, the state game warden needed support from local officials to arrest and prosecute lawbreakers.

During that quail season, Visart also went to Crittenden County, across the Mississippi River from Memphis, Tennessee, the source of many nonresident hunters. Visart appointed thirty deputy game wardens, all that he could find, to hold the line. With those additional troops, the number of deputy game wardens numbered 250 statewide. Visart had appointed them all in just a few short months. As an incentive, deputy game wardens now received half of the fines from those convicted for game violations.[44]

Illegal wildlife sales also concerned Visart. By 1909, laws forbade the sale of wild game and fish inside the state. However, no law existed barring shipping them out of the state to meat, fur, or feather markets if harvested legally. "A citizen who does not hunt has to do without, yet Memphis, Tenn., Joplin, Webb City, Lamar, Mo., and many other foreign markets feast upon Arkansas fish and game," complained the new game warden. The cause of this "slaughter," complained Visart, was the "market hunter and game hogs," whose actions were a "detriment of thc people at large."[45]

Because Memphis contained some of the closest wild-meat markets, Visart traveled across the Mississippi River to investigate and was appalled. "The Memphis market is flooded with game of all kinds," Visart reported, "mostly from Arkansas. The fronts of the restaurants and eating houses are literally lost to the view, so thick are the strings of ducks, quail, turkeys, and venison." The eastern part of the state, he claimed, overlooked game laws, and "conditions in Arkansas [were] about as bad as they can be." Yet, he was powerless to prosecute out-of-state meat markets. So, he tried to stop the deliveries from Arkansas by focusing on the points where the game left the state. Visart enlisted Tennessee game warden Acklen for help during his multiple investigations on the Tennessee border.[46]

Because Mississippi County possessed a well-connected state representative, a few townships were exempted from the state shipping laws. The exemption was so lucrative that a Tennessee game buyer set up headquarters in Mississippi County. From his base, the Tennessean hired hunters from across eastern Arkansas.[47]

Visart determined that the best way to stop game shipments was to go after the shipping companies. He informed the shippers, railroads, and express companies that he would prosecute them for any illegal activity under federal law. Visart tried several times to have five-hundred-dollar fines assessed on companies that shipped wildlife illegally with limited success.[48]

The United States Congress had amended the Lacey bill in March 1909, which explicitly targeted market hunters, making it illegal to ship any "wild animals or birds, where such animals or birds have been killed or shipped in violation of the laws of the State, territory, or district." All shippers must label the containers correctly or face a two-hundred-dollar fine per incident. The receiver, furthermore, could also receive a two-hundred-dollar fine.[49]

Within days, Visart seized nine illicit containers of ducks headed out of state to St. Louis, Missouri; Memphis, Tennessee; and Chicago, Illinois. He donated them to the St. Bernard Hospital at Jonesboro. "I propose to help maintain the charitable institutions of the State of Arkansas so long as people insist upon violating the game laws and attempt to ship ducks outside the State," insisted Visart. The state game warden claimed that nonresidents unlawfully attempted to ship the game to outside markets, and he was hot on their trail. He sent the next seized batch to the "blind school [and] the Old Ladies' Home."[50]

A month later, Jonesboro deputy game warden James Young seized over 1,500 pounds of ducks headed to Chicago. Deputy O. B. Wade of Harrisburg arrested a representative of the Pacific Express Company and consignee W. C. Lande for illegal shipping. Authorities fined the Pacific Express Company one hundred dollars for each count.[51]

After reading an article in a St. Louis sporting magazine about Missouri game wardens seizing crates of Arkansas quail, Visart became determined to discover the shipment's origin. He found that an Ashley County resident, named Louis Myer, had mailed the quail. Piggot deputy game warden J. H. Morgan arrested Myer on Visart's information, and justice of the peace J. F. Watson fined him fifty dollars.[52]

Visart's report for November 1909 listed 109 convictions in fifteen counties. The three primary offenses included selling venison, poaching fish, and other game. The fishing offenses mainly came from unlawful net sizes. Shooting ducks at night made up most of the illegal hunting activity. Fines ranged from twenty-six dollars for selling venison to ten dollars for net violations.[53]

As word spread about the new state game warden, concerned citizens sent scores of letters to Visart, informing him about violations in their area. One day alone, he received complaints about night shooting on Grassy Lake near Conway, illegal fishing in White and Woodruff Counties, and "fish hogs" in Poinsett County. Nevertheless, Visart pledged to investigate them all.[54]

One report stated that the area along the St. Francis River, near the border between Arkansas and Missouri, "offer[ed] a deplorable laxity in the enforcement of the game and fish laws." More specifically, market hunters' fishing traps filled the waters near Hatchie Coon on the river, and sometimes they caught so many fish that "thousands of pounds [were] known to have spoiled waiting for a market." Visart reported a "hearty reception" from the county's sportsmen, who pledged full support for Visart. "There is no possible chance for the stream to survive the present method of taking the bass," complained one member. Visart pledged more deputies for the area.[55]

In another example of local support, Warren sportsmen in Bradley County alerted Visart about illegal fishing on the Saline River. Deputy game warden J. O. Roberts went along to perpetrate any arrests. A local provided a gasoline-powered motorboat, while "several of Warren's enthusiastic sportsmen" accompanied Visart and his deputy and "provided [him] every assistance." The party seized a boatload of nets, carried them back to Warren, and burned them in front of a large crowd. If he could, Visart made a show of destroying these nets publicly. He repeated the public ritual in Cotton Plant later that month. That way, citizens saw the officer's success firsthand, while poachers witnessed what happened to lawbreakers.[56]

While not as common as netting, trapping, or seining, poachers often used dynamite for fishing. The explosives concussed or killed the fish and brought them to the surface. Dynamiters killed all species and sizes of fish, not just the ones they were after, and could quickly ruin a stream in a short time. Arkansas deputy game wardens arrested seventeen dynamiters during that first year.[57]

From his first year as a game warden, Visart learned that many violators and law-abiding citizens did not know the present game and fish laws. He thought that if poachers knew the law, most would not break it. Many citizens asked Visart for information concerning game statutes. In response, he produced an outline of the state laws. Visart's summary ran in several Arkansas newspapers. In this public forum, journalists often pointed out that Arkansas had no state bag limits on game and fish, unlike some other states. Because of the lack of limitations, "game hogs may slaughter indiscriminately without fear of prosecution," claimed one columnist. During this period, individual counties kept county-specific laws. Accordingly, Visart did not list these local laws but encouraged hunting and fishing sporting clubs to establish their own bag limits. The education campaign worked. A few months after his newspaper campaign, Visart met with twenty-five Morrilton fishermen, who claimed they did not want to break the law and would pull their nets from the Arkansas River for inspection. Other commercial fishermen publicly promised Visart that they would thereafter assist him in catching poachers.[58]

The warden also decided that the Arkansas State Sportsmen's Association was too exclusionary. He needed all Arkansawyers on board, not just the sporting gentlemen. He created the Arkansas Fish and Game Protective Association and opened up membership to everyone "interested in the preservation and protection of Arkansas game." No dues or fees. All he asked from members was that they "pledge themselves to give their support to the better protection of the game and fish." He hoped to recruit farmers, housewives, and professionals to the organization. For Visart, the AFGPA and the ASSA could work hand in glove for the betterment of Arkansas.[59]

By the time Visart provided his year-end report to the officers of the ASSA (he was sending them weekly messages from the road, too), he had achieved a decent start. He raised money, made arrests, and appointed several deputy game wardens. Visart had deputy wardens in about twenty-five counties by 1910. Some of his deputies had already made arrests without his assistance. Notably, Deputy Sellers in Lonoke County made arrests for illegal fishing and unlawful venison sales. The ASSA membership approved his actions, considered his work an unmitigated success, and pledged continued support.[60]

Visart explained that reporting violations had decreased since the start of his term, crediting the cause of the increase in public education and respect for the law. However, one of the most significant issues was the game laws themselves. "The laws of this state are so intricate and complicated that I would be afraid of breaking some law myself if I went hunting," complained Visart. Arkansas game and fish regulations had several thorny issues. Some counties, such as Mississippi County, with its exemptions from shipping limitations, ignored state laws. A few counties (or towns) had specific rules for their location. Some of these regulations nullified state laws. This confusion allowed market hunters to work the system. It seemed that the state legislature amended, abolished, and changed the state laws every session. The state government did not publish a guide to these changes to inform the citizenry. A hunting activity that was legal last month was now illegal this month. A legal net in one county was unlawful in the neighboring county. It was frustrating for Arkansas citizens and the authorities. Visart believed all local and county laws needed repealing. They were too confusing. The state legislature needed to pass general rules that were supreme throughout the state and did not change every session.[61]

The confusion over the ever-changing and unclear game laws was apparent in the stature that allowed property-owning nonresidents to hunt and fish on their land. The focal point was Helena. Visart received word that some nonresidents were hunting on the White River in Desha County, and he went to arrest them. Chicago Americans owner Charles Comiskey, president of the American League B. B. Johnson, pitcher William "Bill" Chappelle, catcher Ed Hurlburt, pitcher Charles Shields, and a few others rode on the vessel *White Sox* while hunting. When confronted, the men presented a written invitation to hunt in Desha County, Arkansas. They reminded Visart that authorities had recently amended the nonresident ordinance that allowed guests to hunt with a written invitation from a county resident. A frustrated Visart allowed the celebrity outing to continue. Less than sixty days before, nonresidents could only hunt on land they owned in Arkansas.[62]

Visart and the ASSA believed that all nonresidents needed a license to hunt and should take no more than two days' worth of catches home. These measures would cut out market hunters, relieve the pressure on the wildlife, and bring in license fee revenue. Visart believed Arkansas could advertise in national magazines and allow nonresidents "to come here and spend a few days hunting and fishing" after buying a license. "At present, all other States in the Union but two are working under the license system," he explained. As Arkansas did not have an adequate licensing system, "we are having pushed upon us that class of outlaws that are driven from other States." They come disguised as fishermen or hunters, "but in truth, [they] are robbing our farmers of their cattle, hogs, and even steal[ing] the corn from the fields." They poached the furbearing animals as well.[63]

Visart claimed that fur sales had rivaled the annual Arkansas cotton crop income for the last few years. He insisted that "something must be done to protect the furbearing animals of the state." The early killing of these animals made their fur worthless to sell. It was a waste. The state must pass more protective measures.[64]

In February 1910, the ASSA published the first of many public pleas for assistance, a call to arms. Here, they issued a scathing report on the state government's failure to pass adequate game and fish laws: "As our legislature at its last sitting did not deem it right to give us the necessary relief and protection sought for, in the protection of our game and fish, against the unscrupulous and unlawful destruction of said game and fish, and the pollution of our streams, the trapping and killing for mercenary purposes of the game and birds, we ask that you join with us and help protect the greatest gift God has given the man who loves nature, and enjoys that outing that furnishes the greatest man-exiting and nerve exhilarating sport known to the human race." The ASSA claimed it was up to them to stop the exploitation of the state's wildlife. While they had employed a state game enforcement officer to assist in prosecuting the few game laws, it would only be a short time before the slaughter took all the state's animals. For only two dollars, Arkansas conservationists received an ASSA membership.[65]

In early March 1910, Visart pursued other areas of conservation. He contacted several fish hatcheries outside of Arkansas about rainbow trout. Visart insisted publicly that these new fish would not go to any county that allowed trapping in their waters. By that time, thirty-five counties disallowed the practice. He had seen twenty-five thousand rainbows come to Arkansas in Eureka Springs when Senator Festus O. Butt placed them in the King and Osage Rivers. Visart wanted to add trout to suitable waterways all across Arkansas. Within a few weeks, he had received the National Department of Agriculture forms to request the game fish.[66]

In 1903, the federal government had established a fish hatchery near the Missouri border at Mammoth Springs, Arkansas. They chose the site because of the cold natural spring water and proximity to railroads. In the summer of 1910, Visart received a shipment of seventy-five thousand black bass from the hatchery and placed them in Big Lake, Old River, Rock Creek, Wolf Bayou, Plum Bayou, and Neimeyer Lake. More shipments followed. He stocked water holes across the state.[67]

Visart planned to lease, stock, and manage several lakes near Little Rock "for the benefit of fishermen who are unable to get away from their work for more than a few hours." Two years later, he leased a lake from the Neimeyer Lumber Company (southwest of Little Rock) and stocked it with ten thousand black bass. Visart reserved this lake for sportsmen, and there would be no fishing for the first year until the fish were well established.[68]

The state game warden also attempted to bring pheasants into Arkansas. If he could find a bird that the farmers saw as beneficial to their work, the agrarians would support their protection. The American ringneck "is the most valuable fowl for the farmer," claimed Visart, "being the only known bird that will destroy the potato and chinch bug." Sportsmen also knew the birds were splendid to hunt and the meat delicious. He sent out a notice in 1910 that he had contacted the American Sportsmen's Association in Denver to obtain twenty-thousand pheasant eggs. Unfortunately, he was only able to get one thousand eggs. Visart needed farmers to raise the birds and promised to give volunteers free eggs. He petitioned the state legislature to enact laws to protect them. He also tried to obtain Hungarian partridges and prairie sharp-tailed grouse but had better luck with the pheasants.[69]

According to Visart's plan, landowners would provide the space while he and the ASSA stocked the birds. Subsequently, he would personally manage the birds. His sales pitch expanded, and he also used the opportunity to recruit for the AFGPA. Visart described the immense beauty of the birds' feathers, claimed that raising them was "much more entertaining than the rearing of ordinary fowl," and explained that "industrious and intelligent" women could make money from the venture.[70]

Over five hundred interested parties wrote to Visart within a week requesting further information. Much of the correspondence dealt with questions about the characteristics of the bird. In response, Visart asked the *Daily Arkansas Gazette* to publish an image of the bird. This general description answered most of the questions. When a few eggs arrived, he placed them in the front windows of Sam Henderson's Little Rock cigar store. By the summer, the warden had visited several Arkansas farms to determine whether they suited the birds' needs.[71]

The demand for the free pheasant eggs grew so large that Visart could not meet it. The warden claimed that he could order eggs or birds for people if they paid for them out of their pockets. "I have now something over one thousand application for pheasant eggs, and they continue to come," he said. "There is no question but that a few dollars invested in pheasants will prove to be one of the best investments a farmer can make," he concluded. He bought eggs and birds from Pennsylvania wholesalers Wenz & Mackenson. Visart insisted that if the Arkansas legislature passed bird protection laws, he would also bring in partridges. He also hoped the state government would fund a propagating farm to refill the countryside with animals and birds.[72]

Everything did not go perfectly for Visart that first year, most notably the run-in with the nonresident professional baseball men and their permission slips in Desha County. In April 1910, a judge ordered Visart to return fine money to some wrongdoers in Faulkner County. Their lawyer successfully argued that authorities had not filed a bond with the justice of the peace. Lawbreakers challenged Visart's and his deputy game wardens' authority several times.[73]

A year after the Arkansas State Sportsmen's Association chose Visart as their state game warden, he confidently reported a "marked decrease in the violations of the game and fish laws." Visart expressed his satisfaction with how many Arkansawyers assisted him in enforcing the law and increasing the growth in the general interest of protecting Arkansas fish and game. Because of his progress, the ASSA presented him with an engraved Remington 25–35 caliber rifle. However, Visart believed the state had a long way to go. "It is only a question of a short time, if this wholesale slaughter continues as in the past, until Arkansas will have no game or fish," he lamented.[74]

Visart entered a situation that seemed impossible to change. He operated in a state that had a long history of market hunting. Recent industrialization and urbanization heavily pressured Arkansas's game and fish population, with more destructive killing methods and the growing demand for meat in the cities. When he began his enforcement in September 1909, he had only the backing of the state sportsmen's association, with little actual authority. Within a year, he had built a network of game law enforcement officers who worked to arrest and prosecute poachers across the state. The game warden took it upon himself to educate the public about the benefits of protecting Arkansas wildlife and press the state legislature to make stricter policies. According to Visart, education assisted the law-abiding public spot violators and kept them from accidentally breaking the law. He not only wanted to ensure that Arkansawyers knew the law, but he wanted them to know the benefits of protecting wildlife, and knowing those benefits could cause them to support stricter game and fish laws. When Visart lobbied the state legislature, he asked for a sensible statewide license law to close the shipping codes' loopholes and bag limits. He wanted license fees to pay for a yet-to-be-developed enforcement system with state-employed game wardens with police powers.

Visart had the backing of many hunting and fishing club members, county and state sporting organizations, and the budding conservation clubs—all of them members of the upper and middle class. These folks were the backbone of the push for stricter game and fish laws in Arkansas and the nation. Finally, Visart brought hundreds of birds and thousands of fish into Arkansas. His lake management programs and birds and fish repopulation efforts served as forerunners for today's Arkansas game and fish wildlife management plans.

CHAPTER SEVEN

The Pauper and the Prince of Game Conservation

Earnest Vivian Visart, Edward Avery McIlhenny, and the Political Wrangling of Arkansas Game Law, 1910–1916

500,000 ducks have been sent out of the state during the past two months.
—Edward A. McIlhenny, 1913

I shall not permit myself to think of giving up.
—Earnest V. Visart, 1913

The Arkansas State Sportsmen's Association and Arkansas Fish and Game Protective Association intended to push for a hunting license law in 1910. By then, Arkansas stood alone among all the nation's states for not having a hunting licensing structure. A cadre of Arkansas politicians stood against licenses, viewing it as a tax on property that citizens already owned. Visiting hunters and fishermen did not want to pay a fee either, primarily because of simple economics. On average, neighboring states charged nonresidents from ten to twenty-five dollars, residents a state fee of five to ten dollars, and a county fee of one to ten dollars. Texas recorded the lowest prices at three dollars for residents and ten dollars for nonresidents. Thirty-five years earlier, Arkansas had led the way by passing a nonresident license law. Since that time, they had made little progress.[1]

The ASSA wanted a one-dollar fee for residents and thirteen-dollar fee for nonresidents to hunt in the state. Visart hoped that the licensing fees could fund a state game warden system. He contacted several other state wardens, inquiring about the best plans to protect wildlife. The most common answer involved a licensing program. They claimed that it brought in revenue for enforcement and filtered out undesirables. During the July 1910 session, the Arkansas legislature did not pass a state licensing law for residents or nonresidents, but they did pass a law that required a county hunting license of ten dollars.[2]

The sporting magazine *Forest and Stream* published a scathing report about Arkansas authorities allowing certain nonresidents to hunt, even though a state law barred them. "'No nonresident shall hunt or fish' is as positive a prohibition as 'Thou shalt not kill,'" it declared. The publication claimed local authorities did or did not enforce the law based on the visitor's pocketbook and charged local officials with pocketing some of the fines.[3]

The magazine also pointed out that Arkansas seemed to equate "slaughter" with "non-resident." It argued that a resident gunner could kill more wildlife than three visiting sportsmen during the same time. Despite laws prohibiting shipping game and fish from the state, Arkansas wildlife still ended up in St. Louis and Memphis markets. The growing timber industry destroyed habitats, while recent floods drowned thousands of animals. Indeed, agricultural development and logging efforts in Arkansas had destroyed thousands of acres of wetlands along the river valleys. In 1920, the US Department of the Interior estimated that Arkansas and its two neighbors, Mississippi and Louisiana, contained roughly half fewer wetland acres than previously. "Arkansas game has had a pretty bad year of it, and the years to come offer [a] small promise of better things unless the people of the State awaken to the importance of real game protection by the enforcement of existing laws." the article concluded.[4]

The ASSA continued to lobby the Arkansas legislature and supported Visart's work. In 1911, they redoubled efforts, raising thousands of dollars for game protection, supporting game introductions and fish restocking in the state, and continued promoting outdoor life through education programs and shooting competitions. The ASSA fought to get more Arkansawyers to support wildlife protection, especially among the farmers and young people.

Edward Avery McIlhenny, heir to the Tabasco Company on Avery Island, Louisiana, and early bird conservationist, purchased thousands of acres in Louisiana and created avian sanctuaries. He started with a refuge on his family's island. He partnered with similar-minded preservationists like Charles W. Ward and Margaret Olivia Slocum Sage, a railroad tycoon's widow, to acquire more than 175,000 acres for bird conservation in Louisiana through purchase and lease. The hot-sauce businessman had assisted the state governments of Louisiana and Mississippi in developing game protection legislation. McIlhenny soon became interested in Arkansas bird protection and the possibility of avian sanctuaries there. Arkansas is on the Mississippi Flyway, a corridor where migrating birds travel south and back north. In 1912 and 1913, the Louisianan visited Arkansas several times, met with Visart to discuss legislation, and continued to carry on a lengthy correspondence with the newly made game warden. He planned to establish a series of safe havens for migratory fowl along the Mississippi Flyway, from Canada to the Gulf of Mexico, including Arkansas. McIlhenny wanted Arkansas to have a game law enforcement arm before he spent money to buy the land.[5]

McIlhenny believed that Arkansas contained some of the best ground in the Mississippi Valley for waterfowl, and the state government needed to pass laws that protected those birds. He explained that although he had a long interest in Arkansas, several people had warned him that Arkansawyers did not want stricter game laws. They claimed he should not actively assist in passing them since he did not live in the state. However, during his trips over those six months, Arkansas sportsmen had welcomed him and urged him to lobby the state legislature. Those meetings had changed his mindset.[6]

In January 1912, Visart informed the Tabasco heir that the death of Arkansas Senator Jeff Davis had thrown Arkansas politics into "turmoil." The secretary believed Governor Joseph "Joe" T. Robinson sought to replace Davis, so Robinson's attention remained focused on that issue, not wildlife conservation. A conservation bill thus might have to wait, claimed Visart. "Never in the history of Arkansas has the political situation been as it is today," Visart explained. He urged McIlhenny to send a conservation statement to the Arkansas people immediately so that it might hit print before the legislature met.[7]

McIlhenny authored a conservation bill for Arkansas and asked Visart what the governor thought about it and whether he needed to make a solid push to get it passed at the time or to wait. McIlhenny immediately sent several copies of an article arguing for a general game bill and stricter protection laws. He told Visart to send copies to several Arkansas newspapers and ensure they "appear simultaneously in all papers and should appear as a direct contribution from each Editor of each paper." McIlhenny informed Visart that he wanted to arrive after all the political wrangling over the appointment ended because it made little sense for him to waste time during that tumultuous period.[8]

Previously, Robinson expressed support for game legislation. Several supportive Arkansas politicians, allied with Visart, hoped the governor would read the bill's latest version and provide recommendations and improvements. Visart informed the Louisianan that he hoped Robinson could send those suggestions "in time that they might be submitted to [him] in time for [him] to prepare the final bill." A few legislators had offered some backing, but Visart claimed there "will be [a] line up against us."[9]

Somehow, Visart managed to get to Governor Robinson's office. He wrote excitedly to McIlhenny that the governor "highly favor[ed] the bill and [would] select men to introduce it and do all in his power to secure the passage of it." McIlhenny wrote for more details, claiming with more information, he could "advise [the governer] more definitely."[10]

McIlhenny's plea made it into the *Daily Arkansas Gazette* and the *Arkansas Democrat* before the state legislature met. The *Democrat* printing had no signature but ran with the headline "An Appeal for Wild Life to the People of Arkansas." The *Gazette* published "Makes Plea for Game Protection: Great Wrong Is Being Done in State, Writes an Arkansan" under the "From the People" column, so the piece appeared that an Arkansawyer had penned it, just as McIlhenny and Visart planned. The article argued that Arkansas wildlife, the property of the state's citizenry, faced extermination: "It is the duty of our people to demand" that the legislature repeal all of the old rules and enact meaningful, efficient ones. It argued that birds kill boll weevils and that the insects cost Arkansas millions of dollars annually. "Wanton human butchers," it continued, "murdered the deer, turkeys, and quail" for "there were no State officers to protect our game!" Half a million ducks shipped out of Arkansas from three counties in October and November 1912. On one October day in 1911, 90,600 ducks shipped in one order. "Arkansas needs more up-to-date conservation laws," and the state also "needs a conservation commission," the article concluded.[11]

Most Arkansawyers remained skeptical. Several nonresident club members supported new game laws, which worried some locals. Why do nonresidents want a game law that "will operate to the disadvantage of the non-resident hunter?" they questioned. The law should keep nonresidents from entering the state and shipping as much game and fish out as they desire. One critic believed the present conditions were "unjust, unfair and discriminating" and "against the peace and dignity of the state of Arkansas."[12]

Politically, the sides shaped up over supporters of Robinson as Davis's replacement and anti-Robinson forces. The latter could not postpone the first vote on replacing Davis, which damaged the governor's chances of taking the seat. Indeed, Davis's public support of the game bill remained a calculated risk for those supporting the legislation. Adversaries might torpedo any bill that Robinson might support, like a new game bill, simply because they opposed his senatorial bid. Also, Robinson might withdraw his support to curry favor with the game bill's opponents.[13]

Several political opponents stood prepared to battle once more. State representative A. G. Little from Mississippi County, an outspoken friend and lawyer for market hunters, had led the opposition to the 1909 game enforcement bill. He corralled enough support to kill it. Little again led the opposition in the State House, along with Green County representative Grant Love and L. Clyde Goings from Poinsett County. Goings owned a large percentage of a fishing operation based in his hometown of Marked Tree, which reportedly made more money than the "cotton crop in any five counties of the state." Allegedly, boats illegally ran out of Poinsett County carrying eight to ten thousand dollars of fish per week to Rosedale, Mississippi, for further transport. By 1913, Goings, an ally of Little, was elected to the state senate, representing Poinsett, Jackson, and Mississippi Counties. They stood ready to fight any and all game legislation.[14]

Visart called upon politicians for support and recruited grassroots advocates for the new laws. The Phillips County Rod and Gun Club voted to send representatives to Little Rock to lobby lawmakers to support the Visart-McIlhenny legislation. Visart had friends in the club and reminded them their support was critical to success. Porter Lake Hunting Club in Helena also agreed to send lobbyists to the capital on the same day. A group of Paragould sportsmen also visited the capital to lend support.[15]

McIlhenny finally made it to Arkansas on January 20, 1913. The hot-sauce tycoon met with Arkansas governor Robinson, receiving a friendly reception. In a preemptive move, Visart also arranged for McIlhennny to meet with the State House's Committee on Natural Resources to urge support for the bill that surely would come their way. Visart and McIlhenny decided to move forward with their legislation during the discussions.[16]

Pulaski County state representative Robert Martin formally introduced the conservation bill on January 23. After passing through its second reading, lawmakers sent it to the Committee on Natural Resources. This action illustrated that the State House defined game and fish as an official Arkansas natural resource. In the past, legislators, if they did send game bills to a committee at all, sent them to the Committee on Agriculture, which thereby defined game and fish as just a food source. This change in perception is essential because wildlife moved from a foodstuff to a resource that needed management and protection.[17]

Visart combined portions of McIlhenny's latest letter with his own explanation of the Martin Bill and submitted it to the *Daily Arkansas Gazette* for publication. The article urged support for Martin's Bill No. 152, "An Act to conserve the game, wild quadrupeds, birds, fish, mineral and forestry resources of the State of Arkansas," claiming that passage of the proposal would "place Arkansas in the front ranks of wild life conservation." The bill established a governor-appointed three-member conservation commission for fifty years. One commissioner had to possess demonstrable knowledge of wildlife and hunting. Another had to fish and have an understanding of aquatic life. The final member was required to possess experience with forestry and minerals. The bill contained a budget, salaries for secretarial staff, "four conservation inspectors," a game and fish propagator, and "conservation agents." The agents served as the police arm of the commission and enforced the game and fish laws. It also would establish a licensing system. Most remarkably, the bill banned all market hunting. McIlhenny held so much confidence in the passage of the Martin Bill that he obtained a lease on thirty-five thousand acres in northeast Arkansas with an option to purchase. Privately, McIlhenny told Visart he thought the bill was fair to residents and nonresidents. He also held confidence that his national partners in purchasing land for preserves would buy Arkansas land to establish refuges if the Martin bill passed. He had already spotted two other locations his group wanted to purchase as soon as the bill became law.[18]

Visart had also sought land for a game propagation farm. As Visart traveled around the state bolstering support for stricter game laws a couple of years before, he became acquainted with several sportsmen in Helena. He assisted them in developing the Phillips County Rod and Gun Club, which worked to end poaching in the area. Visart asked the club to search for roughly twenty thousand acres for use as a game propagation farm. However, the game bill had to pass to fund such an operation, and the situation had soured.[19]

Many lawmakers balked at the bill's one-dollar resident license fees. Seeing that Martin's bill struggled to gain traction in the House, Senator Futrell introduced Senate Bill No. 162, a state game warden bill. It called for creating the position of game enforcement officer with an annual salary of three thousand dollars. Funds from market hunting, trapping, and nonresident hunting licenses would fund the operation of an office and staff based in Little Rock. Now, the pro–game and fish protection forces had two chances of success.[20]

Everyone understood that Visart actively lobbied for and wanted Martin's game bill to pass. If citizens did not want more stringent game laws, they generally did not like Visart. If someone supported Visart's work as an unofficial state game warden over the past five years, they generally supported stricter game laws and the Martin bill. Therefore, anyone with a personal issue with Visart condemned this new game legislation because its passage practically guaranteed Visart the official state game warden position. Visart had enemies, too, even if they had seemed allies initially.[21]

Visart claimed that the ASSA attorney Baldy Vinson sent negative letters about him across the state and had "poisoned the minds of not only the Governor but many of the Representatives" against him. Vinson and another ASSA officer wanted a man they could control in the state game warden position. Visart thought Vinson wanted the job for himself.[22]

On the contrary, Visart explained, he was the best man for the new position. The game warden argued that he knew the present situation in the state better than anyone. State game warden was a vital office, and it took a delicate touch, one that he possessed. The person on the job had to use caution in the first few years to build public confidence in the office. He certainly did not want to see "it in the hands of grafters, or men who would not conduct it on the high plain in which it should be conducted." Visart insisted that he had learned the power of money and that men like Vinson could wield significant influence in Arkansas. They worked in the shadows to discredit him for personal motives.[23]

For Visart, opposition to the bill was personal. He asked friends to secretly poll some of the representatives. They found that the politicians claimed to support the bill when one group asked and admittedly opposed it in other circles. When he learned of this information, Visart exploded: "I never was as disgusted in my life," he later said. In a self-pitying funk, he wished he was on a deserted island with birds and animals and could escape what he called the "bunch that I have to contend with here." He exclaimed that the only reason he remained in the fight was because of "the love that [he had] for the wildlife of Arkansas." Visart explained that he enjoyed educating the public about game and fish, but he could not handle crooked politicians. "I want to see this thing settled, and then I am ready to leave Arkansas forever," he concluded.[24]

Publicly, the game warden continued to show unwavering confidence in the bill's passage. Privately, he thought it did not have a chance without McIlhenny's presence. His friends had polled the representatives again and reported that about forty officials supported the bill. However, Visart believed that at least half lied about their support.[25]

Visart ultimately lost total confidence in Martin, calling the representative "indifferent." He was losing faith and had repeatedly fought back the feeling of quitting. Visart now claimed that a "dangerous element" led the opposition, and they remained hidden. "It has always been said that there was a dirty set of politicians in Pulaski Co., and I have now begun to believe it," he confessed.[26]

Visart and McIlhenny called upon their strongest ally to tip the scales in their favor. As the vote neared on Martin's bill, they pushed Governor Robinson to make a statement, and he agreed. On March 3rd, 1913, Robinson gave a special announcement to the Arkansas House of Representatives that basically repeated McIlhenny verbatim. Visart claimed the governor's heart was not in it, and the message sounded insincere and rehearsed.[27]

McIlhenny left for Louisiana after Robinson's speech, not waiting for the vote to occur. Visart, however, continued to lobby for the bill throughout the day. Opponents defeated the measure later that afternoon. Visart became convinced that no wildlife protection bills could pass because if McIlhenny could not accomplish the task, no one could. While McIlhenny, the millionaire, had given money to the cause, Visart borrowed money for living expenses, spending all of his funds toward lobbying efforts. Initially, the ASSA had set aside money for his salary through 1911. The warden had limped along on donations for the last two years. He owed $150 to debtors and started selling all the furniture and some of his other household goods to settle it. A month later, Visart owed $500 and prepared to sell his automobile to cover the debt.[28]

Visart became despondent. He wanted a job in conservation and was willing to move away from Arkansas to get one. He believed, however, that his last five years had convinced many Arkansawyers about the importance of wildlife protection. If a grassroots movement for stricter game laws forced measures to appear on a general ballot, Visart believed that the public would vote for animal protection. Yet, he was tired, willing to step aside to allow someone else to lead the fight. The previous eight years had worn him out.[29]

Nonetheless, Visart propositioned McIlhenny to develop a wildlife breeding business in Arkansas, a stock company that purchased land, raised animals, and sold them. Visart would work for room and board only at any job that McIlhenny provided, his salary going to the Louisianan for the repayment of the money that he had spent trying to get better game protection in Arkansas. Visart apologized to McIlhenny, claiming that the people of Arkansas owed him a great deal for his efforts and that he was ashamed of how the Arkansas politicians treated him.[30]

For conservationists, the 1913 legislative session was a bust. Opponents defeated Futrell's game warden bill in the state senate. Although heavily supported, Martin's bill fell by thirty-nine to fifty in the state House of Representatives.[31]

When Visart wrote McIlhenny on March 14, he seemed a changed man. He realized that although the Arkansas legislators had failed to pass any meaningful game protective laws, the general attitude toward these laws had changed from two years ago for the better. The game warden believed the battle in Arkansas was "a continued fight [they would] finally win." McIlhenny, according to Visart, had "laid a foundation that [would] bear fruit in the future."[32]

McIlhenny contacted the editor of *Illustrated Outdoor World and Recreation* about making Visart an agent for their company in Arkansas. At least Visart could draw a small paycheck until something else opened up. The impoverished game warden prepared to try to sell some magazine subscriptions if he could.[33]

In mid-April 1913, an environmental event created an occasion for Visart and other conservationists to show the need for a state-sanctioned and funded game protection organization. Spring rains had caused severe flooding in eastern Arkansas. The newly appointed Governor Futrell called out the state militia, and they traveled to the site to assist flood victims and provide temporary tent housing. Conservationists worried about the displaced wildlife. This opening was what Visart had awaited.[34]

If the heavy rains broke levees in eastern Arkansas, Visart would rush to those areas to protect the wildlife from market hunter slaughter. If this occurred, he needed funding. "I shall not permit myself to think of giving up," Visart concluded. Governor Futrell suggested that he, Judge William F. Kirby, state senator Lee Miles, Visart, and maybe some other of Arkansas's leading conservation advocates conduct a fundraising tour through speeches about wildlife, expecting to raise at least five hundred dollars. The governor also promised to write a letter to the people, which newspapers presumably would publish, urging the citizens of Arkansas to assist Visart in his efforts to protect wildlife. Visart informed Futrell of McIlhenny's idea of appealing to other conservations outside of Arkansas for assistance, and the governor approved. Visart made the call through the state newspapers, and McIlhenny requested assistance through his national connections. If Arkansawyers saw that nonresidents committed to conservation efforts outside of their state, Visart believed, residents might be more apt to donate too.[35]

Several levees ultimately broke, causing people and animals to flee to higher grounds. Millions of acres flooded across several states along the Mississippi River, causing billions of dollars in damage. "The greatest opportunity for conservation work is now being offered in Arkansas," Visart wrote McIlhenny from Jonesboro. Visart claimed that thousands of deer would drown if someone did not intervene to save them and he estimated costs at one hundred dollars per day for the efforts. Many Arkansawyers donated, but Visart considered the problem too large for just residents to fund. He needed access to a larger audience of donors. The game warden suggested naming the fund the "McIlhenny Fund for the Preservation of Wildlife of Arkansas." Visart claimed that with such a public effort, the Arkansas legislature would have no choice but to listen to McIlhenny during the 1915 session.[36]

The AFGPA voted to use all remaining funds in their accounts to save the stranded animals. Visart now had the means to travel to the flooded areas and coordinate efforts there. McIlhenny told Visart that he would assist in any way that he could. The governor informed Visart that he could order the state militia to help with wildlife protection and rescue if he did not receive the local officials' cooperation. Futrell claimed he wanted to give Visart the police power to enforce the game laws during this crisis, but he did not have the authority to do so.[37]

Visart jumped into the battle. On April 9, he visited Jonesboro to recruit more assistance. Three days later, the warden pursued two Cross County men who had killed two deer that had floodwaters had pushed into town. Two days later, fifty men arrived in Helena with boats to save as many trapped animals as possible from the islands.[38]

Visart intended to follow the rising waters. As a levee broke, he planned to rush to that area because of the immediate danger to the wildlife there. If he was not on site, the state game warden told his deputies to submit any evidence of game violations to the sheriff. If the officer did not prosecute the criminals, the wardens should gather enough evidence for a grand jury. Visart and his deputies were not the only Arkansawyers fighting to protect the helpless wildlife. Judge A. B. Grace of Pine Bluff contacted Desha County sheriff George R. Lacy, telling him to prevent the slaughter of trapped animals. During the previous year's flooding, poachers slaughtered several stranded deer out of season, and Grace did not want this to happen again. He promised to convene a grand jury to assist in the efforts if needed. Governor Futrell issued a decree not to kill the stranded deer. Visart did not receive assistance from local authorities in every county, though. He complained that law enforcement officials in St. Francis and Poinsett Counties refused to help him protect the stranded wildlife. However, the slow rise of the water had allowed many animals in those counties to get to higher ground. Plus, citizens from other counties assisted Visart where they could.[39]

"I am making a great hit with the people of the state in this effort to save the wildlife," Visart excitedly told McIlhenny. He thought that the event's work would pay off in the future and convince more people to support stricter game laws. "You never saw such a change in sentiment of the people in your life," Visart chirped. He enthusiastically concluded that the small amount of work "strengthened our cause more than you can imagine." While making his rounds, he even managed to sell some subscriptions to the *Illustrated Outdoor World and Recreation*.[40]

McIlhenny apologized that he could not come to assist Visart during the floods. He promised to return to Arkansas "to have suitable laws passed" and "attempt to establish . . . game refuges within the borders of your state." The philanthropist confided that he had just closed a land deal for one hundred thousand acres on the Mississippi gulf coast, saying that it could have happened in Arkansas.[41]

By the end of April, the floodwaters started receding, which allowed Visart to assess the level of success of his efforts. He believed that his actions had saved hundreds of animals. During the 1912 floods, several market hunters had killed hundreds of deer, but this time reports indicated that the number of deaths in the flooded regions remained low. "Patrolling the flooded districts" made the difference, according to Visart. "All we need is cooperation of the sportsmen, and the game of Arkansas can be saved until we have adequate laws to protect them," Visart concluded.[42]

Visart provided Governor Futrell with a "Game Protection Report" as part of his flood investigation. He informed the governor about McIlhenny's land purchase. Arkansas had missed an opportunity. However, the game warden argued, by the time the Arkansas legislature reconvened in 1915, McIlhenny would have more funds raised, possibly for Arkansas land.[43]

After his success in the delta, the Arkansas conservationist seemed reinvigorated in his chosen vocation. "My work is now becoming a pleasure," Visart admitted. Wherever he traveled, locals greeted him, appreciated his work, and wanted to hear his message. He possessed enough funds to print hundreds of Arkansas Fish and Game Protective Association membership blanks and intended to crisscross the state, establishing new chapters in every town. He expected to add five thousand members over the next year, bringing the total to ten thousand by the next legislative session. He sent a lifetime membership card to McIlhenny. "I am going after this thing to win," he excitedly told McIlhenny. Although his mood had brightened, Visart sank deeper into debt. Visart had hoped his work with the magazine company might fund his interstate travels and living expenses. He asked McIlhenny for assistance securing a job somewhere, anywhere.[44]

During the summer of 1913, Visart pushed the United States government to set aside some land in the National Forest in Arkansas for a game preserve. He claimed that "if set aside as a National Game Preserve, no doubt [it] would soon be as noted as the Yellow Stone." In July, he asked McIlhenny for assistance securing a position with the Department of Agriculture as an enforcement officer for the Weeks-McClean Act. The law allowed the federal government to set seasons, ban spring shooting, and stop plumage and meat sales of migratory birds nationwide. It served as the first national law regulating the shooting of migratory birds and had passed in March 1913.[45] Assistant chief Theodore S. Palmer and William F. Bancroft of the United States Biological Survey, a division of the Department of Agriculture, assured Visart that he had a good chance at the position.[46]

All of his hard work had paid off, along with the proper connections. On October 11, 1913, the United States government named Earnest V. Visart a United States game inspector under the Weeks-McClean Act. Finally, he could climb out of debt. He became one of thirteen head inspectors and oversaw between twenty-five and thirty deputy inspectors covering Arkansas, Mississippi, Oklahoma, Texas, Louisiana, Missouri, and Tennessee. Critics pointed out that he got the job, "now we will see if he protects the game or the nonresident hunter." The editor of the *Jonesboro Daily Tribune* defended Visart, claiming that his work had saved countless animals from slaughter and continued, "He does not deserve the criticism and abuse heaped upon him." The unkind remarks are not appreciated, the editor claimed, and "Mr. Visart should be encouraged, not vilified."[47]

The following month, the Arkansas Supreme Court made a ruling that Arkansas conservationists had advocated for several years. In November 1913, a crucial legal case about local Arkansas law occurred over nonresident hunting, and it had a rippling effect across the state. Visart had caught nonresident

George Lewis hunting unlawfully in Grant County a few weeks before. When his case finally came to court, local officials found him guilty and fined Lewis ten dollars. He appealed the case to the Arkansas Supreme Court and won. Based on the case, the justices declared the local game law that Lewis had violated unconstitutional because it gave "privileges and immunities to people of Grant County that were not given to people living outside of that territory." The decision invalidated "every county, local or sectional game law that discriminates between the privileges of resident and non-resident hunters." Almost all Arkansas counties had their own game and fish laws. Until then, county officials commonly passed local game and fish laws that did not pertain to any other areas than their own. This decision caused state legislators to work on statewide regulations, invalidating all county-specific game laws, the good, the bad, and the ugly.[48]

Most importantly, the Arkansas Supreme Court ruled in the Lewis case that the state owned the animals and fish living in the state. Because all Arkansas citizens owned the wildlife, county laws could not require people residing outside the county to obey a rule that the county citizens were not required to follow. One newspaper declared, "It is now possible for Arkansas nimrods to hunt where and when they please, so long as they respect the federal laws that recently were made effective."[49]

By 1913, national sporting organizations and the federal government took more interest in migratory waterfowl. A National Game and Fish Commissioners report noted that Arkansas and Texas had no waterfowl season, while four states had no bag limits on waterfowl. They recommended that Arkansas and Tennessee adopt a closed season and bag limits (twenty-five birds) for migratory birds. The commissioners also recommended that each state have its own game and fish commissioners oversee wildlife protection, that spring shooting should end nationwide, aigrettes and heron plumage markets end, and "that politics and game wardens be kept beyond reaching distance." In other words, political infighting and change should not influence the state game law enforcement structure. Finally, the organization recommended that all states support stricter federal game and fish protection laws.[50]

As it stood, twenty-five states each had one dedicated officer enforcing their wildlife laws. Eighteen states had a commission. Five states, including Arkansas, left enforcement to county officials. Thirty-six states had state resident licensing laws. Arkansas, Mississippi, and Florida were among the states that did not require residents to purchase a state license. By 1914, Arkansas was the only state with deputy sheriffs still enforcing state game and fish regulations.[51]

In January 1914, newly minted United States game inspector Visart had contacted Arkansas attorney general W. L. Moore, asking him to provide an opinion on Mississippi County's exemptions on the state law forbidding the shipping of game and fish from the state. Moore claimed the 1909 measure that allowed the shipping exclusion discriminated against other Arkansas citizens. Possibly taking a cue from the Arkansas Supreme Court ruling in the Lewis case, the attorney general ruled the 1909 act void. Although challengeable in court, this opinion went a long way in stopping any game and fish shipments from Arkansas. It took another Supreme Court ruling to kill the exemption finally.[52]

In February 1915, the Arkansas Supreme Court ruled on *Adams et al. v. Jonesboro, Lake Charles, and Eastern Railroad.* Big Lake market hunters had consigned the railroad company to ship a few barrels of ducks to Chicago. When the ducks arrived in Chicago, the Illinois game wardens confiscated the ducks, acting on behalf of the United States government. The market hunters sued for losses and a Mississippi County court found in their favor. Still, the railroad company appealed the case, claiming that shipping the ducks contradicted the Arkansas law (Whitley Law 1905) banning shipping wildlife from the state.[53]

J. N. Adams and his market-hunting colleagues sued the railroad company for damages, claiming that the business was responsible for the cargo until it reached its destination in the Chicago market. The Supreme Court ruled the special Mississippi County township exemption unconstitutional because it discriminated against other Arkansas citizens. Sure, other Arkansawyers could ship to Mississippi County first and then out of the state, but such actions proved inconvenient for people living in distant counties. Therefore, according to the Arkansas constitution, the government cannot grant special privileges to the few, and consequently, the Little exemption was null and void. The case closed the Mississippi County shipping loophole permanently. The exemption had ended, striking a massive blow against market hunters, especially those around Big Lake. With the shipping loophole closed, conservationist organizations like the Arkansas Fish and Game Protective Association could focus on tightening waterfowl bag limits and hunting seasons.[54]

The Arkansas Fish and Game Protective Association members worked to develop a bill based partially on the earlier failed Futrell and Martin bills. The new Futrell legislation (rather the AFGPA's) included provisions to set up a three-man governor-appointed game and fish commission board: the commission "shall have management of all matters pertaining to bird, game and fish protection, and shall receive all funds arising from the sale of hunting and fishing licenses and from fines imposed." They had complete control over this money and could hire wardens, publish regulation books, set seasons, restock fish, mammals, and birds, make the laws (including licenses), enforce the rules, and buy land for wildlife propagation. A petition containing signatures from hundreds of sportsmen went along with the bill. Senate Bill No. 128 was the state's first complete game and fish bill. Legislators attempted to exempt thirty-six different counties but failed. No county exemptions this time.[55]

The ever-present Visart, even though a United States game inspector, remained the AFGPA secretary and had assisted in developing the bill. Visart believed that Futrell's legislation had promise because of the groundswell of citizen support that he had worked to cultivate over the past few years. State and county conservationist organizations like the Phillips Gun and Rod Club in Helena, the Arkansas Fish and Game Protective Association, and various ladies' clubs lobbied toward creating a game commission. Speaker of the House L. E. Sawyer had promised to support the bill when it reached the floor. Big Lake Shooting Club secretary Green Benton wrote to McIlhenny to offer the club's assistance in passing conservation laws. Benton claimed that Mississippi County representative Edward Alexander would support any measure at the club's request, stating, "I believe we can bring pretty strong pressure to bear on him to get him to do anything within reason for us." This support was quite the change when Representative A. G. Little fought tooth and nail from Mississippi County against all game laws only a few years before in the interest of the market hunters who slaughtered on Big Lake.[56]

The bill blew through the Arkansas Senate in early February with a vote of twenty-eight to four. When it passed the House, fifty-four to thirty-two on February 26, 1915, Visart wired McIlhenny a four-word message: "General Game Law Passed." The Louisiana bird conservationist called the passage a "great achievement for those who have so faithfully worked in Arkansas for the passage of this law." Furthermore, the philanthropist would begin work on creating a bird refuge on Big Lake in the sunken lands of northeastern Arkansas, just as he had promised. He already had commitments from many Big Lake Shooting Club members to turn over their stock to him.[57]

One observer claimed that the legislation was clear and concise, unlike all before it. The act stated that no new laws could take effect until they were posted in at least two state newspapers and had the governor's signature.

State officials had financial oversight. Wardens would have the police power to enforce the rules. The bill removed all previous game laws from the books. With a long history of convoluted, misleading, repealed, exempted protection laws, they needed to start with a clean slate.[58]

The year 1915 brought massive changes nationally in wildlife protection. State governments passed 240 new game laws, and 1915 served as Arkansas's turning point in protection laws. Arkansas added bag limits for bear, deer, quail, turkey, and waterfowl for the first time. Big game seasons shrank to ninety days, the shortest ever. The shipping of wildlife out of Arkansas had ended. Resident hunters now purchased a license to hunt certain animals. They paid one dollar to hunt deer and one dollar to hunt quail with dogs. To incentivize owners to license dogs, the bill made stealing a licensed dog a felony. One of the most significant changes was the establishment of the Arkansas Game and Fish Commission. It had a governor-appointed five-member panel, and they had the powers, as mentioned earlier, to protect Arkansas wildlife. This panel appointed game wardens for two-year periods. The commission also had to provide a yearly report to the governor.[59]

Newspapers announced the passage of the new game bill, and several citizens responded with confusion about what it meant for them. A few residents wrote complaint letters to Arkansas governor G. W. Hays, thinking the bill ended fishing and hunting. Other Arkansawyers questioned why the state government protected the wild animals and fish as the people went hungry.[60]

Clark County resident E. Pickins claimed that he spoke for the "sitisins of this county" when he argued, they "don't want no laws to stop fishin. . . . no bills . . . against the sal of fish." He was angry that representatives of other counties were making laws that pertained to his county. If Hays did not sign the bill into law, Pickins claimed, "you will earn hundreds of frinds." Director of the Farmer's Supply Company J. S. Gibson of Black Rock told the governor, "If you sign this bill it will put hundreds of people out of employment. It [is] bread and meat to their families & they are not able to pay such a heavy tax these hard times. The Lord put the fish here to Ketch & eat & it matters not how much they trap & hoop-net."[61]

Waco resident W. J. Johnson wrote, "Mr Govner Hays Dear sir to day I finde in the ComerShell apeel [Commercial Appeal] news paper where you have made it a law for no game to Been killd." If the Arkansas government wanted to protect something, argued Johnson sarcastically, why not protect his children from chiggers. The government should first help the Arkansas citizens. He continued, less facetiously this time: "We aint got the Money and we cant get any thing to eat and our faimlys are naked for Cloths and Bare footed." There is no record that Hays answered any of these letters.[62]

Hays did appoint the first five commissioners, however. They served with no salary, but they did receive expenses. The list resembled a who's who of Arkansas elite, but Hays carefully chose representatives from different parts of the state. D. G. Beauchamp from Paragould became the commission's first director. Beauchamp was an attorney in Greene County and a quail hunter who raised Llewellyn setters. Dr. Horatio Wells from Monticello had worked as a surgeon at the state mental asylum and with the Second Arkansas Regiment in the Spanish-American War. He had retired from the medical field and worked with Arkansas zinc mines. Texarkana attorney George Webber raised registered hogs and pointers. He had served on several railroads and banks boards. Senator C. C. Calvert from Fort Smith operated a printing company with partner John P. McBridge. He was an officer in the Democratic Party Committee in western Arkansas and the Arkansas National Highway Association. Calvert also raised pointers and hunted quail. Lee Miles from Little Rock, who became the commission's first secretary, had served as Pulaski County's state senator from 1911 to 1913. He continued as a county and probate judge in the capital city. Reports called him a "thorough sportsman and a lover of birds."[63]

11. First Arkansas Game and Fish Commission, 1915. (*Left to right, standing*) Wardens --Foster Rogers, Med Donaldson, Joe A. Galvin, Herschel Neely, Clyde Marsh, Wash Clibourne, James Fernandez, George Rison Jr., Ed Huddleston; (*left to right, sitting*) Commissioners --Dr. Horatio Wells, George Webber, C. C. Calvert, Lee Miles, D. G. Beauchamp. Courtesy of the Arkansas Game and Fish Commission.

During their first meeting, after choosing a chairman and a secretary, they all agreed that if they were determined to achieve any results in their work, the members had to remain outside of politics. They did not want undue influence placed upon them, and each commissioner pledged to stay politically objective. Members agreed that education sat at the top of the to-do list. The public needed to know the law. The legislature had ordered one thousand copies of the new game laws to be distributed statewide and several state newspapers published the new law in its entirety. One newspaper cautioned readers, "Do well to study this law carefully, not only that you may be guilty of its violation, but that you may be enabled to advise its provisions to the less informed so that they may keep out of trouble." The commission expected the wardens to spread the word as well.[64]

Enforcement, however, proved the most challenging issue for the commission. The first issue was funding for warden salaries. The law designated that fines and license fees would fund the warden system, but how could the AGFC hire wardens to track down violators who paid these moneys unless they had adequate funding to start? Most licenses were only one dollar, and filling an enforcement chest at that rate took a while, which led to the second issue.[65]

They needed to establish a way to distribute licenses. At first, the commission announced that all license seekers should send their fees to the state treasurer, who would provide two receipts. One of the vouchers went to the state auditor and the other to the customer. The license seeker then sent a complete AGFC-supplied application form and the receipt to the Arkansas Game and Fish Commission secretary, who would then issue the license. The commission did not even have the money to print blank applications. This convoluted system would never work anyway. The commission needed locations in each county to sell licenses instead, and they needed start-up money.[66]

Additionally, and most crucial, the Futrell bill did not include appropriations. The commission realized they had a few hundred dollars from license fees in the treasury but no way to access those funds. In the past, the Arkansas Supreme Court ruled that any withdrawals from the state treasury needed specific appropriation, and the Futrell game law did not contain any such allocation. State auditor M. F. Dickson agreed that the law forbade the AGFC from withdrawing funds. Regarding the fiasco, the *Monticellonian* remarked, under "the new game law, the Commissioners work without pay; and from a reading of the Bill, the wardens are to be fed on thin air and respectability."[67]

Senator Futrell, however, argued that the AGFC could access the treasury funds through a governor proclamation and AGFC voucher. The law required that all license fees go to the state treasury and not into a direct commission-controlled account, he contended, allowing for proper handling of the funds through accounting. He claimed that the legislators who passed the bill did not make an appropriation because they did not know how much the fund would contain, and any allocation required specific amounts, something they had no way of knowing. "The failure to make an appropriation was not an oversight," he concluded.[68]

Within five weeks, the fee account contained $1,200. By that time, the commission issued approximately two hundred licenses per day. The AGFC could still not access the money despite what Futrell had claimed. At one of their April meetings, the commission decided to petition the Arkansas attorney general, W. L. Moose, for his opinion about their access to the funds. While they awaited the attorney general's decision, the commission went forward with gathering information about potential game wardens. They set the annual salary for a warden at one thousand dollars plus travel expenses. They had five men in mind for the warden positions and asked a couple to work part-time until they heard from the attorney general. The commissioners decided to wait for the attorney general's opinion before permanently appointing anyone.[69]

At the end of April, Attorney General Moose announced that the commission did not need a "specific legislative appropriation" to access the money. After receiving the news, the AGFC submitted warrants to the state treasury for a little over five hundred dollars for office and commissioner traveling expenses. More importantly, they started officially hiring game wardens.[70]

The AGFC had hired Texarkana chief of police Foster Rogers as the first warden, but he did not start his job immediately because of the funding question. In April 1915, he left his career with the city. Quickly disillusioned with wardening, Rogers ran unsuccessfully for Miller County sheriff in July. By 1917, Officer Rogers had returned to his old position as Texarkana's chief of police.[71]

Warden Meriwether "Med" Donaldson came from a family who owned several large farms near Paragould and Lake City in northeast Arkansas. Forty-five-year-old Hershal Neely sold dry goods in Paragould. Joseph "Joe" A. Galvin worked as a salesman in Stuttgart. Little Rock deputy constable George B. Rison, Jr., son of Little Rock sheriff George B. Rison, Sr., remained a game warden until 1920.[72]

The other four appointed wardens were Marshal J. Clyde Marsh from Prescott, Faulkner County deputy sheriff Washington E. "Wash" Clibourne Sr.

from Conway, Fort Smith chief of police James "Jim" Fernandez, and McGehee city marshal Ed Huddleston. Visart had appointed Huddleston while he was state game warden. Huddleston never left his job as McGehee City Marshal while working for the AGFC. He ran unsuccessfully for sheriff in November 1915, a few months after his appointment. Huddleston left the commission and became a Desha County deputy sheriff. These men had full-time jobs when the commission hired them. Some quit their old positions to game warden, while others worked both.[73]

More than half of the new wardens had law enforcement experience. One was a former game warden. Only three stayed with AGFC after 1919, most likely because of the poor salaries and the extensive travel. Of the originals, only Clibourne and Rison remained on the payroll until 1920.[74]

Only time could determine if the current funding program could support more wardens and if the commission needed more men to enforce the law. "We are all beginners in this line of work," wrote Secretary Miles to McIlhenny during that summer, "and would very much appreciate a call from you [and] any fatherly advice that you may feel disposed to give us."[75]

As the commission had outlined, the new wardens' first job was educating the public. During the first summer, Warden Clibourne spent a month in eastern and southern Arkansas, "inform[ing] the people of the provisions of the law and to warn them against future violations." The Arkansas Game and Fish Commission reported that courts had only convicted seventy violators of the new laws four months later. These violations mainly constituted illegal seines and netting. However, the regulations included a multitude of new and revised rules.[76]

Section 19 of Futrell's new law banned the shipping or carrying of any game or game fish out of Arkansas. Letters flooded the governor's office concerning this statute, not from commercial hunters or fishermen, but nonresident sportsmen, most from Memphis. A St. Louis businessman asked if he owned land in Arkansas, could he carry the game back to their home state? Shelby County, Tennessee, sheriff J. A. Riechman asked, since he was a member of a club in Arkansas, if he could carry home fish. But the AGFC did not consider nonresident sportsmen a top priority.[77]

According to the commission, the wardens' most difficult day-to-day issues were finding the owners of illegal nets, seines, and traps. The owners set the traps, then left the area and periodically came back to check the devices. No names appeared on the contraptions. Wardens could not sit in one location and wait hours or days while other sites needed patrolling. Burning the illegal equipment proved the only recourse for the wardens. Illegal fishing exemplified the most extensive violations in the commission's early days. The nets seemed to fill every Arkansas waterway. For example, during one trip along the White River in the spring of 1916, two wardens pulled out and burned eighty-two illegal nets. When word reached the surrounding area, other fishermen rushed to pull their nets from the water to save them from burning.[78]

A warden's job was also dangerous. During the first year of operations and waterway patrolling, four AGFC wardens had their boats stolen or damaged. On the White River, two vessels disappeared, and hooligans smashed another boat's motor. On the Cache River, vandals shot another warden's boat full of holes when he left it unattended.[79]

Fish meant money. The commission argued that market fishermen shipped approximately twenty million pounds of fish out of Arkansas during a twenty-month period. The average price registered five cents per pound or one million dollars. The lower White River alone produced 126 carloads of fish at twenty-four thousand pounds per car. With other types of shipping, the commission believed that nearly four million pounds of fish came out of the lower White River annually and went to market. The new laws were meant to curtail such activity, but the job proved difficult. Luckily, Futrell's law provided some tools for the wardens to fight these crimes.[80]

One of the most remarkable sections of the new game law had to do with licensing. Under the provisions, game wardens had the right to revoke a license immediately if the holder refused to provide information about law violators, and a court could fine the individual for the same. So, commercial fishermen could lose their licenses and livelihood if they did not inform on fellow fishermen for illegal activity. Another clause forbade drunkards from getting any permits.[81]

The fines from these various infractions remained low, however. By November 1915, the total fine money in the state treasury reached a paltry $83.78. On January 1st, 1916, the total read only $165.48. One of the most apparent reasons for the small amount had to do with the amount of the individual penalties. For example, when Warden Med Donaldson caught Paragould residents R. P. Marley and H. N. Henricks shooting quail out of season in November 1915, justice of the peace A. B. Hays found them guilty and fined them only one dollar. Another reason was that many of the violators appealed their convictions from the justice of the peace to the circuit courts, and the process took time.[82]

The licensing fees, however, were a different story. Since the law had passed, license money poured into the state treasury, sometimes at $200 per day. During the first nine months, the state collected $6,352 from commercial fishing licenses, $4,810 from artificial bait licenses, $3,853 from hunting licenses, $2,358 for dog licenses, and $320 from trapping permits for a total of $17,466. They issued 11,293 different permits. Expenditures included salaries, travel expenses, office supplies, and purchasing boats and motors at $17,220.26. During the first six months of 1916, licensing fees brought in $7,401, with fines lagging at only $229.75. The license fees during those first six months did not equal or surpass the same period during the first year because the AGFC did not have enough blank 1916 license forms.[83]

The new year also brought more financial woes for the embattled commission. Acting on behalf of H. F. Jackson of Lonoke County, attorney Hal Norwood filed a mandamus suit attempting to halt the commission from withdrawing its funds from the state treasury. The lawsuit claimed that since the legislature made no appropriation for the AGFC during its last session, it was illegal for the commission to access the funds with warrants. The board members announced that they had dismissed the current nine wardens until the Pulaski County Chancery Court decided the matter.[84]

Commission members and several state citizens claimed that the move originated either from nonresident sportsmen who wanted to stop the enforcement of Arkansas game and fish laws or political opponents bent on embarrassing some commission members. Nonresidents could not obtain licenses, nor could they ship game or fish from Arkansas. These clauses angered many of them, especially if they owned land in the Bear State. Several Tennessee men, members of the Big Lake Shooting Club near Manila, received fifty- and one-hundred-dollar fines for nonresident hunting, although Governor Hays had waived their penalties.[85]

In pretrial discussions, the commission and Norwood agreed that the AGFC could access funds to pay all the current outstanding bills, so at least the wardens and secretarial staff could receive their salaries. The editor of the *Batesville Daily Guard* suggested two possible outcomes if the court ruled against the AGFC, both rather dire. The wildlife would have no protection because the commission dismissed the wardens or hunting and fishing in Arkansas would end. After all, no one could obtain licenses without forms and no way to issue them. Sure, local authorities could enforce the laws, but that had already been a problem in the past and one of the causes for the commission's creation in the first place.[86]

The lawsuit and warden dismissal could not have happened at a worse time. Winter rains had driven several eastern rivers from their banks, pushing fleeing animals to higher ground, where they concentrated. The Ouachita, White, Cache, St. Francis, Arkansas, and Mississippi Rivers rose above flood stage, flooding thousands of acres. AGFC game wardens calculated that nearly one thousand deer were stuck on higher ground. In response to the wildlife emergency, the AGFC appointed ten extra wardens and asked the others to return temporarily without pay. Sheriff's offices in Desha and Phillips counties employed special deputies to stop poaching.[87]

According to the AGFC, white farm owners forbade African American farmhands from carrying firearms during the flood. They inferred that this proved the only method to halt Black workers from breaking the law, as if the temptation of stranded animals was too great for their self-control. If farm owners did not take away the guns, they at least ensured African Americans did not carry buckshot, so theoretically, small animals were the only prey they could kill. In its first annual report, the commission published a picture of approximately twenty African Americans of various ages who had fled the floodwaters. The caption under the photo read, "More hazardous to the game driven from its refuge by the overflows, than panthers. They are the real menace."[88]

During another flood a few years later, AGFC secretary Lee Miles reported that the wardens had arrested and convicted eight African American men for killing one deer. He claimed, "The negroes have been more of a menace during the floods than the white people." Miles insisted that no African Americans had assisted efforts to protect the wildlife, while "practically every white person [had] lent his assistance."[89]

But what Miles later claimed in 1922 was not reported in 1916. The newspapers actually reported the opposite. When information arrived in Little Rock about illegal poaching, most violators were white. Wardens hired locals, including several African Americans, to monitor the wildlife, report any violations to authorities, and provide animal sustenance. As the AGFC efforts continued, two events brightened the situation. Many local citizens assisted officials in reporting violators, animals' locations, and general cooperation, a successful turn from past experiences. Second, Pulaski County Chancery Court Judge John E. Martineau ruled for the Arkansas Game and Fish Commission. For now, they could access the funds through warrant draws, but the legal battle did not end.[90]

When the floodwaters receded, and the danger had passed, the AGFC issued a public report. The following fall, hunters claimed the deer had increased in population and pointed to the commission's protective efforts.

The commission made a special effort to thank the citizens of the eastern counties for their cooperation, and they credited the public assistance in saving many animals' lives. During the emergency, the state wardens, special deputies, and even locals had protected 180 deer and twelve bears. Prosecutors had convicted fifteen deer poachers. The monetary costs totaled a little over one hundred dollars for the extensive effort. But there was a problem.[91]

Although the Pulaski County judge had ruled that the AGFC could access their money, Arkansas state auditor M. F. Dickson and state treasurer R. G. McDaniel refused to release their funds and assets from seven other accounts, "unless ordered by a court." In May 1916, the Arkansas Game and Fish Commission, through game warden Wash Clibourne, brought mandamus action against the pair to force them to issue the moneys. Clibourne asked the court to compel the officials to pay his salary of thirteen dollars. At noon on May 8, circuit court judge G. W. Hendricks confirmed the writ of mandamus, ordering the two Arkansas officials to pay Wash Clibourne. But Wash did not get paid in May.[92]

Arkansas state auditor M. F. Dickenson appealed the AGFC funding case to the Arkansas Supreme Court. Although two county judges had ruled that the AGFC did not need a specific legislative appropriation to access their funds in the state treasury, the Arkansas Supreme Court declared they did. It ruled that "a specific appropriation is an absolute prerequisite of the drawing from or payment out of the State treasury any money therein required to be appropriated." Therefore, the AGFC commission could not access the $12,900 in their account until the Arkansas state legislature met in 1917.[93]

"Game and Fish Wardens May Have to Quit Job," read the bold headline of the *Batesville Daily Guard* that July 5. AGFC wardens had already suffered through funding issues, political attacks on the commission, attacks on equipment, and lawsuits. Now, the commission could not pay their salary for six months. According to the newspaper, the only way to continue was if the wardens were "financed by private individuals." Luckily, they did not have to take such action.[94] In response to the court ruling, Governor Hays issued a proclamation. He released the AGFC funds to pay the salaries and expenses through a voucher.[95]

The commission estimated that Arkansas contained approximately two thousand deer statewide in 1915, a stunning few. They blamed the paltry number on several factors, including overhunting, loss of habitat, and waste. They argued, however, that the most critical factor for the significant decrease in the deer population was the killing of does. For example, one eastern Arkansas hunting party harvested twenty-one deer one fall, and eighteen of those were does. Female deer can birth many fawns over their lifetimes. Therefore, killing one doe is the equivalent of slaughtering multiple deer. If sufficiently protected, the commission argued, the deer population remained high enough to repopulate the state. They suggested that the legislature ban deer hunting for five years to give them time to rebound, a measure that was an impossible ask. Most lawmakers believed that the opening deer hunting season for only sixty days and closing it the rest of the year was enough. At the very least, the commission thought officials should pass a measure outlawing the killing of does and the chasing of deer using hounds. Despite concerns, during the 1916 fall, the commission issued over five hundred deer licenses to Arkansas hunters. They did not have the power to halt license sales.[96]

However, the deer, bear, turkey, and prairie chicken populations were so low that market hunters no longer targeted them regularly. They became a target of opportunity. Instead, the migratory birds that passed through Arkansas twice a year became the primary objective for market hunters. For example, in November 1916, Warden George Rison Jr. arrested market hunter John Morton for killing more than the twenty-five-bird limit allowed for ducks. Rison confiscated the waterfowl and distributed them to charitable organizations around Little Rock.[97]

Since the commission's inception, its wardens had gone after market fishermen, and now, they declared war on the duck shooting market hunter. With access to pump and semiautomatic guns, ice, cheap cartridges, and more efficient shipping methods, market hunters continued slaughtering migratory waterfowl. "This type of citizen is beyond all question the most detrimental to the existence of game in this State," argued the commission. Market hunters used any means necessary, illegal or not, to butcher as many birds as possible and did not fear prosecution. They do not care how many, what kind, or when they kill, just as long as there is a market.[98]

In closing their first annual report to the governor, the commission urged more protection laws. They wanted a five-year ban on prairie chicken hunting, a squirrel season, prohibition on squirrel sales, reduction of duck bag limits from twenty-five to fifteen, a bag limit on geese, a ban on killing does, bag limits on bass, licensing for each net or seine, allow the AGFC to ship fish for scientific and stocking purposes, enable the sale of fish and game raised on private lands and waters, permit the commission to create bird hatcheries, a gun license, make dynamiting fish a felony, and create a furbearing season.[99]

Most importantly, however, the AGFC suggested a licensing system for nonresidents. They pointed out that most other states had nonresident permits, and Arkansas should not be an exception. Five nonresident hunting clubs existed in Arkansas, one controlling five thousand acres of land and water. On these lands, deer, turkey, and other animals thrive with the assistance and expense of club members who are forbidden to kill the wildlife. Although not directly stated, the commission indicated that the state rewarded these nonresidents with the ability to hunt the game that they had a direct hand in conserving and increasing.[100]

Why did the initial law not contain many of these essential elements? Commission secretary Lee Miles explained behind the shortcomings was that "no one anticipated that it would pass." Multiple bills had failed in 1913 and during earlier 1915 legislative sessions. Supporters introduced the Futrell bill "in order to keep up the fight for the influence it would have." Therefore, the law's passage shocked many people, and as Miles would later say privately, the commissioners did not know what they were doing once the governors appointed them to their positions.[101]

CHAPTER EIGHT

The Attempted Destruction of the Arkansas Game and Fish Commission, 1917–1920

> The game laws are well enforced now, and the hunter who exceeds his bag limit or who transgresses any of the section of the laws is pretty sure to find a game warden on his trail.
>
> —*Daily Arkansas Gazette*, 1917

> The general game law, as declared by leaders of the movement, does not provide sufficient protection to wildlife in the state.
>
> —*Arkansas Democrat*, 1920

Heading into 1917, commissioners hoped the Arkansas legislature would pass a perpetual appropriations measure during their next session. Concerned conservationists and the AGFC also lobbied for more game legislation, laws that streamlined and clarified those already in existence, and new laws to cover vital areas that the AGFC had identified during their first year. During that 1917 session, newly elected state representative Baldy Vinson and state senator Lon Slaughter led the conservationist-minded lawmakers to pass measures to aid the newly created Arkansas Game and Fish Commission.

The 1919 legislative session, however, proved disastrous to the commission and, by extension, the protection of Arkansas wildlife. With Vinson, Slaughter, and many of their allies gone from the state assemblies, opponents such as state representative James A. Choate of White County and C. A. Holloway of Lonoke County fought the few commission advocates that remained in the legislature. The opposition forces, believing the commission represented government control over individual rights, proceeded to remove as much power as possible from the commission and its wardens. After an unsuccessful attempt to remove all of the commission's funding, they passed legislation that halved the AGFC budget and removed the wardens' police powers. In response, the AGFC relied more upon like-minded Arkansawyers, sportsmen's clubs, local law enforcement, and the federal government to assist them in their duties. Since its creation, the AGFC consistently faced opposition from within the state. The 1919 legislative session represented the zenith for opponents in their efforts to destroy the commission.

During a joint session of the Arkansas legislature in January 1917, the outgoing Arkansas governor George W. Hays presented his departing message. Hays cautioned the policymakers not to spend Arkansas into debt. He admitted that he issued funding proclamations in 1916, including $12,900 to the fish and game fund. But those represented emergencies. Hays nonetheless urged lawmakers to fulfill the Game and Fish Commission's requests in their first annual report, which he included for the record. He went no further to promote continued conservation law development.[1]

At the onset of the 1917 legislative session, newly elected Arkansas governor Charles H. Brough delivered his inaugural address. Besides the call for whiskey taxes and better roads, he congratulated the legislators for creating the Arkansas Game and Fish Commission in 1915. Brough described the action as "valuable, constructive work for our State." He claimed that the lack of permanent funding for the AGFC had severely hindered their work in protecting the considerable number of wild animals and birds. The governor pointed out that more work was needed, especially concerning a nonresident license program. Brough asked the assembly to strengthen the laws, "which would greatly conserve the game of our rich hunting and fishing preserves." Finally, Brough recommended that legislators remove the one-dollar license on bird dogs and place the tax on the individual hunter because several hunters could use the same dog to hunt.[2]

State senator Lon Slaughter, a lifelong hunter who represented St. Francis and Crittenden Counties, and seventeen legislative allies introduced "An Act to provide for a State Game and Fish Commission and to protect the game and fish and fur-bearing animals of the state and to regulate the killing and taking of same" on February 8, 1917. Slaughter and his cosponsors intended to correct all of the issues in the Futrell law. The Arkansas Game and Fish

Commission authored the legislation and included all the requests outlined in their first annual report. The new bill's sections added the commission's ability to import, raise, and distribute game and fish and develop propagation farms. It also included a measure that allowed circuit clerks and the state treasurer to sell licenses, improving the current structure where the AGFC's secretary had to issue them. More importantly, the legislation contained a nonresident licensing system. Nonresidents could hunt for fifteen dollars or fish for five dollars. Noncitizens could even take home one day's catch. Fish dealers had to maintain records, including poundage purchased, place, and person of origin. They paid an annual tax of one-sixteenth of one cent to the AGFC budget. The bill set minimum-size limits of fish that dealers could sell. Commercial fishermen would also have to purchase licenses. Hunters could not harvest turkey hens, doves, robins, woodcock, pheasant, grouse, or prairie chickens until 1922, a five-year ban. Hunters could only harvest deer with five-inch antlers or longer. No night shooting would thereafter be permitted. Hunting seasons covered more animals and were shortened. Bag limits remained but were reduced. Those who dynamited fish could receive a year in prison. Finally, the sale of venison, bear, turkey, pheasant, grouse, quail, and partridges would be thereafter banned. Rabbits had no protection in the new game law. Hunters could harvest and sell them without special fees other than a hunting license.[3]

By the time the Slaughter bill came through the Arkansas legislature, the atmosphere surrounding game conservation laws had changed. More politicians supported the protective legislation, but a vocal minority of opposition remained. Arkansas State Sportsmen's Association attorney Baldy Vinson, also a state representative for Chicot County, urged passage of the Slaughter bill in the House. Vinson pointed out that the commission's wardens saved hundreds of deer during the latest flooding in eastern Arkansas. Vinson noted that several surrounding states collected hundreds of thousands of dollars in fees and fines during the enforcement of wildlife conservation laws. He expected if lawmakers passed the Slaughter bill, the state could collect between $75,000 and $100,000 annually. For Vinson, Arkansas hemorrhaged money as dealers sent millions of dollars in game and fish from the state. Over six million ducks shipped from Mississippi County in 1914 and 1915, worth $1 million. At a value of $2 million, Vinson argued, twenty million pounds of fish left Arkansas during one twenty-month period. The AGFC thus needed the tools to enforce the laws and, by extension, the rights of the Arkansas people. Vinson's leadership proved successful. The Slaughter bill made it through the Senate and passed the House by only one vote. On February 23, Governor Brough signed it into law.[4]

On February 22, 1917, Slaughter also introduced "An Act to appropriate the receipts of the game protection funds for the use and benefit of the Game and Fish Commission." It passed twenty-two to five in the state senate. However, the bill met with concerted opposition when it went to the Arkansas House. This bill dealt with funding, and some opponents proved stingy.[5]

State tepresentatives William J. "Bill" Waggoner of Lonoke County and E. Newton Ellis of Randolph County led the opposition against Vinson and the second Slaughter bill. The *Daily Arkansas Gazette* called the fight over the funding bill "the hardest fight and stormiest session" they had observed in years. Witnesses described the raucous debate on the State House floor. Supporters and opponents battled one another until "toward the last the wildest scenes were enacted, with every member on his feet, many standing on chairs and tables and yelling at the top of their voices for recognition." Observers claimed that name-calling and "bitter remarks" filled the room during the debate, and "common decency" disappeared.[6]

The most contentious issue in the debate was over warden salaries. The original document listed twenty-five wardens with an annual salary budget of $85,000. Waggoner argued that they only needed five officers at $22,500. He believed that sportsmen did not fund the game fund, but instead argued, "It is paid by the farmer and the poor country boy." State representatives J. B. Harris of Madison County and Louis Josephs of Miller County argued that the state needed all twenty-five wardens with appropriate salaries. For these legislators, $85,000 was a small amount compared to the millions of dollars worth of game and fish that left Arkansas annually.[7]

Josiah "Joe" Hardage of Clark County called the proposed legislation "the most outrageous bill offered in this house." He argued that the majority applied the "steam roller" to the minority and that supporters of the bill had mistreated him. Josephs rebutted that no one had poorly treated anyone else. Ellis jumped in, angrily stating, "We are doing deeds this day that will smash this administration into smithereens in the next two years, and you will have to answer for this nefarious measure!" Sebastion County representative T. S. Osbourne said the bill's appropriation amount shocked "his sense of justice and right."[8]

Amendments reduced much of the line-item amounts. Opponents slashed the $85,000 salary budget to $50,000. The boat budget went from $2,500 to $1,500. They cut the $3,500 fish hatchery and game preserve budget completely. After a $52,500 cut, $73,000 remained in the allocation. The State House approved the amended bill in March, and the Senate followed soon afterward. In reality, the amount proved much lower. From April 1918 to April 1919, the amount paid to warden's salaries registered at $15, 873.10 with another $11, 747.57 in expenses. The following year, during the same period, wardens received only $15, 123.24 for salaries and expenses, a loss of over $12,000. The entire AGFC budget dropped from $35, 495.44 to only $25, 250.47 some 29 percent.[9]

With the funding out of the way and the creation of several new laws, the AGFC and its wardens continued their work. Because of damage and theft, they had lost several of their patrol boats during the first two years. They decided to spend some of the $2,500-allocation on a new vessel. A ten-by-thirty-four-foot craft christened the *Mary B* launched into the White River near Des Arc on March 26. The *Mary B*, named after Judge Lee Miles's secretary Miss Mary Blakeney, became the houseboat for the wardens on the lower White River and allowed them to use it as a base of operations thereafter and on the greater Arkansas River watershed.[10]

Nonresidents who were previously barred from hunting or fishing in Arkansas unless they owned land now rushed to obtain permits. Within a month of the law's passage, two hundred nonresidents mailed license requests. The majority of these requests came from Memphis club members. By April 21, resident license seekers could purchase permits from circuit clerks, but nonresidents had to acquire theirs from the commission's office.[11]

The AGFC employed just eight game wardens during the first two years, but with an increased budget, they thought they could hire more help. At least they might employ day laborers. Flood season was upon them. Past experiences indicated that the commissioners needed all the workers they could muster.[12]

Illegal market fishing continued to plague Arkansas. Unlawful seines in Arkansas waters proved too easily used for many unscrupulous fishermen. Game wardens confiscated nets, seines, and other types of equipment on nearly every water patrol. Unless the warden actually observed them launching or checking the traps, they could not prove the violators owned the fishing tools. In the White River, wardens discovered fishermen using two gasoline-powered boats pulling a twenty-foot-deep and eighteen-hundred-foot-long net, scooping up all of the fish in the river as they traveled. They continued to arrest and prosecute fish dealers throughout the state. Reports of dynamiting continued, now a felony offense. The general attitude of many of these men was, "The fish will soon be gone, and I want to realize out of them all I can while they last," claimed the AGFC.[13]

Between 1909 and 1913, several poachers had sued Arkansas State Sportsmen's Association game warden Earnest V. Visart for destroying their nets. Yet, they never won against Visart, proving that the Arkansas judicial system backed his actions. During the summer, R. L. Prestridge of Lake Village sued the AGFC and Warden W. A. "Dutch" Jones for seven hundred dollars in a civil suit for destroying his dismantled seine net. In a desperate act of local resistance, Chicot County deputy sheriff Norman Friedman arrested Warden Jones for trespassing and forced the officer to make a five-hundred-dollar bond, which Commissioner Lee Miles and Secretary A. H. Little signed. Now, could the authority of the Arkansas Game and Fish Commission officers stand up in court?[14]

Warden George Rison claimed that he had warned Prestridge of the illegality of the net several weeks beforehand. The owner agreed to dispose of it but had not. Jones claimed the length of the net stretched to a staggering 3,500 feet, over 1,000 yards. The wardens claimed they had evidence of several illicit fishing activities, and they planned to return to Lake Village to prosecute the perpetrators.[15]

When Warden Jones appeared before the Lake Village justice of the peace J. R. Spragins, he received the maximum one-hundred-dollar fine for trespassing. Jones appealed the ruling, and the case went before the Chicot Circuit Court. Arkansas Game and Fish commissioner Lee Miles, a Pulaski County judge, and commission secretary A. G. Little, observed the trial when this court met. Because of the general attitude of the Chicot County citizenry, Miles and Little decided to hold the next meeting of the commission in Lake Village. They also announced that they would conduct a town hall meeting with residents.[16]

By fall 1917, the AGFC had five members. D. G. Beauchamp from Paragould continued as commission chairman. Texarkana attorney George Webber, Dr. Horatio Wells of Monticello, and Pulaski County judge Lee Miles remained original members. Only one commissioner replacement occurred. Secretary Albert H. Little had replaced state senator C. C. Calvert of Fort Smith.[17]

This group faced the problematic task of educating the public about the newest laws. Most Arkansawyers had just started to accept stricter wildlife protection laws. Despite the best efforts from the ASSA, Visart, and local enforcement officials, the few laws enacted before 1915 went unenforced. But, with more manpower, stricter regulations, and police authority, the Arkansas Game and Fish Commission made more arrests, successful prosecutions, and added progress toward ending most wildlife slaughter. "The game laws are well enforced now," explained one observer, "and the hunter who exceeds his bag limit, or who transgresses any of the section of the laws, is pretty sure to find a game warden on his trail."[18]

In weekly and sometimes daily reminders to the public about seasons and bag limits, educational efforts from Secretary Little assisted ethical outdoorsmen to stay lawful. If Arkansas citizens did not have access to one of the countless newspapers that printed the game laws on multiple occasions, they could write to the AGFC office in Little Rock and receive a free printed copy of all wildlife laws. Ignorance of the law was not an excuse. Secretary Little even corrected federal and local officials on the rules in some cases.[19]

In preparation for World War I, the United States Food Administration under Herbert Hoover began a propaganda campaign for Americans to reduce their food consumption. "Meatless Meals," along with other slogans, asked citizens to choose a day to go without certain types of food. When Hot Springs businessman and Arkansas state food administrator Hamp Williams announced meatless and wheatless days for the state in November 1917, he encouraged restaurants to serve wild game on the meatless days instead of beef and pork. The United States Food Administration expected state officials to overlook wildlife protection laws because of the urgent situation. If this action occurred, the nation could save significant amounts of meat for the military. In response to William's message, the AGFC declared that selling turkey or waterfowl broke game laws. Secretary Little added that if hotels or restaurants bought or served these birds, they violated the law. The AGFC had no plans to suspend the regulations. The editor of the *Arkansas Democrat* agreed, stating that citizens should cooperate as much as possible, but the federal government should not promote the breaking of state laws.[20]

In a self-assessment for their third year, the AGFC claimed that because of their efforts, reports indicated that "game and fish are more plentiful at this time than they were three years ago." The AGFC claimed that when they had started, Arkansawyers fought against them at every step to protect their "ancient rights" of unrestrictive hunting. This resistance may have existed in some locations, but Arkansas contained enough wildlife protection advocates to convince lawmakers to pass conservation laws and create an enforcement structure to impose them. In their annual recommendations for the future, the AGFC spoke directly to Arkansas citizens. They asked a series of rhetorical questions. Did they miss the old days when billions of passenger pigeons flew the skies and millions of bison roamed the plains? Do you want to be responsible for the extinction of other Arkansas game, like the pigeon? Do you want to keep tough game and fish laws or remove them and drive the wildlife to an end? Instead, would you like to see more wildlife in Arkansas in the future? We must protect them now for that to happen.[21]

Interestingly, the AGFC suggested the creation of a state sportsmen's organization with county-based membership that could inform wardens "where the laws have been violated and to have the cooperation of that county's members in prosecuting violations of the law." They expected the county organizations to suggest new game laws and ways to enforce them. But the ASSA had begun before 1900, while the Arkansas Fish and Game Protective Association existed for nearly as long. They were both statewide organizations. Did the commission believe these groups were too politicized, or did they not care for certain leaders of these associations? Why they did not call upon these existing sportsmen's groups is unknown.[22]

The AGFC pointed out that their funds did not allow them to hire enough wardens or provide proper equipment to police the state properly. Nonetheless, the AGFC wardens arrested 660 violators between June 15, 1916, and March 31, 1918, and added $3,450.84 in fines to the state treasury.[23]

With a newer technique in wildlife conservation using biological scientific study, the AGFC determined the spawning seasons of different fish species in the state. Using that information and federal scientific assistance, they quickly developed more efficient fishing seasons. To allow fish to hatch the most offspring possible, the AGFC closed fishing during those periods. In addition, the state fish hatchery needed expansion, as well as a wildlife propagation farm established. With these facilities, the AGFC could restock overfished waterways and refill the forests and prairies. In an earlier statement surrounding the war's necessity to use wildlife as a beef and pork substitute, Hoover acknowledged that wildlife numbers had decreased, some to near extinction. "The decrease in game became so serious that a universal demand throughout the country persuaded Congress to pass a law placing the jurisdiction of migratory game birds under federal supervision," he explained. Hoover said that a national effort existed to save the wild animals. By then, several Arkansas locations already received an infusion of fish.[24]

Claiming that dynamite and illegal market fishing had "cleaned out" some of the Arkansas rivers, the AGFC restocked them as best they could on a limited budget. Besides their own fish hatchery, the AGFC worked with the Federal Bureau of Fisheries in obtaining stock for several Arkansas waterways. Several individual citizens submitted applications for shipments from the federal hatchery. The AGFC had approved those requests. In 1917 and 1918, over fifty separate Arkansas streams, creeks, and rivers experienced restocking from nearly one hundred individuals. Restocking fish was not a new activity in Arkansas. Several fishing clubs and many Arkansawyers had obtained black bass from the United States government hatchery in Mammoth Springs since it had opened in 1903.[25]

Several years prior, E. A. McIlhenny offered to create a preserve along the Mississippi Flyway in Arkansas, but only on the condition that the Arkansas government pass protective measures. Yet, by the time Arkansas finally passed a general game law, McIlhenny had already purchased ground elsewhere. The AGFC believed that the state needed a game preserve. Arkansas wildlife could live in an area without competition or encroachment from humankind. Specifically, they hoped such a place could save the prairie chicken, who might nest there unmolested.[26]

Arkansas Game and Fish secretary A. H. Little claimed that prairie chickens once covered the Arkansas prairie counties of Lonoke, Prairie, and Arkansas. The birds numbered so many that hunters challenged one another to determine who could come back with the most bird heads in one day. In 1918, Little claimed that he knew of only one group of these birds living in the entire state, and those existed on a private propagation farm, closely guarded from poachers.[27]

Additionally, the AGFC requested that legislators end the sale of any game currently under protective laws, tighten net-fishing regulations, create a squirrel season, and regulate market fish size. They also included a recommendation for a general gun license. That is, a permit to hunt with firearms. The regulation of firearms seems like a controversial issue. However, the most troublesome request concerned hunting dogs.[28]

The AGFC wanted to end the chasing of deer or bears with dogs. They admitted that "this [had] been the greatest sport ever enjoyed by the hunters of Arkansas," who had participated in running dogs "from the time of the first settlers." Unfortunately, the commission explained that they expected the state's deer and bear would quickly be extinct if this practice did not end. In 1915, a dog-running party in eastern Arkansas killed eight bears, an incredible number for the time. The AGFC claimed "few bears left" in 1918. The previous year, members of just one hunting camp harvested twenty-three deer using this method. The AGFC believed that if the party had not used hounds, they would not have killed nearly as many deer.[29]

When legislation appeared in 1919 banning hunting bears and deer with dogs, Arkansawyer W. I Varner argued that nature built deer for the chase and that hunters pursuing them had a fun experience, but they also trained as soldiers. Practical training at the time. He disagreed that dog hunting reduced the deer population, claiming that "the axe and the torch" were to blame. Varner proclaimed, "It grows to be a matter of wonder how far such ridiculous legislation can be carried."[30]

Secretary Little, a disciple of early conservationists like William T. Hornady, proffered the most significant reasons for the continued wildlife decline in Arkansas. Little blamed the game hogs and pot hunters, not the market hunter whom sportsmen had targeted for generations. But the legislation had ended much of the independent market hunting. Closing the shipping loopholes and adding new animals to the do-not-sell list caused most professional hunters to find other employment. However, new technologies allowed game hogs and pot hunters to kill more efficiently and wholesale. A game hog kills as many animals or catches as many fish as possible, even if they cannot use them. A pot hunter continues to hunt as the sole means of feeding himself or his family. Sometimes, pot hunters and market hunters meant the same. None of these are mutually exclusive. Little claimed, "The greatest aids to the [game hogs' or pot hunters'] slaughter of game are good roads, automobiles, automatic and pump guns." No one can reason with these men, he argued, because they "possess no sentiment or conscience." Little continued: "They use any means to capture or kill any animal, regardless of age or sex, without fear of prosecution. They sell their surpluses to anyone" and "create the greatest howl when caught," sometimes claiming they are conservationists or sportsmen. Little proclaimed they were not.[31]

Sportsmen followed the rules. They did not kill does, hen turkeys, or young animals. Secretary Little offered examples of "true sportsmen," listing Governor J. M. Futrell, Conway hunter William Cole, DeVall's Bluff resident William H. Wheeler, and Englander T. B. Goldsby. Little claimed that they are the true sportsmen of Arkansas and deserve a spot in the "hall of fame as leaders in the work of conservation of game."[32]

Besides the attempted ban on using dogs to chase bears and deer in 1919, state representative George A. Hillhouse of Jackson County tried to toughen wildlife protection regulations. Unfortunately for conservation advocates, however, the previous leader of prowildlife legislation, Baldy Vinson, died in March 1917. His principal allies, furthermore, Lon Slaughter, J. P. Scrimshire, and Louis Josephs, no longer served in the State House. Instead, anti–game commission forces had grown.[33]

In 1919, Hillhouse went to war with state representative James A. Choate from White County and his supporters over the Arkansas Game and Fish Commission and the general game laws. The Hillhouse House bill amended the 1917 general game and fish law, adding measures the AGFC had requested in their annual report in 1918. Besides the ending of running deer and bears with dogs, the changes included adding or altering licenses for general hunting, trapping, and all hunting dogs. The bill called for more protection for game fish, banning the killing of does and insect-eating birds. It allowed for the propagation of wildlife with a proper permit. All state bag limits would conform with federal limits.[34]

When Hillhouse introduced his amendment to the 1917 game and fish bill, he "encountered violent opposition from some members." Observers called the room "an atmosphere of antagonism." Choate called the bill "an infernal fool law" and claimed the state wardens "discriminate in favor of wealthy sportsmen." Game laws sometimes favored the affluent because they could afford the license fees and did not need the meat to supplement their tables. Therefore, some of the poorer residents had to break the law to feed their families because they did not have enough money to pay the fee or they needed to harvest more game and fish than the rules permitted. State representative C. A. Holloway of Lonoke County argued that AGFC wardens had committed "flagrant abuses of personal rights" in his county. Choate got the Hillhouse bill tabled, sealing its demise.[35]

Seeing his opening, Choate introduced a bill to abolish the AGFC and the general game law. With a vote of fifty-seven to twenty-one, the Choate legislation passed the House of Representatives, proving that most representatives supported the destruction of the relatively new AGFC. If this bill made it through the Senate, decades of work from countless advocates would come to naught. Such an action could set back Arkansas conservation efforts for years and destroy the few remaining wildlife and fish the state had, including sending some into extinction.[36]

When Choate's bill made it to the Senate, Johnson County senator Lee Cazort immediately moved to postpone it indefinitely. Seconds and "ayes" followed, carrying the bill into oblivion. "The bill which had caused so much commotion and strife in the House passed into the archives of the Arkansas Senate," an observer concluded. The Senate had averted a catastrophe for Arkansas's conservationists and wildlife.[37]

However, Choate, Holloway, and their backers were not finished challenging any favorable legislation for the AGFC. State representative Robert L.

Kendrick of Franklin County submitted an appropriation of $100,000 drawn from the fish and game fund to support the AGFC. The funding issue had long plagued commission supporters, and political clashes had occurred over the last four years over the matter. Choate, referred to by many as "an avowed enemy of the commission," oversaw the debate from the committee chair.[38]

Holloway opened up with a salvo against the AGFC, claiming it "had caused more criticism than any body ever created by the legislature." State representative G. B. Knott of Monroe County argued that the commission's wardens did nothing more than "go up and down the state causing trouble." He maintained that if the game officers worked half as hard at doing their job as they did try to get this appropriation, "they might do a durned sight more good." He wanted the AGFC to operate their organization with no money and see how they did. "We will not be responsible for it," Knott concluded. A few other members followed suit, attacking the commission. If the opponents could keep from funding the AGFC, it could not operate unless the governor stepped in with special appropriations. Earlier, the Senate ended Choate, Holloway, Knott, and the other anticommission politicians' efforts to abolish the commission when they permanently tabled Choate's bill. Now, these forces attempted another maneuver to destroy the AGFC by withholding operational money. By the end of the day, the House had tabled the Kendrick appropriations bill.[39]

Known as "one of the hardest fighters against the Fish and Game Commission," Representative Holloway then submitted a bill of his own, amending Act 133 of the Acts of 1917 game and fish. His amendments had fifty cosponsors, more than half of the House membership. Holloway took Hillhouse's defeated legislation and altered it, hoping that the bill could make it through both houses as a compromise. It included all of the significant components of the previous proposal and made tremendous changes regarding wardens. House Bill No. 508's amendments stripped wardens of their police powers and suggested they hold managerial roles. They could arrest only if deputized in the county where the poaching occurred. Wardens had to report violations to law enforcement officers who had jurisdiction in the area where the crime occurred. They had no unique search and seizure powers for illegal shipments of contraband wildlife or fish. The commission could employ no more than eight wardens. These measures alone gutted the efficiency of the game wardens' work. However, the bill weakened the game and fish laws too.[40]

The Holloway amendments passed the House overwhelmingly and the Senate with a twenty-five to six vote. It went into effect as Act 276 on March

17, 1919, committing a significant blow against wildlife protection and conservation. Market fishermen could sell game fish now. If they paid fifty dollars annually, these men could catch any fish species in any quantity they wanted using any tackle, except nets under two-and-a-half inches. In addition, any violation fine money went to the counties and not the general fish and game fund, meaning that licensing became the sole source of revenue for the game and fish fund. Bird dogs required a license, but no other hunting dogs did.[41]

Satisfied with the damage they had inflicted on the AGFC, Choate, Holloway, and their supporters allowed a funding bill through. However, instead of $100,000 for the commission's operation, including salaries and expenses, they successfully reduced the total to $50,000.[42]

With the ink barely dry on the newly amended game and fish law, Pulaski County brothers A. L. Roberts and Jadie Roberts decided to dynamite Parlarm Creek under the main highway bridge. Still a registered felony offense, dynamiting could get the men thrown into the penitentiary. The Roberts had also heavily damaged the bridge's struts during their fishing expedition. This action carried an additional property destruction charge. Warden Wash Clibourne investigated and gathered together enough evidence to arrest the two men. However, because the local authorities had not deputized him, Clibourne had to provide the information to deputy sheriff Clifton Evans, who made the arrests. This incident provided the commission with an idea about how the wardens now had to operate. When Arkansas State Sportsmen's Association game warden E. V. Visart carried out his work starting in 1909, he did not have the authority to arrest either, an incredible hindrance.[43]

Although the AGFC previously requested assistance from Arkansas citizens to stop poaching, these requests increased once the new legislation removed their policing powers. The AGFC claimed that the legislation had "handicapped the work of enforcing the laws to such an extent that there have been many violations where the Commission could not prevent it." The commission needed citizens to help with protection and conservation, but they also needed them to pressure their lawmakers to correct the current system. In November 1919, game warden Arthur G. Steadman, operating in Marion County in northern Arkansas, asked for help from anyone who could provide evidence toward a conviction of a group of Missourians who continued to come into the county and illegally kill deer and turkey. He even offered twenty-five dollars in gold as a reward, which had never occurred prior.[44]

The game wardens also assisted in creating more county sportsmen organizations, another tactic that Visart had used during his early fight for wildlife protection. During the spring of 1919, game warden W. C. Cochran helped organize the Bentonville Fish and Game Protective Association. The new members pledged to protect the newly arrived fish allotment from the commission and to build ladders in county streams as fish pathways during spawning season.[45]

Around the same time, sportsmen from Marianna and Lee Counties organized together to "see that the laws for the protection of game and fish are rigidly enforced." These men intended to print copies of the laws and distribute them so that everyone understood the regulations. "The officers [had] agreed to be extra alert, and with the help of the members of the organization," the sportsmen agreed that they could stop violations in their area. They claimed that many Arkanswayers believed the new game law allowed market hunters and fishermen to resume their unchecked poaching.[46]

Commissioner Lee Miles attempted to restart the effort for a statewide sportsmen organization, thinking it might be a powerful lobbying group in the capital. It had worked with the Arkansas Fish and Propagation Association in 1915. He sent word throughout the state for a meeting in Little Rock of all those interested in wildlife protection to "formulate measures for preservation and propagation of game and fish in Arkansas." This statement is a strong indicator of just how much damage Holloway, Choate, and the other opposing Arkansas state legislators had done to the AGFC. Grassroots efforts had driven Arkansas legislators to pass the general game bill in 1915, and now, the AGFC needed another movement to push the lawmakers to fix the enforcement system. The commission also required help in the field. Miles lined up a slate of speakers for the first meeting, including president of the American Game Protective and Propagation Association John B. Burnham, chief United States game warden George A. Lawyer, secretary of the American Fisheries Association John P. Woods, along with Arkansawyers former governor J. M. Futrell, Joe Belloue, and J. V. Walker.[47]

With only eight wardens, patrolling the entire state proved impossible, especially during the rainy season. During the flooding, all of the AGFC wardens went to eastern Arkansas. Heavily reported in the newspapers, this information opened up the remainder of the state to anyone who wanted to break the law. Again, the AGFC had to lean on conservation-minded Arkansawyers to help save the animals in the inundated areas and stand vigil in other parts of the state.[48]

The AGFC also attempted to team with federal authorities. In January 1920, chief of the Bureau of the Biological Survey E. W. Nelson in Washington, DC, sent a letter to the AGFC asking their opinions concerning a proposed fifty-cent federal hunting license. The BBS could use the collected money to fund the enforcement of federal migratory bird laws, particularly on bird reserves like the Big Lake National Wildlife Refuge near Manila, Arkansas. Chairman D. G. Beauchamp, Secretary Dick Brundidge, and Commissioner Lee Miles all wrote letters in support and offered cooperation in implementing the fee. Miles thought such a fee might "break the ice" for the AGFC's long-sought-after general Arkansas gun license, which had failed to pass during the last two legislative sessions.[49]

Federal game warden Visart patrolled Arkansas to enforce the Migratory Bird Treaty as part of his duties. He teamed with state game wardens on several occasions where their work overlapped. For example, Visart and AGFC warden Wash Clibourne arrested restaurateur Billy Bale in his café on Main Street in Little Rock after an investigation proved the owner served wild duck. In March 1920, Visart and state warden Arthur Thomas arrested five men at Clarendon for buying and selling ducks. Hotelier R. M. Merritt, the purchaser of the waterfowl, Thomas Bolin, Thomas William, R. W. Williams, and Dave Moss appeared before a federal grand jury in Helena facing federal charges, not state. In both cases, as a federal warden, Visart had the authority to arrest the men when they broke federal law.[50]

The embattled AGFC continued to receive negative publicity. In a *Gazette* article, an observer complained about Arkansas's cost for "protection," including figures for 1919. He pointed out that the state spent $33,909 for fish and game wardens but only $19,721 for the state militia, $24,127 on banking regulation enforcement, and $11,957 for insurance. The $41,000 spent on public service was the only amount in this category higher than the AGFC tally. Opposing politicians could use information like this in future debates on funding for the commission, but why did opposition against game and fish protection remain among some Arkansas citizens?[51]

It took decades for budding conservationists to compel Arkansas legislators to pass meaningful wildlife protection laws. Although lawmakers introduced hundreds of minor game and fish regulations before 1915, the regulations were often confusing, contradictory, ineffective, quickly amended, or repealed. Different counties had specific laws, and those changed every two years. Then, the first general game bill passed in 1915, establishing statewide regulations and the AGFC. However, within four years, opposition forces had weakened the agency tremendously, hindering its protective efforts. Conservationists realized they needed to reinvigorate their efforts to influence state lawmakers to correct the Choate-led destructive activities of the 1919 session.[52]

In their 1920 biennial report to Governor Brough, the Game and Fish commissioners explained that Arkansas contained more hunters than in its history. At the same time, game such as deer and turkey had reached their lowest number. Only about four hundred prairie chickens existed in the state, and they survived on private land. To build back the wildlife and fish population, the AGFC needed revenue to fund propagation sites. They also needed money to hire more wardens.[53]

The commission intended to continue their educational programming, teaching Arkansawyers that they "were not being deprived of rights belonging to him as an American citizen when he is denied the right to use the methods to take fish" with dynamite and illegal seines. The AGFC wanted to educate the public about the conservation of the state's natural resources and bring them to the realization that the individual caused long-term damage to his community and state when he poached, an activity that brought him only temporary benefits.[54]

Arkansas sportsmen continued to pass information to the AGFC about illegal game sales. Just as prohibition had ended legal alcohol sales, however, a fellow could find a drink at the right location with the correct password. The same went for a duck dinner. Several Little Rock restaurants served waterfowl if the patron spoke the secret phrase. The state legislature's restrictions on the AGFC made it impossible for adequate enforcement. The commission's eight wardens had to use their time in the woodlands and fields during the fall because of the opening of hunting season and did not have the time to investigate cities and towns for these violations. "The demand for our wardens is such as to need them in a dozen places at once," explained the commission. Some counties, claiming that the general game laws were not strong enough, attempted to pass their own stricter game and fish laws. Lee County even considered hiring county game wardens. Concerned conservationists looked to the next legislative session to determine whether the continued violations could spur enough support from Arkansas citizens to force state legislators to improve general game laws.[55]

CHAPTER NINE

Battling Back

Arkansas Conservationists and the Game and Fish Commission, 1921–1925

> I want to express the appreciation of Arkansas sportsmen and Conservationists [for] the interest you have manifested in this important work.
> —Lee Miles, chairman of the AGFC, 1921

In 1921, the new Democratic Arkansas governor, Thomas C. McRae, declared war on government corruption and waste. With his plan to streamline the administrative structure, the Arkansas Game and Fish Commission again came under attack. Luckily, the ever-growing number of Arkansas conservationists fought off the challenge and managed to strengthen the AGFC during the coming years. However, the commission struggled with a lack of manpower. The AGFC continued to cover thousands of square miles with only eight wardens.

According to historian Sam Hilliard, the deer population reached a low point in the South between 1900 and 1910, during the same time that agriculture acreage peaked. Although shrinking alarmingly, fish numbers remained high enough to continue commercial sales. Many local sporting clubs, individuals, and the AGFC continued to stock Arkansas waterways with thousands of fish. However, the fish came from federal sources.[1]

When outgoing Arkansas Democratic governor Charles H. Brough presented his message to the Arkansas General Assembly on January 12, 1921, he claimed that Arkansas needed a proper game propagation farm and a fish hatchery to restock the state's forests and streams. Brough pointed out that Arkansas remained one of the last states in the nation without these facilities. Hunters possessed greater firepower than ever, thus the greater need for tougher restrictions to protect wildlife. Brough also supported game preserves where animals might find rest and public shooting grounds where outdoorsmen who could not afford land or belong to a club could hunt and fish. He reiterated that the state's game and fish belonged to the state. Harvesting them was thus a privilege, not a right. He thought Arkansawyers who wanted to hunt should purchase a firearm license to pay for the opportunity. Arkansas joined Mississippi and Florida as the only states with no general hunting license.[2]

Nevertheless, the AGFC soon became caught up in the incoming governor's crusade to destroy corruption and political machinery in Arkansas. A banker and lawyer, Thomas C. McRae, the new Arkansas governor, campaigned to eliminate government waste. A postwar recession produced economic uneasiness, creating a significant need for government fiscal responsibility and thrift. Furthermore, McRae contended that political insiders, such as the Little Rock political ring, had long developed ways to control the government through bureaus, commissions, and other offices. McRae intended to abolish many superfluous commissions and boards and change the membership of others. Several Arkansas state senators heeded the call and abolished several commissions, including the Penitentiary and Corporation Commissions. The Arkansas Tax Commission, Boiler Inspection Department, Highway Department, State Inheritance Tax Collector, and the Oil Inspection Department also faced challenges to their continued existence.[3]

At McRae's direction, Democratic state senator James R. Woods introduced Senate Bill No. 159, a proposal to eliminate the AGFC and replace it with an honorary commission of the governor and two governor-appointed members. As he submitted the bill for consideration, Woods claimed that the governor's aide had handed him the bill, calling it an "administrative measure." Although the legislation moved to end the AGFC, it made no alterations or removal of any current game law. All general enforcement moved from AGFC wardens back to the state's county sheriffs if passed. Some policymakers even hinted at launching investigations into the AGFC and others. State senator David C. Arnold alleged that one of the wardens worked for the AGFC yet moonlighted for the Rock Island Railroad, which was prohibited. The charge proved unfounded, but commission opponents nonetheless used it for political ammunition.[4]

By February 1921, the AGFC needed immediate funding to continue operations. Senator Woods submitted an emergency appropriations bill to pay salaries and expenses through the end of the state fiscal year. Woods got it quickly passed twenty-seven to two. Unfortunately, the appropriation did not cover all of the AGFC expenses. By the end of June 1921, all the wardens, plus AGFC office secretary Miss Nellie Patton, had not received their entire salaries. A few weeks later, Woods ensured the AGFC had the funding to operate the following year. The legislation provided an annual $40,000 appropriation for warden and secretarial salaries and maintenance expenses for 1922 and 1923.[5]

United States chief game warden George A. Lawyer wrote Lee Miles congratulating the AGFC on receiving their appropriations. Miles had asked the top federal warden to visit Arkansas during the legislative battle over the AGFC and its funding. "The situation looked very gloomy when I was there, and I expect you must have put some good hard licks after I left," Lawyer commented.[6]

To implement McRae's anticorruption agenda, Arkansas lawmakers ordered all special funding deposited in the general revenue fund, from dog licensing fees to the cotton rag tax. The legislature argued that this action allowed them tighter control over the state's revenue and would stem corruption. For the AGFC, it meant continued political wrangling each legislative cycle to obtain their appropriations, something they had experienced since 1915.[7]

Seeking to hold down additional costs, the state legislature passed a resolution calling for the federal government to create game reserves in the Arkansas national forests. That meant no land costs for the state. Thousands of acres lay in several Arkansas counties that could hold and manage wildlife in protected areas. Arkansas lawmakers suggested that current federal forest rangers could patrol these areas if established. That meant no enforcement costs for the state. Some Arkansawyers, however, did not want reserves or preserves and resented the federal foresty authorities.[8]

Several Arkansas sporting organizations joined with the AGFC and other conservationists from around the nation to push the United States Congress to create public shooting grounds in their states. United States Senate Bill No. 1452 and United State House Bill No. 5823 attempted to establish public shooting grounds, refuges, and nesting areas for migratory waterfowl. Commissioner Lee Miles wrote to secretary of Agriculture Henry C. Wallace to express the AGFC's support of the bills. If the laws passed, Miles explained to Wallace, "the enforcement of it will depend very largely on the cooperation of your department." The judge claimed that if the conservationists and the federal and state governments had cooperated in the past, the bison and the passenger pigeon might remain. "I want to express the appreciation of Arkansas sportsmen and Conservationists [for] the interest you have manifested in this important work," Miles concluded.[9]

When the United States House of Representatives Committee on Agriculture held hearings to discuss bird refuges and public hunting grounds, supporters asked representatives from every state to pen support letters to add to the Congressional Record. Miles wrote on behalf of conservationists in Arkansas. He explained that hunting and fishing clubs owned thousands of Arkansas acres, but a few spots remained that could serve as refuges and public shooting grounds. "I fear that within a few years more there will be practically no good shooting ground left open to the public. I have talked with a great many sportsmen in Arkansas, and all of them with whom I have talked favor this bill," he explained. The *New York Times* printed a quote from Miles's statement that summarized the argument's crux: "I cannot understand how a man could be a sportsman and not support this law." The American Game Protective Association claimed that migratory bird populations would never decrease if one of the bills passed.[10]

While state lawmakers asked the United States government to establish preserves, one AGFC warden appealed directly to the people. Warden W. H. Cochran campaigned to create a game preserve in northwestern Arkansas, meeting with sportsmen in Fayetteville and Springdale. He claimed that the AGFC needed only another $30,000 to add to the $70,000 they had already in their coffers to buy the preserve. He encouraged anyone interested in wildlife to support the operation.[11]

Although the AGFC had suffered setbacks during the 1919 legislative session, these threats did not deter Arkansas conservationists. The exact opposite happened. Soon, Arkansawyers joined and founded more conservationist organizations and sportsmen's associations, and already established groups became more involved in wildlife protection and education. Many Arkansas conservationists believed the only way to save Arkansas's game and fish from annihilation was through concerted efforts. Calhoun County Democratic representative J. F. Plunkett explained, "As soon as the people of Arkansas who are interested in wildlife conservation get together on what they want and are to fight for it, the Legislature is ready and will [*sic*] to help them." These local groups worked to educate their populace, assisted the AGFC, lobbied for changes to the state laws, and restocked county waterways and forests with propagated wildlife. Local protective associations and rod and gun clubs started or grew across the state. In 1924 alone, Arkansasawyers organized fifty of these societies. In some cases, county sheriffs deputized a few members of these organizations so that they could arrest poachers.[12]

For example, Arkansas conservationists started the first chapter of the Izaak Walton League in 1924. Founded in Chicago, Illinois, in 1922 as a fishing conservation organization, the league is named after early sports fisherman Izaak Walton (1593–1683), known as the "Father of Flyfishing" and author of *The Compleat Angler*. The first annual state convention met in Mena on November 21, 1924, with fifty Arkansas delegates. By 1926, there were eighty-one chapters with nearly three hundred members of the Izaak Walton League in Arkansas. Proponents claimed that "the birth of this statewide division of a great national sportsman's organization is a notable step in conservation work."[13]

While these new organizations formed and joined the conservation battle, some older clubs redoubled their efforts. When word leaked in 1920 of a possible attempt to abolish the AGFC, the Twin City Fish and Game Protective Association of Fort Smith sent a letter of support to the commission, claiming they "would not support any candidate for governor who is not in sympathy the game and fish laws." The eighty-member Bentonville Fish and Game Protective Association lobbied for a fish hatchery on Lake Bella Vista. Commissioner E. L. Boyce and game warden Stedman helped organize the Jackson County Sportsmen's Club in Newport. The Phillips County Protective Association asked the AGFC to appoint a special deputy just for their county to whom they could submit violation reports. The Arkansas Federation of Women's Clubs chairwoman Mrs. Edwin Bevens submitted a request for game conservation pamphlets to distribute at their meetings. Bevins intended to form a district women's group to advocate for enforcing state and federal fish and game laws. Several schools started Audubon Clubs, first at Rightsell School in Little Rock, with Miss Leona Chapline as the sponsor.[14]

Like in the schools, other wildlife educational programs compelled young citizens to develop a protectionist attitude toward fish and game. In a recruiting plug for the Boy Scouts, the writer pointed out that a man is not a sportsman because he is a good shot or angler. Instead, a sportsman does not break the law and does not injure young animals or kill wildlife or fish just because he can do so. A sportsman "considers the rights of others and those of the rising generations," he said, and "Scouts believe in true sportsmanship." So, the Boy Scouts and the schools pushed for conservation. But Arkansas needed a new statewide sportsmen's organization to help coordinate these renewed conservation efforts.[15]

The Arkansas State Sportsmen's Association had splintered during the legislative fights of 1915. In its place, the Arkansas Fish and Game Protective Association had become the leading statewide organization, with Earnest V. Visart at the helm, to fight for creating the Game and Fish Commission and general game legislation. By 1920, with Visart gone to the United States Biological Survey and several other leaders in both organizations deceased, the conservationists needed a new state organization to fight for game and fish protection. The new group needed to include both sportsmen and other conservationists for maximum effect. In the summer of 1920, forty-two attendees formed the State Conservation Association in the Hotel Marion in Little Rock. They elected E. L. Boyce as the organization's first president, J. M. Futrell as vice president, Lee Miles as secretary, and W. E. Lenon as treasurer. The SCA's mission included protecting Arkansas game and fish, promoting cooperation between local and state authorities and conservationists, and assisting in establishing preserves in the state. When the public learned of the SCA's creation, local conservation organizations started a letter-writing campaign, sending hundreds of mailings to their fellow Arkansawyers requesting that they support the organization. These regional associations even mailed some letters in Italian so the newly arrived immigrants in Chicot County could read the appeal. Unfortunately, the SCA dissolved shortly after the 1921 legislative session ended, the last statewide sportsmen's association in existence at the time.[16]

The attempts to abolish or alter the commission failed in 1921. Arkansas conservationists were able to breathe a sigh of relief. Also, Democratic senator Grover Owen introduced Bill No. 154, amending the 1919 statutes that had removed police powers from the wardens, arguing that criminals had escaped while the game wardens searched for a law enforcement official to arrest them. The bill's passage would have returned police powers to the AGFC's wardens. When Owen's bill passed the Senate, hopes rose within the AGFC. When it reached the House, however, opponents defeated it. Therefore, the wardens continued to rely on local authorities to arrest poachers.[17]

In a short time, the AGFC had a new member, Judge E. L. Boyce from Newport, whom McRae appointed to replace Monticellonian Dr. Horatio Wells. Chairman D. G. Beauchamp from Paragould, C. C. Calvert from Fort Smith, Judge Lee Miles from Little Rock, and J. Vol Walker of Fayetteville rounded out the remainder of the commission in 1921. These men immediately hired three new game wardens to replace those who had left. Grady native John F. McQuiston was a former member of the now-abolished Arkansas penitentiary board.

J. C. Cowell from Jasper had worked as the Arkansas Senate head janitor. John Hinton from Paragould was the last of the hires. A. G. Stedman, A. E. Denton, W. S. Cochran, H. V. Oldham, and Arthur Thomas were the remaining five game wardens in 1921. The AGFC needed a full roster as spring spawning season started to warm up, and poachers continued to break the law.[18]

Dynamiting remained a continuing problem for Arkansas waterways. Besides killing thousands of fish and destroying lake and stream structures, explosives were dangerous to handle. In June 1921, Warden A. G. Stedman investigated the death of Camp Pike soldier Byron Carr near Greenbriar in Faulkner County. Carr had procured dynamite and intended to use it for fishing in Blackfork Creek. The inexperienced soldier knew nothing about handling explosives or fuses. He unknowingly obtained a fast-burning fuse, and when he lit it, witnesses claimed "that fire flashed almost instantaneously, and the [charge exploded] before he could throw it in the creek." A month later, Warden Joe Cowell found six men blasting a stream in Pope County. After investigation, Cowell discovered some stolen explosives from the Forestry Department, which had used the dynamite for road building in the Ozark National Forest. Because of the continuing problems with these infractions, the AGFC, a few municipalities, and several sporting associations issued twenty-five- to one-hundred-dollar rewards for information. In September 1921, a judge sentenced Ben Fore of Algoa in Jackson County to a six-month prison term for dynamiting the Cache River. His actions nearly killed Dennis Richards, a fellow dynamite poacher.[19]

Frequent explosive and firearm use and heavy illegal seining in many Arkansas waterways denuded significant fish areas. When bass spawn in the spring, they move to shallower waters, making them easy targets for riflemen. The shock kills the bass and the eggs when a shooter fires into the water at the fish. The shockwave drives other fish from their nests, leaving them unprotected. The AGFC believed that only one shot could potentially kill thousands of fish. To replace the reduction, many individual citizens and groups continued to order fish through the AGFC and restock their local lakes, streams, and ponds. These activities increased the importance of expanding the state propagation program.[20]

In the fall of 1921, as part of a new educational program, Lee Miles wrote an op-ed in the *Arkansas Democrat* about the history of the fight for game and fish laws in Arkansas. According to Miles, New York contained much less land area than Arkansas (not true), but because of their conservation program, hunters harvested over twenty thousand deer there in 1919. Arkansas had more than eight thousand deer hunters. He explained that if they killed one deer in 1921, the deer would be extinct in Arkansas. The AGFC desperately needed money to restock depleted game and fish populations and protect the wildlife remaining in the forests.[21]

Despite the cautionary lesson, Arkansas continued to have a deer hunting season. The reasons varied. For instance, the backlash from a closed deer season would inevitably turn many citizens against the AGFC. Indeed, so many Arkansawyers killed deer indiscriminately that eight wardens could not prosecute all the poachers. In 1921, the AGFC did not determine the season dates as they do today, which resulted from an act of the Arkansas state legislature. Although Miles did not raise the issue, Arkansas also continued to have an open bear hunting season. By that year, black bear numbers had fallen to a critical level, yet hunters continued to hunt down the remaining bruins.[22]

Around the same time that newspapers published Miles's article, the annual meeting of the Arkansas Sheriff's Association occurred in Little Rock. The main topics of the day's discussion included auto theft and illegal whiskey. AGFC warden Stedman spoke during the morning session of the meeting, encouraging cooperation between local law enforcement and wardens. Stedman explained to the sheriffs how vital their assistance could be to the game officers. In several instances, traffickers simultaneously used the same avenues to transport illegal alcohol and illicit game.[23]

Spring flooding deluged the state's eastern portion the following year, exhausting the AGFC's limited resources. Federal warden Visart, all eight AGFC wardens, Phillips County sheriff J. D. Mays, and Phillips County Game Protective Association members went into action in March. Once again, heavy flooding drove wildlife to higher ground, concentrating them in isolated areas. Federal, state, and local authorities had long worked to educate the public concerning fish and wildlife regulations. As such, the poachers no longer deserved the benefit of the doubt. Visart, the AGFC, and Sheriff Mays announced that "the educational campaign and warnings will be supplanted by legal action," covering the entire state, not just Arkansas's flooded regions.[24]

With the continued cooperation of the local Protective Association, the backing of local judicial officials, and the support of many local authorities, the AGFC progressed toward stopping most of the poaching in the flooded areas that year. Thirty days into the flood, authorities arrested and charged only three game cases. When they concentrated their efforts, wardens and their assistants experienced success in sending the message that they would not tolerate poaching.[25]

Arkansas employed the fewest wardens of any state in the nation. "The Game and Fish Commission has needed more than 150 wardens at one time," explained AGFC secretary Jonathan W. Allen. But the budget nor the law allowed more than eight full-time wardens. In the summer of 1922, each time a warden patrolled an area, they found violators. In remote areas like Cotter in the Ozarks, poachers plied their trade with little fear of arrest. After observers heard explosions for ten straight nights and saw dead fish floating down the White River, they claimed that "this county has no game warden, and the fish and game laws are being violated with impunity." The remote regions were not the only locations that experienced illegal activity. That summer, dynamiters or poisoners killed thousands of fish in Barber's Lake in White County, a popular fishing spot.[26]

Commissioner Miles continued his efforts to educate the public about the AGFC and Arkansas's wildlife. In 1922, the AGFC created a complex display for the Arkansas State Fair in Little Rock, which included taxidermy wildlife, a list of the commission's duties, and other educational information. Accordingly, many people inquired about the purpose of the AGFC. In response, Miles delivered another report for newspaper publication. Indeed, the organization produced an unprecedented amount of studies into wildlife management. Their objective was to determine how many animals lived in Arkansas and how to sustain and grow that number so that citizens could continue to hunt and fish. Miles encouraged the creation of more local clubs to assist in those efforts and again called upon the state legislature to allow the AGFC funding for a propagation farm.[27]

The Arkansas press widely supported the efforts of the AGFC. The state's leading newspaper, the *Arkansas Gazette,* claimed that the AGFC served the greater good. Editor J. N. Heiskell, one of the South's leading Progressive thinkers, explained that sportsmen operated the organization and worked to preserve fish and game so future generations could hunt and fish. "If all the boys in Arkansas knew what this commission is doing for the benefit of the rising generation in this state, they would realize their interest in it and be active supporters of it," Heiskell concluded."[28] Two weeks later, *Arkansas Democrat* editor W. T. Sitlington expressed similar sentiments. He claimed that "game hogs and unscrupulous so-called sportsmen [were] flagrantly violating the state and federal game laws daily." Poachers killed the protected game, shot more waterfowl than allowed, and left dead carcasses behind. Sitlington hypothesized that either the AGFC was not working correctly or did not employ enough officers but that "the state should take vigorous action to put a stop to the shameful practices." Forrest City journalist Ed Landvoigt reported that some shooters in his county continued to violate the law. St. Francis County wanted a game warden to end illegal game sales, doe killing, and quail poaching. Landvoigt agreed that the AGFC did not have enough wardens to patrol the state, so judicial officials needed to throw the book at any convicted offenders. Since St. Francis County contained so many poachers, they could provide one or two men as examples to other criminals. "Unless the game hogs are stopped by the fear of the law, the big effort for game conservation now being made by the Arkansas Game and Fish Commission will go for naught," Landvoigt contended.[29]

In the state's northern section, Lead Hill residents were fed up with poachers and game hogs and claimed they were "making preparations to make war on them." They insisted that the Boone County Sheriff's Office place a deputy sheriff in Lead Hill to round up the lawbreakers. In nearby Hardy, farmers banned all quail hunting on their lands, claiming that the birds ate invasive insects. The now legendary Arkansas State Sportsmen's Association game warden, now federal game warden, Visart started an educational campaign that included the agricultural benefits of birds over twenty years prior. Avian conservationist McIlhenny had traveled all over Arkansas, presenting information about birds and their advantages for farmers. Several women's clubs, including the Audobon Society, had continued children's education about birds. After all this time and effort, the agriculturalists finally accepted the information as truth.[30]

The AGFC also noticed that more nonresidents had purchased licenses in 1922 to hunt and fish in Arkansas than ever. Reports indicated that out-of-staters bought between six hundred and seven hundred fifteen-dollar permits. "Good roads and an unusually large amount of game are bringing" them into the state, explained the commission. Most of the sportsmen stated that they mainly sought waterfowl. American Game Protective and Propagation Association vice president and editor of their bulletin, Ray P. Holland from New York, was among them. In December 1922, Holland wrote to APPA secretary George M. Fayles, asking him to send him all the Arkansas laws about game shipping. Holland wanted to hunt for a day or two in Arkansas, travel to Memphis, ship his quarry home, and then return for another hunt in Arkansas. Holland explained that if he had to ship it through a resident, he could send his game to his friend, Judge Lee Miles, in Little Rock. Fayles replied that Holland could send only one day's harvest from Arkansas to Tennessee and ship it out to New York. No provision barred Holland from making more than one trip, carrying one day's bag out. He could then accumulate several days' worth and ship them in one package from Tennessee. Federal law prohibited exporting more than two days' limit of migratory birds at once. So, they could not be waterfowl. Nonresidents could carry or ship out their harvests, but under specific restrictions.[31]

In their 1922 biennial report, the AGFC assessed the Arkansas wildlife situation. In the section "What Was Here Fifty Years Ago," Secretary Allen explained that Arkansas wildlife and fish once flooded the meat markets as the forests and streams teemed with thousands of animals and fish. Few railroads traversed Arkansas. Instead, boats carried the harvests to market. Arkansas farmers stored enough wild meat to feed their families through the winters. Arkansas lawmakers failed to protect the animals and the fish for so many years that few remained.[32]

One of the most interesting aspects of the AGFC's biennial report concerned the crime of dynamiting fish. Since 1919, dynamiting convictions carried stiff penalties, including penitentiary time. Despite the increased punishments, dynamiting continued unabated. Secretary Allen claimed that few juries would send a man to prison for fishing violations. Therefore, the AGFC asked that the penalty for dynamiting change from a felony to a misdemeanor, which included a heavy fine. With this change, the wardens believed they could obtain more dynamiting convictions.[33]

Two years later, United States Bureau of Fisheries pathologist H. S. Davis visited the Ouachita River near Arkadelphia to investigate why the river contained so few fish. Reports claimed that pollution had killed the aquatic life, but Davis found little evidence. Nonetheless, Davis did agree that both the Ouachita and Caddo Rivers held few fish, whereas, during previous years, both abounded with aquatic life. Davis claimed that the two most significant factors for fish depletion in United States waterways were population growth–induced environmental changes to habitat and the rise of automobiles. Deforestation and similar issues also polluted streams. Automobiles allowed access to previously isolated waterways. However, neither was the case for Arkansas. Davis argued that overfishing, particularly dynamiting, destroyed fish populations in Arkansas streams and rivers. For Davis, the only way to reverse the present course toward fish extinction in Arkansas was through more regulation and propagation.[34]

1923 brought new troubles for the AGFC game wardens. Garland County municipal judge Verne S. Ledgerwood dismissed a case against Dr. E. L. Thompson and Tom Garen for hunting without a license on the grounds "that no warning had been given by the state authorities that there were no days of grace" for buying a new permit for the following year. Ledgerwood incredulously summarized that hundreds of hunters had not purchased their licenses, so they did not know they had broken the law. Ledgerwood wanted the new law publicized, as he believed the public would buy permits if they knew they needed to. Ledgerwood declared that he would penalize these men only after they did not heed the warden's warning. The judge's logic proved true when the Garland County clerk sold fifty-five hunting licenses within two days.[35]

In January 1923, the *Arkansas Gazette* sporting editor, Henry "Heinie" Loesch, proclaimed that Arkansas remained the "Paradise for Fishermen and Hunters." Arkansas waterways brimmed with fish while her forests teemed with animals. Besides touting the wonders of the Bear State, Loesch believed that the AGFC continued to stop game hogs, ended the commercialization of fish and wildlife, and instilled a sporting ethic. He insisted that hunters might even find an old bear during one of their hunting trips. It remains unclear how Heinie could make this determination as every Arkansas conservationist collectively claimed the contrary.[36]

In early 1923, based on AGFC observations in the waterfowl regions, Commissioner Miles wrote to the chief United States game warden at the Biological Survey, George A. Lawyer, requesting that the federal government reduce the daily bag limit on ducks from twenty-five to fifteen. Miles claimed that he found "the sentiment in Arkansas so universal" to reduce the limit, as evidenced by Arkansas United States senator Joe T. Robinson supporting the reduction.[37] Lawyer replied that he needed clarification. Arkansas waterfowl hunting was some of the best in the world. Did Miles mean to push for limit reduction for the entire country or just Arkansas? Plenty of states did not hold waterfowl like Arkansas, and few shooters achieved a limit in most places during their hunts. Miles told Lawyer that he would bring up the reduction at the next advisory meeting. At the time, Miles sat on the Biological Survey's shooting advisory board. "I do not want to be cranky about something of this kind," but if Arkansas hunters continued to take twenty-five ducks per day for the entire open season, they would reduce the chances for shooters in other states. The AGFC announced that they thought a reduction in Arkansas while the rest of the nation had no reduction was the wrong decision. The AGFC continued to argue for the reduced bag limit two years later.[38]

A few months later, Miles wrote chief of the United States Biological Survey E. W. Nelson that he intended to lobby the advisory board to designate certain days during the waterfowl season as "rest days or non-shooting days." In Miles's opinion, the board could set the number and dates for these rests. Miles wanted to know Nelson's opinion about what days the chief thought best to ban shooting. Nelson immediately sent letters to all state game wardens asking for their opinion on rest days and a reduction in daily bag limits. Lawyer also interviewed state game wardens and sportsmen during his travels. The decision to reduce bag limits or set rest days would take months.[39]

Before the outset of the 1923 hunting season, the Arkansas state legislators passed a five-year moratorium on the bear, beaver, otter, pheasant, and prairie chicken seasons. Those seasons remained closed until December 1, 1928. By that time, bear, beaver, otter, and prairie chicken numbers remained so low that they were practically extinct in Arkansas. They banned pheasant hunting in hopes that five years would provide enough time for the stocking efforts to build a self-sustaining state population of these birds. Lawmakers did not ban quail hunting but still reduced the season period.[40]

The Arkansas state legislature finally allowed the AGFC to establish a propagation farm in Rogers. The facility purchased broodstock English ring-necked pheasants in 1923–1924 with hopes of raising nearly two thousand birds annually. Kansas and Texas also had plans to raise ring-necked pheasants too. According to the AGFC, their stocking efforts' most challenging problem was Arkansawyers illegally shooting the pheasants. Until then, most Arkansawyers had never actually seen a pheasant. The AGFC decided to send birds to every county for exhibition in their county fairs and instructions that Arkansas law barred killing the birds until 1928. By the following year, the AGFC began to raise wild turkeys at the Rogers facility.[41]

Another significant development to come out of the 1923 session surrounded funding. Arkansas lawmakers provided the AGFC with double their previous appropriation, $160,000. Although promising, the increased financing did not do much for the AGFC and its work because of the limitations the law placed on it. Legislators defeated a bill that would have increased the number of wardens from just eight. Additionally, the AGFC could spend no more than they took in from licensing fees. So, they could not spend all the money the legislature had appropriated.[42]

Local conservation organizations continued to form in different parts of the state. It was not just the sportsmen who took a role in their development. In 1924, two articles appeared in a pair of Ozarks newspapers concerning the decreasing number of fish in their streams. Two decades prior, *Baxter Bulletin* editor Tom Shiras wrote that a man could catch twenty fish in a day, none under two pounds from the North Fork River. The lack of fish makes that feat impossible now, he explained. "It's up to you and your children, and that means everybody and everybody's children, whether we are going to enjoy this little game of fishing anymore," Shiras argued. He explained that the Arkansas Game and Fish Commission was not getting results. Citizens needed to take care of the fish. Shiras concluded that Arkansas needed more fish hatcheries to rebuild the lost population. The *Mountain Echo* in Yellville, a competing paper in nearby Marion County, agreed with Shiras. "The game fish in [the] White, Buffalo, and North Fork Rivers are getting scarce," editor Don Matthews argued. Some Baxter County residents had decided to form a fish conservation organization, and other Arkansas counties should follow their example. The Baxter County contingent planned to purchase fish from the federal hatchery at Mammoth Springs and lobby the Arkansas legislature for stricter fish laws. Superintendent of the United States Hatchery at Mammoth Springs Dell Brown echoed the same sentiments. Brown explained that ten years earlier, on the Ouachita River, a man could catch a fine string of black bass in an hour, which now took more than a week. That was the case if the angler proved lucky.[43]

Consequently, the Yellville Booster Club and the Chamber of Commerce stocked pheasants and fish. They purchased brood pairs for volunteer farmers to raise the birds. Arkansas had a five-year ban on pheasant hunting, and if the group worked hard enough, they might have a good crop of pheasants when the season opened again in 1928. Farmers should also look after the wild quail on their farms to ensure the birds did not starve in the winters and poachers did not destroy entire coveys.[44]

The Mountain Home Chamber of Commerce developed a fish and game committee inside their offices. They met with Baxter County representative R. C. Love and pressed him to introduce protective measures for the county's fish population. The committee also decided to lobby the rest of Arkansas's legislators to allow the AGFC to build state fish hatcheries. "Unless this is done, and the [Arkansas Game and Fish] Commission looks carefully after the [fish] plantings, Arkansas counties can kiss its game fish good bye," concluded Shiras.[45] In 1924, Arkansas game wardens estimated that hunters harvested only two hundred deer and two hundred turkeys statewide. No one had reported a bear sighting. Editor Shiras argued that Arkansas conservationists must support the AGFC. "WHO WILL BUILD ARKANSAS IF NOT HER OWN PEOPLE DO NOT?" he asked sternly.[46]

Baxter County conservationists took the first step toward establishing a state fish hatchery in 1925 when a group purchased five acres with a spring and two ponds. The AGFC teamed up with National Fish Hatchery System superintendent Del Brown to advise the Baxter County men during construction. Brown believed the facility could simultaneously raise two fish species, crappie and black bass. According to Brown, they might be able to produce as many as one hundred thousand fish annually.[47]

Unfortunately, as some new local conservationist organizations bloomed, others dissolved. It is unclear why this was the case. It is possible that they felt that the AGFC was enough to do the job or, to the contrary, the AGFC was not worth supporting. In other instances, members might have joined other conservation organizations or become apathetic over the entire effort. The Twin City Fish and Game Protective Association of Fort Smith, who had claimed in 1920 that they "would not support any candidate for governor who is not in sympathy the game and fish laws," ceased meetings four years later. In Helena, the Phillips County Protective Association saved many animals' lives during the 1922 flood. For over five years, the PCPA proved instrumental in supporting conservation efforts in Phillips County and had submitted evidence to the AGFC against many poachers during their prosecution. They, too, disbanded in 1924. AGFC secretary Guy Amsler hoped that when the legislative session began in 1925, some defunct groups would reorganize to join the fight.[48]

Amsler argued that these inactive clubs had no idea how substantial the conservation fight continued to be. Arrogantly, he hypothesized that the dissolution of these clubs lay in the erosion of the members' resolve. Amsler claimed that newly formed organizations had lofty, unrealistic ideals, such as their own AGFC game warden to patrol their county and unlimited access to the AGFC's fish and game restocking supplies. Because some new club members anonymously supplied the names of a few poachers, they expected a special warden for their area. The secretary explained that the AGFC informed the public that state law forbade more than eight wardens statewide and that all of the restocking pheasants were all shipped out, they had no hatchery or more money to purchase more restocking fish and to please take these matters up with their legislator, the governor, or some other political official. Accordingly, these clubs become angry at the AGFC. Some of these clubs declared the AGFC corrupt. A few argued that they would join the poaching ranks before all of the game and fish were gone. Amsler demurred, admitting only a few conservationist organizations go through these steps while the majority "keep on keeping on."[49]

As part of their educational and public outreach program, the AGFC began publishing a magazine in April 1924. Anyone who purchased a license received a free copy. Using this forum, the AGFC touted successes, answered questions, advocated for amending or adding game and fish laws, educated the public about current regulations, and provided how-to instruction in conservation and propagation techniques. Additionally, the AGFC created a questionnaire for statewide distribution, preparing for the next legislative battle. With "Education Means Conservation" as its slogan, the *Arkansas Deer*, quickly renamed the *Arkansas Conservationist*, became the AGFC's official organ.[50]

In the *Arkansas Conservationist*'s second edition, Chairman Lee Miles listed new game and fish laws that the Arkansas legislature needed to pass in the 1925 legislative session. Demonstrating just how low wildlife numbers were by 1924, during the previous legislative session, lawmakers had banned hunting bear, beaver, otter, pheasant, and prairie chickens for five years. Miles suggested the legislature also end the sale, shorten the season, and reduce the bag limit on squirrels. He also wanted to reduce the raccoon season to twenty days, prohibit spring turkey shooting, ban all hen turkey shooting, reduce the quail limit from twenty to twelve with only a sixty-day season, end all hunting of deer with hounds permanently, end game fish sales, enact a twenty-fish daily limit on all fish, establish more propagation facilities, create a statewide gun license, and allow the AGFC to hire as many wardens as funding permits. Most importantly, he sought to return police powers to the wardens so that they could execute their work more efficiently.[51]

AGFC secretary Amsler again called to establish a new statewide sportsmen's organization. Amsler argued that if conservation efforts in Arkansas were to progress, the sportsmen should unite to press their government officials. He believed Arkansas contained enough sportsmen to ram through sensible game legislation, but they had to get organized. The state's sportsmen needed a plan, a direction, and leadership.[52]

In 1921, Governor Thomas C. McRae's plan to streamline Arkansas's state infrastructure to eliminate waste and corruption had placed the AGFC on the chopping block. Four years later, he played a different tune. During his 1925 farewell address to the state legislature, McRae claimed that "Arkansas, originally, was a great game and fish preserve. No State, in proportion to her area, had more, but she has permitted the most of it to be destroyed." He implored lawmakers to "strengthen the work of the Fish and Game Commission."[53]

Secretary Guy Amsler used the governor's change of stance to call out to all Arkansas sportsmen. The new governor, Tom J. Terral, "and the assemblymen should hear from every sportsman in the State relative to conservation legislation," Amsler urged. Making a telephone call, wiring a cable, mailing a letter, sending a petition "from YOU may be the means of saving our game and fish," Amsler exclaimed. They thought Terral would join the battle to save Arkansas game and fish. "We feel confident he would heartily favor such a program," Amsler insisted.[54]

They misjudged Terral. When the new Arkansas governor presented his speech to the joint session of the Arkansas legislature in January 1925, it appeared that the AGFC faced real trouble again. Regarding the $160,000 appropriation that the commission had received as an operating budget for the 1923 to 1925 period, Terral claimed that that money could have paid the salaries of two hundred first-grade teachers for eight months instead. Teachers could have taught over twenty-six thousand children for that price. Too many Arkansas children were left untaught while commissioners and wardens received their salaries, traveling expenses, furniture, stationery, and printing. In the best political spin of the day, Terral claimed that the state could use the revenue from licensing "for the protection of child life instead of using [it] for the protection of wild life."[55]

Terrel wanted the AGFC abolished but not the laws that protected the game and fish. On the contrary, Terral claimed that he wanted to strengthen those regulations. He saw "no reason why there should be a special set of officers to enforce these game and fish laws." "Are human lives less important than animal life? There are no special officers to investigate a murder, so why do special officers investigate poaching? Sheriffs, deputies, and constables can enforce these laws," Terral argued. Local law enforcement had not enforced these laws in the past because the state provided "an army of game wardens" to do the job for them.[56]

In early 1925, Terral set out to restructure how the state government operated. Like the previous governor, Terral had campaigned on reform, efficiency, and ending government waste. He believed that special bureaus, departments, and honorary commissions should be abolished or consolidated under fewer established departments, thus reducing costs. Terrel contacted federal game warden Visart, asking for a meeting. Visart claimed that during their phone conversation, Terral contended that he did not want to do away with the AGFC, seemingly a complete change of heart. Instead, Terrel supposedly informed Visart that Arkansawyers wanted stricter game and fish laws, so he intended to give them. The governor told Visart that he was in the middle of having a bill written that "had teeth in it" and that "a man caught violating the game and fish law would pay a penalty that he would not forget." Terral wanted Visart to suggest any needed alterations to the bill. The federal warden claimed that he could not participate in state legislative activities without permission from Washington. The governor told Visart to wire immediately for permission. Visart contacted chief game warden of the Biological Survey W. R. Dillon in Washington, explaining that he had no intention of assisting in writing the bill because he felt "sure it is the Governors intention to abolish the present game commission." Visart believed if Terral's legislation did not pass, the governor would veto any subsequent bills that benefited the AGFC, including appropriations bills. The AGFC would cease operations without funding, giving Terral what he wanted anyway. Visart had known Terral for several years and worried that once the governor made up his mind, it might prove difficult for anyone to change it. The federal warden resolved to stay away from Little Rock.[57]

Ultimately, Terral dissolved a few targeted commissions but failed to destroy the AGFC. Additionally, the legislature provided the same level of funding as the previous session despite Terral's efforts to reduce it. The governor did not have as much legislative support as he expected. However, a few AGFC opponents remained.[58]

As 1925 closed, editor Shiras again asked his Baxter and Marion County neighbors the problematic questions. The byline read, "Who Pays for Conserving Our Game?" The editor argued that a stream of disinformation continued to swirl around Arkansas that the rising property taxes supported the AGFC and other state bureaus and departments. These rumors incorrectly claimed that Arkansawyers support the AGFC with their own tax money. As he explained the state law concerning the funding of the AGFC, Shiras illustrated that the law did not allow tax revenue to finance the commission. Instead, the law required the AGFC to self-fund. "No taxpayer has ever contributed one cent to the upkeep of the Arkansas Game and Fish Commission except by procuring a license to hunt, fish, trap, or use a dog to hunt," he concluded. Shiras hoped his fellow citizens would educate his neighbor and give the AGFC a fair assessment.[59]

As 1925 closed, the AGFC pushed for wildlife refuges and predator bounties. The state needed fewer open seasons and more daily game and fish limits. Most importantly, the state required additional game wardens. From 1921 to 1925, the conservationist movement grew in Arkansas. Arkansawyers founded hundreds of local and state organizations committed to protecting and propagating the state's game and fish. By 1925, national organizations like the Bird Lovers Society, the Isaak Walton League, and the Audubon Society had chapters in Arkansas. Individuals and clubs reported nearly two hundred poaching violations in 1924.[60]

Although Governors McRae and Terral attempted to destroy the AGFC, they did so because of their quest for governmental efficiency. The two politicians did not stand entirely against game and fish protection. Instead, they thought they knew a less expensive way to achieve Arkansas conservationists' goals. Terral believed children's needs came before wildlife. He could not see that he could simultaneously accomplish wildlife and child life protection. McRae thought the commission was too political and part of the Little Rock political machine that he wanted to end. Neither achieved their goal because the growing number of Arkansas conservationists placed enough pressure on their legislators to end the gubernatorial assault.

More Arkansas newspapers promoted conservation efforts. These informational outlets became essential to the AGFC efforts to get more citizens to support protective efforts in Arkansas. Secretary Amsler claimed, "The fact that our press is taking up the conservation problem encourages us to believe that an effective solution will be reached." AGFC gains proved small during this period, but progress did occur in tightening some laws, creating a game propagation farm in Rogers, and statewide restocking efforts of both game and fish. Unfortunately, by 1925, game and fish amounts had reached such a low point that the only way for Arkansas to obtain suitable fish and wildlife levels was through an enormous restocking effort, which took Arkansas another twenty years to achieve.[61]

Law forbade the AGFC from employing more than eight wardens, a constricting prohibition that plagued the agency for years. In addition, the law allowed the AGFC to spend only what license and penalty fees total for the previous year, which was not uncommon at the time. So, if the legislature appropriated $80,000 annually, which they did from 1921 to 1924, but the AGFC only collected $61,822.77, which they did during the 1923–1924 fiscal year, the department could only receive the latter. Neighboring commissions in Texas and Louisiana spent triple that amount annually. In the final biennial report before 1925, Secretary Amsler added this ominous conclusion: "The future of conservation work in this State is rather problematic."[62]

Epilogue

I have learned that the "Natural State" is a manufactured construct. The deer, bear, elk, and wild turkeys currently inhabiting the Arkansas woods and the trout in her streams began with human restocking and introduction. Trout are not indigenous to Arkansas. Hunters killed most of the other animals by the late 1920s. Born at the foot of the Ouachita Mountains on the Logan-Scott County line in 1936, my father did not see his first wild Arkansas bear until he was seventy. His family free-ranged hogs in the mountains, and he rarely saw a deer or turkey in his youth. However, during his lifetime, the Arkansas Game and Fish Commission and the federal government rebuilt the natural state, but it proved a complex and "problematic" task.

The AGFC remained underfunded and did not reach the $200,000 mark until 1930. By 1925, Arkansas's big game and much of the fish were gone, and the small game were headed in that direction. Humans had made war on the game and fish for decades, and they had driven more than one species into extinction. The fight for only passing stricter game laws to protect Arkansas's game and fish had ended. Conservationists had lost. Therefore, after 1925, Arkansas, with additional assistance from the United States government, entered into a new stage. It was game and fish management. The Arkansas Game and Fish Commission funded propagation and reintroduction as part of that plan, which had started slowly with fish before 1925. In 1927, the commission finally received legislative permission to build a game propagation farm, which it placed in Benton County. However, by 1930, funding issues because of the Great Depression forced its closure. The agency continued its game-restocking efforts with mixed success, mainly because of the lack of funding. During the 1949 turkey season, hunters harvested only 195 statewide, but only because the season was closed the previous three years. Instead of shipping deer meat out of state, Arkansas imported live ones in 1926 and 1927. The state built a deer farm in 1939 near Dierks.[1]

Although the AGFC had acquired a few small ponds, most had proved too small for adequate fish production. However, two places in northern Arkansas had produced a fair hatch in their first year. With significant public interest, the commission created more fish hatcheries in subsequent years. In one of the most extensive projects, construction crews completed multiple rearing ponds totaling twenty-seven acres near Lonoke by the summer of 1930. The facility produced half a million black bass during that year.[2]

12. Established in 1903, the Mammoth Spring National Fish Hatchery is one of the oldest in the United States. Courtesy of the Arkansas Game and Fish Commission.

The Great Depression affected the AGFC negatively, reducing their budget to one-third by 1932. The 1927 Great Mississippi River flood and the 1930 drought produced catastrophic results for AGFC fish propagation efforts. Just as it seemed the commission had a plan and began to experience some success in the propagation field, the national economic tragedy and a series of yearlong weather extremes destroyed their accomplishments, and the situation worsened. It struggled over the next few years until a 1937 law abolished the 1915 Game and Fish Commission and created a new one. Over the subsequent three legislative sessions, the Arkansas General Assembly returned to its previous practice of passing county game and fish legislation, much to the chagrin of many conservationists. State politics continued to hinder the AGFC in the execution of its duties. Just as the Arkansas Sportsmen's Association had done around the turn of the twentieth century, the newest sportsmen's organization, the Arkansas Wildlife Federation, took it upon themselves to push for a change.[3]

The AWF determined that the only way for the AGFC to conserve Arkansas wildlife adequately was to remove it from the political arena, which Missouri had previously done with their commission. By 1944, the AWF had authored a constitutional amendment and obtained enough signatures to place it on the fall ballot. Like in the past, local sporting organizations publicly supported the effort. The amendment passed in November 1944, "divorcing" the Arkansas Game and Fish Commission from legislative control. This passage served as a watershed event for the agency, giving them more power over conservation efforts. However, securing appropriations from the state legislature plagued the AGFC for decades. The agency continued its reintroduction efforts for fifty years, growing, transplanting, and importing birds, fish, and mammals. Success with those programs ranged widely.[4]

By the late 1940s, the agency increased efforts to bring back quail, and by the early 1960s, their labors were paying off. State game farms and federal government incentives for farmers who left a portion of their lands untilled caused a rebound in quail numbers. By the 1980s, a few parts of the state contained a significant population. However, quail numbers had dropped to near previous lows within a decade due to too little habitat. The AGFC was just growing quail for hunters to shoot. They had much better luck with wild turkeys and white-tailed deer. Hunters harvested a fifty-year low 145 turkeys in 1950. However, thirty-seven years later, they took over 10,000. In 1938, hunters killed 203 deer. By 1987, they harvested over 100,000. Both are examples of the AGFC's continued wildlife management efforts.[5]

Throughout the first fifty years of the twentieth century, Arkansas farmers continued to clear and drain fields and swamps to plant cotton, rice, and soybeans. In massive public works projects, the government built drainage ditches and levee systems to combat the annual river flooding. In 1950, Arkansas contained 18.8 million acres of farmland out of 33.7 million total. By 1960, public efforts had drained nearly 90 percent of Arkansas wetlands. Between World War II and 1980, 2.5 million acres of Arkansas forest fell before the saw.[6]

Some lessons are difficult to learn. In 1950, the Arkansas General Assembly passed a law mandating the extermination of several species of predators. Professional trappers went after bobcats and wolves, thinning their already scant numbers. Officials declared the state absent of panthers after an extensive study found no evidence in 1988. Arkansas currently protects most predators from annihilation, knowing they serve a role in a balanced ecosystem. However, one invasive species has proven incredibly destructive lately.[7]

Over the past twenty years, the wild hog has been one of the most significant issues to plague Arkansas and other southern states. They destroy crops, wildlife, and habit. They carry diseases that can pass to domestic animals. The Arkansas Department of Agriculture created the Arkansas Feral Hog Eradication Task Force to deal with the menace, and the agency has eradicated nearly thirty thousand hogs between January 2020 and April 2023.[8]

The federal authorities, like the AGFC, have protected Arkansas game and fish over the past one hundred years. The United States government passed laws regarding fish and wildlife, and those laws had an impact on Arkansas. The Migratory Bird Law and, later, the Migratory Bird Treaty were designed to protect migratory birds, which Arkansas contained millions of during certain times of the year. Starting in 1915, federal authorities arrested and prosecuted Arkansas bird poachers. During these court proceedings, the Arkansas judicial system played a significant role in establishing the constitutionality of federal bird law. Furthermore, in northeastern Arkansas, Big Lake served as a battleground for market hunters and authorities, even after becoming the Big Lake National Wildlife Refuge. As my next project, I intend to tell the federal side of Arkansas conservation before the Great Depression.

Arkansas, the Natural State, has a checkered past regarding the Arkansas Game and Fish Commission. For decades, opponents, both publicly and privately, fought against the creation of conservation legislation. Despite years of opposition, sportsmen and fellow conservationists finally persuaded lawmakers to create the commission. Then, for years, adversaries attempted to destroy the agency. Even today, few Arkansawyers seem completely satisfied with the agency. In 1990, newspaperman Paul Greenberg complained that the AGFC was "too independent." In 2010, the commissioners had a tussle over a perceived power grab from three of its board members. In March 2023, several Arkansas hunters complained about the setting of duck season for the following year. As we have seen, the state has a long hunting and fishing legacy. It is only clear that if any government entity attempts regulation of an activity that many residents consider their right, not only as state and national citizens but as human beings, it will not please most people. Most Arkansawyers do not realize that they would have no game or fish to pursue if it were not for the Arkansas Game and Fish Commission and their efforts over the last 108 years.[9]

Appendix A

Synopsis of Act 124 Creating the Arkansas State Game Commission

The act creating the Game and Fish Commission of Arkansas defines the open season in which dear, bear and turkey may be hunted, to be the last 20 days of November, December and to January 10, inclusive; Also provides that Turkey gobblers may be hunted from the 15th day of April to May 31st of each year. License for hunting deer is $1, it being a violation of the law to kill a deer without being licensed.

The act provides for a license for bird dogs at $1 and it is a violation of the law to use a dog to hunt birds without its being licensed. It is a felony to steal a dog that has been licensed, whether a bird dog or any other kind.

The act provides that no non-resident shall be licensed, and also provides that no person shall be licensed who has been convicted of violations of this law twice, and that no person who has been convicted of larceny or is an habitual drunkard shall be licensed.

Under this act it is the duty of the prosecuting attorney to file information against persons violating the law, and on conviction he receives a fee of $25. In the event of an appeal to the Circuit Court, the prosecuting attorney receives an additional fee of $10.

It is the duty of the prosecuting attorneys to make semiannual reports to the Commission.

Bag limit for season is one bear, two deer, and four turkeys. The act fixes the open season on quail, the months of December and January, and the bag limit not to exceed 15 per day.

The law prohibits hunting on enclosed premises without written permission of the owner. It provides that no person shall fish for the market without paying a license fee of $25, and no one is permitted to use a seine or net with mesh is less than three inches square.

Users of artificial bait are required to pay a license fee of $1. Persons using seines for their family or picnic purposes are permitted to use one not exceeding 60 feet in length with mesh is not less than three inches square. It is also a violation of the law to ship or export game or game fish out of the state.

Trappers using more than ten traps are required to pay a license fee of $10. Beavers and otters are protected for a period of five years.

The law gives any warden or deputy warden the right to search any package, express, freight or otherwise, in violations of the law.

The bag limit on ducks is 25 per day.[1]

Appendix B

List of Arkansas Game and Fish Commissioners, 1915–1925

1915[1]
D. G. Beauchamp, Chairman
Dr. Horatio Wells
George Webber
C. C. Calvert
Lee Miles, Secretary

1917[2]
D. G. Beauchamp, Chairman
Dr. Horatio Wells
George Webber
C. C. Calvert
Lee Miles
A. H. Little, Secretary

1919[3]
D. G. Beauchamp, Chairman
Dr. Horatio Wells
C. C. Calvert
Lee Miles
J. Vol Walker
Dick Brundidge, Secretary

1921[4]
D. G. Beauchamp, Chairman
C. C. Calvert
Lee Miles

J. Vol Walker
E. L. Boyce
John Allen, Secretary

1923[5]

Judge Lee Miles, Chairman, Little Rock
Hon. Charles H. Thompkins, Prescott
Hon. Wm. F. Jeffett, Helena
Hon. J. Vol Walker, Fayetteville
Hon. Harry Neely, Searcy.
Hon. Guy Amsler, Secretary.

1925[6]

Judge Lee Miles, Chairman, Little Rock
Hon. Charles H. Tompkins, Prescott
Hon. William F. Jeffett, Helena
Hon. Harry Jones, Hot Springs
Hon. Roy L. Thompson, Little Rock.
Hon. Guy Amsler, Secretary.

Notes

INTRODUCTION

1. A sportsman hunts or shoots wild animals as a pastime or for enjoyment. A market hunter is someone who harvests wild animals for an occupation or for income. While this study uses "pot hunter" and "market hunter," most people did not distinguish between them at the time. Indeed, most Americans used the terms interchangeably. Pot hunters hunted for their own consumption, while market hunters relied on their game for money. Both categories of hunters and fishermen tried to secure as many animals or fish as possible. The pot hunter wanted enough food to eat for the year. Market hunters wanted the most cash. Both categories disregarded the law in most cases. In many cases, the line often blurred between market hunters, sportsmen, and pot hunters. Many hunters who sold game to the market considered themselves sportsmen for one species or another. Men would hunt for the market on a part-time basis, the majority having other professions such as farming. They might hunt for food and pleasure but also chase a particular quarry, like beaver or wolves, for the market. Another term used at the time and in this study is *game hog*, which means to kill more than one person's share or need. "Pot hunter," Dictionary.com, and "market hunting," Lexico.com, accessed August 4, 2021.

2. Conevery Bolton Valencius's *The Lost History of the New Madrid Earthquakes* is an excellent study covering the New Madrid earthquakes and their effect on northeastern Arkansas, the sunk lands, and the surrounding area, and her *The Health of the Country* provides an insightful look at how American settlers related to their environment and how that relationship affected their health. Conevery Bolton Valencius, *The Lost History of the New Madrid Earthquakes* (Chicago: University of Chicago Press, 2013); Conevery Bolton Valencius, *The Health of the Country: How American Settlers Understood Themselves and Their Land* (New York: Basic Books, 2002).

3. There are many studies covering the American conservation movement. Some cover individual players in the movement, such as *The Green Roosevelt* or studies about John Muir or Aldo Leopold. Other works deal with larger concepts, like Dorceta E. Taylor's *The Rise of the American Conservation Movement*, David Stradling's *Conservation in the Progressive Era*, and Samuel P. Hays's *Conservation and the Gospel of Efficiency*. There are many others. These studies do not include Arkansas because although it teemed with wildlife, the state was not at the conservation forefront. On the contrary,

in most cases, Arkansas remained the last state to enact game and fish protective measures. Nonresidents flocked to Arkansas for many years because of the lack of wildlife protection laws and their enforcement. For more on Progressive environmentalism in the United States, see Zachary Michael Jack, ed. *The Green Roosevelt: Theodore Roosevelt in Appreciation of Wilderness, Wildlife, and Wild Places* (New York: Cambria, 2010); Dorceta E. Taylor, *The Rise of the American Conservation Movement: Power, Privilege, and Environmental Protection* (Durham, NC: Duke University Press, 2016); David Stradling, ed., *Conservation in the Progressive Era* (Seattle: University of Washington Press, 2004); Samuel P. Hays, *Conservation, and the Gospel of Efficiency: The Progressive Conservation Movement, 1890–1920* (Pittsburgh: University of Pittsburgh Press, 1999); Mark David Spence, *Dispossessing the Wilderness: Indian Removal and the Making of the National Parks* (New York: Oxford University Press, 1999); E. Donnal Thomas, Jr., *How Sportsmen Saved the World: The Unsung Conservation Efforts of Hunters and Anglers* (Guilford, CT: Lyon, 2010); Robert Dorman, *A Word for Nature: Four Pioneering Environmental Advocates* (Chapel Hill: University of North Carolina Press, 1998); Stephen Fox, *The American Conservation Movement: John Muir and His Legacy* (Boston : Little, Brown, 1981); John F. Reiger, *American Sportsmen and the Origins of Conservation* (Corvallis: Oregon University Press, 2001); David Petersen, *Heartsblood: Hunting, Spirituality, and Wildness in America* (Washington, DC: Island, 2000); Scott E. Giltner, *Hunting and Fishing in the New South: Black Labor and White Leisure after the Civil War* (Baltimore: Johns Hopkins University Press, 2008); Louis Warren, *The Hunter's Game: Poachers and Conservationists in Twentieth-Century America* (New Haven, CT: Yale University Press, 1997); Karl Jacoby, *Crimes against Nature: Squatters, Poachers, Thieves, and the Hidden History of Conservation* (Berkeley: University of California Press, 2001); Thomas R. Dunlap, *Saving America's Wildlife: Ecology and the American Mind, 1850–1990* (Princeton, NJ: Princeton University Press, 1988); Daniel Herman, *Hunting in the American Imagination* (Washington, DC: Smithsonian Institution Press, 2001).

4. Morris S. Arnold, *The Arkansas Post of Louisiana* (Fayetteville: University of Arkansas Press, 2017). Judge Morris Arnold has written extensively about the Arkansas colonial period, particularly concerning hunters and fur traders in and around Arkansas Post. His work *Unequal Laws unto A Savage Race* studies Arkansas's early French, Spanish, and American administrations, Native Americans, market hunters, European traders, colonial administrators, and soldiers. According to Arnold, Brooks Blevins, and other historians, these early European hunters gave the Arkansas River valley a reputation as a rough, wild, and savage land. In *The Rumble of a Distant Drum*, Arnold explored the relationship between the Arkansas Quapaw and Europeans, especially their trading relationship, which hinged heavily on market hunting, trapping, and fishing. Joseph P. Key also writes about the Quapaw-European relationship, including how the natives manipulated their environment for easier hunting. Kathleen DuVal's *The Native Ground* expands those Native American and European relationships through the Arkansas River valley, covering the Osage, Quapaw, Cherokee, and

European colonials during their struggle for power and dominance in the regions. Market hunting played a crucial role in that struggle. Pelts, meat, and oil trade served as the day's currency, and whoever controlled them controlled the Arkansas River valley. Morris S. Arnold, *Unequal Laws unto a Savage Race: European Legal Traditions in Arkansas, 1686–1836* (Fayetteville: University of Arkansas Press, 1985); Morris S. Arnold, *The Rumble of a Distant Drum: The Quapaws and Old World Newcomers, 1673–1804* (Fayetteville: University of Arkansas Press, 2000); Joseph P. Key, "Masters of this Country: The Quapaws and Environmental Change," (Ph.D. diss., University of Arkansas, 2001); Joseph P. Key, "Indians and Ecological Conflict in Territorial Arkansas," *Arkansas Historical Quarterly* 59, no. 2 (Summer 2000): 127–46; Joseph P. Key, "An Environmental History of the Quapaws, 1673–1803," *Arkansas Historical Quarterly* 79, no. 4 (Winter 2020): 297–316; Kathleen DuVal, *The Native Ground: Indians and Colonists in the Heart of the Continent* (Philadelphia: University of Pennsylvania Press, 2007). See also Jeannie Whayne, ed., *Cultural Encounters in the Early South: Indians and Europeans in Arkansas* (Fayetteville: University of Arkansas Press, 1995); and Brooks Blevins, *Arkansas/Arkansaw: How Bear Hunters, Hillbillies, and Good Ol' Boys Defined a State* (Fayetteville: University of Arkansas Press, 2009).

5. Retired physician L. Wayne Capooth has written extensively about the golden age of waterfowl hunting, market hunters, and sportsmen. His self-published manuscripts contain some of the best Mississippi Flyway waterfowling photographs in existence. He has collected much information about late nineteenth-century and early twentieth-century hunting in the South. He has donated some of his rare collections to Ducks Unlimited and the University of Memphis. Unfortunately, Dr. Capooth has no citations in his publications or analysis, so they are of limited use to historians. I recommend his works to anyone who loves hunting and fishing, especially waterfowling. Luther Wayne Capooth, *The Golden Age of Waterfowling*, (Germantown, TN: self-pub, 2001); Luther Wayne Capooth, *The Golden Age of Hunting: Arkansas, Mississippi, Tennessee* (Germantown, TN: self-pub, 2004); Luther Wayne Capooth, *Waterfowling America*, vols. 1 and 2 (Germantown, TN: self-pub, 2008); Luther Wayne Capooth, *Arkansas Duck Hunting* (Germantown, TN: self-pub, 2018).

6. Samuel Dickinson, ed. *New Travels in North America by Jean-Bernard Bossu, 1770–1771* (Natchitoches: Northwestern State University, 1982), 44.

7. Edward Howe Forbush, *A History of Gamebirds, Wildfowl, and Shore Birds*, issued by the Massachusetts State Board of Agriculture (Boston: Wright and Potter, 1912), 510–57. There was also a group within the wildlife conservation movement known as the "more game" movement or the "more game and fewer game laws" movement. New York lawyer and sportsman Dwight W. Huntington first advocated game farming around 1910, a call that became a conservation philosophy. Supporters of this concept did not believe that passing stricter game laws alone would save the animals. Instead, men like Bronx Zoo director and conservationist William T. Hornaday, president of the National Audubon Societies T. Gilbert Pearson, outdoor magazine editor George

O. Shields, and sportsman George Shiras believed in game farms, controlling predators, game preserves, and controlling fish and game diseases. Some supporters advocated the commercialization of wildlife farms and the limited use of market hunters to harvest meat. In 1917, the American Game Protective Association pledged to provide print space to advocates of the movement. In the same year, Cornell University established a game breeding as a "farm enterprise" and conservation program that supported the work of this movement. Dwight W. Huntington, *Game Farming For Profit and Pleasure* (Wilmington, DE: Hercules Powder, 1914); John Huntington, "The Game Conservation Institute," *American Game*, April/May 1929, 48–49; *How to Insure the People of New York An Abundance of Fish and Game*, pamphlet issued by New York State Conservation Department, Albany, NY, 1912, 5–18; "T. Gilbert Pearson," *Game Breeder* 18, no. 3 (December 1920): 87–88; John B. Burnham, "Association and Magazine Combine in Interest of More Game," *Bulletin of the American Game Protective Association* 6, no. 4 (October 1917): 1–8. Daniel Herman argues that when sportsmen asked the government to become involved in saving game animals, they laid the groundwork for the government to protect nongame species as well. Herman, *Hunting*, 12. Louis Warren's *The Hunter's Game* examines the battle between budding urban-based conservationists and established hunters, primarily studying the American West. He argues that the fight stems from social divisions between different peoples: the wealthy, immigrants, Native Americans, and the local pot hunters. American sportsmen served as the driving force behind American conservation and sought to ban other classes from hunting while protecting vanishing wildlife populations for their sport. Warren, *Hunter's Game*. Sports writer E. Donnal Thomas Jr., in *How Sportsmen Saved the World*, agrees with Warren's observation about the importance of amateur sportsmen leading the American conservation movement. He argues that these men and women have led each significant advance in wildlife conservation during the last one-hundred-plus years. Because of these efforts, Thomas claims that North America remains the best-managed location globally because of these efforts. Thomas does not mention Arkansas, however. E. Thomás, *Sportsmen Saved*. See also Tara K. Kelly, *The Hunter Elite: Manly Sport, Hunting Narratives, and American Conservation, 1880–1925* (Lawrence: University of Kansas Press, 2018).

8. Arkansas contained 781,530 improved acres in farmland in 1850; 1,983,313 in 1860; 1,850,821 in 1870; 3,595,603 in 1880; 5,475,043 in 1890; 6,953,735 in 1900; and 8,076,254 in 1910. These numbers do not include the unimproved acreage of farms. United States Census, Statistics on Agriculture, 1850, 1860, 1870, 1880, 1890, 1900, 1910.

9. Lynn Morrow has written about early hunting clubs, game wardens, sportsmen, market hunting, and some about the Big Lake War. Her work primarily covers Missouri and the Missouri-Arkansas border. Along with James F. Keefe, Morrow edited *The White River Chronicles of S. C. Turnbo*, an incredible compilation of early Ozark hunting and fishing stories. Before the appearance of databases, Morrow created an

index on early hunting and fishing mentions in the St. Louis Globe. This index resides in the Rolla branch of the State Historical Society of Missouri. James F. Keefe and Lynn Morrow, eds., *The White River Chronicles of S. C. Turnbo: Man and Wildlife on the Ozarks Frontier* (Fayetteville: University of Arkansas Press, 1994); Lynn Morrow, "St. Louisans at the Knobel: Sport Tourism and Conservation in Arkansas," Southeast Missouri State University Press website, accessed August 12, 2021; Lynn Morrow, "A 'Duck and Goose Shambles': Sportsmen and Market Hunters at Big Lake, Arkansas," Southeast Missouri State University Press website, accessed May 5, 2019.

10. Initially published in 1975, John F. Reiger's *American Sportsmen and the Origins of Conservation* provides a top-down look at how notable sportsmen like President Theodore Roosevelt, George Bird Grinnell, New York Zoo director William T. Hornady, Forester Gifford Pinchot, and other like-minded individuals used their influence, the press, and sporting organizations to push for stricter wildlife protection and the American conservation movement. Reiger does not mention Arkansas. Reiger, *American Sportsmen.*

11. Sam B. Hilliard, *Hog Meat and Hoecake: Food Supply in the Old South, 1840–1860* (Carbondale: Southern Illinois University Press, 1972), 71.

12. Scott Giltner's *Hunting and Fishing in the New South* investigates the growing American conservation movement in the South during the last half of the nineteenth and early twentieth centuries. He argues that white conservationists passed many wildlife protection laws to control the game and remove African American access to wildlife. White attitudes toward Black hunters and their hated dogs caused some southern state politicians to pass stricter game and fish laws to deny African Americans access to natural resources. They wanted to save those wild resources for upper- and middle-class white folks. White employers also wanted their workers in the fields, not hunting through the nights for raccoons and possums. During my research, I found cursory evidence of institutional racism toward African Americans in Arkansas but no documented evidence that Arkansas conservationist organizations targeted African Americans or any other minority group based on race or religion, as other studies suggest. I am not arguing that racism did not motivate some individuals. There are several uses of the phrase *lower class* in a negative way when describing poachers. However, *lower class* referred to criminals in these examples, not specifically white, Black, poor, or nonnative. I discovered evidence that some Arkansas government officials held racist attitudes, a common stance for the time. Giltner, *Hunting and Fishing.* Also see Steven Hahn, *A Nation Under Our Feet: Black Political Struggles in the Rural South from Slavery to the Great Migration* (Cambridge: Harvard University Press, 2003).

13. I am using the term "Progressive" as part of the Progressive-movement mentality of the period. A part of a national reform concept to end wanton waste and to conserve the nation's fast-dwindling natural resources.

14. Daniel Herman argues that sportsmen are different from farmers because they identified with the "wilderness rather than the cultivated field" and "took command

of nature—and the continent—not by clearing and plowing but by doing noble battle with wild beasts." He points out that farmers and Gilded Age sportsmen clashed over who was closer to the land. Herman, *Hunting*, 10.

15. In 1990, Ted Ownby published *Subduing Satan: Religion, Recreation, and Manhood in the Rural South, 1865–1920*, a study about the life of the southern male. As part of the section about recreation, Ownby discusses hunting and the importance of that activity to southern men, both white and African American. My research suggests that Arkansas society often interlinked hunting and manhood. Not only was someone considered healthy and robust if they hunted and fished, but young men who did these activities had also reached a point where they could provide for their families. These actions were steps toward full manhood. Ted Ownby, *Subduing Satan: Religion, Recreation, and Manhood in the Rural South, 1865–1920* (Chapel Hill: University of North Carolina Press, 1990).

16. Many outdoor authors wrote about Arkansas during the earliest days of sporting magazine publications. The state remained a popular hunting and fishing location at the turn of the twentieth century, and many new sportsmen wanted to read about the wildlife, terrain, waterways, and hunting and fishing there. These articles perpetuated Arkansas's reputation across the nation as one of the last best places to chase bear, deer, turkey, or the big bass. *Forest and Stream*'s western editor, Emerson Hough, visited Arkansas several times during the fifteen years he held the post. He worked under George Bird Grinnell, whose office was in the East. In March 1894, Hough hunted quail with Little Rock sportsman and hotel manager John Irwin. Hough was influential in his push for game protection and conservation. In his article following one of his Arkansas hunts, Hough remarked, "Arkansas is one of the very best game States in the Union today. The West is out of it now. The West is done, and its game is gone." Hough became nationally known for his works of fiction, many of them made into films. Hotel manager John Irwin and Little Rock attorney J. M. Rose took the Chicago sportsman and writer to their Prothro Hunting Club in the cypress swamps outside of the capital city in January 1897. After viewing the hunting grounds, Hough remarked, "Sometimes I think Dante, or Dore, must at some time have hunted mallards in a cypress swamp. Here you have an Inferno on earth, desolation absolute. Silent, gray, cheerless, almost malignant in its forbiddingness, it lay before us, mile after mile." Once he was placed in a hollowed-out cypress for a hunting blind; the Chicagoan peered out across his dreary surroundings, observing, "I never was in a lonesomer place in my life." Irwin and Rose began calling the area "Hough's Inferno" after Hough's remarks. Emerson Hough, "In Dixie Land-IV," *Forest and Stream*, March 31, 1894, 269; Delbert E. Wylder, *Emerson Hough* (Boston: G. K. Hall, 1981); "Arkansas Club," *Forest and Stream*, October 23, 1897, 327. Joseph Irwin also became the manager of Hotel Richelieu in Little Rock after leaving the Capital Hotel.

17. Karl Jacoby's *Crimes Against Nature* studies how American conservationists criminalized previously accepted practices such as hunting, fishing, gathering, and

cutting wood. He explores this outlawing after the establishment of three US national parks. What I found in Arkansas supports Jacoby's findings. A similar situation occurred with Arkansas's new game and fish protection laws. Families who had hunted and fished to provide food for their tables or men who had shot, trapped, and netted for the markets could no longer conduct these activities. Many Arkansawyers cried government overreach even if they did not hunt and fish. Jacoby, *Crimes against Nature.*

18. Keith B. Sutton and the Arkansas Game and Fish Commission published a history of the organization in 1998. He pulled the first few years together with the published AGFC annual reports and the Commission's magazine, the *Arkansas Conservationist,* which started publication in 1924. Part scientific and part historical, Sutton covers the history of the Arkansas Game and Fish Commission and early state conservation efforts. The AGFC meant this publication for public audiences who wanted to know more about Arkansas wildlife and the work of the Arkansas Game and Fish Commission. Sutton's book contains little historical interpretation, although it is a good outline of the overall history of the commission. Keith Sutton, *Arkansas Wildlife: A History* (Fayetteville: University of Arkansas Press, 1998).

CHAPTER ONE

1. For an excellent environmental study of the Yazoo-Mississippi River floodplain, see Mikko Saikku, *This Delta, This Land: An Environmental History of the Yazoo-Mississippi Floodplain* (Athens: University of Georgia Press, 2005). For a more thorough environmental study of the Mississippi River, see Christopher Morris, *The Big Muddy: An Environmental History of the Mississippi and Its Peoples from Hernando De Soto to Hurricane Katrina* (Oxford: Oxford University Press, 2012).

2. Stuart A. Marks, *Southern Hunting in Black and White: Nature, History, and Ritual in a Carolina Community* (Princeton, NJ: Princeton University Press, 1991), 25–27. In this work, farmer-hunters are hunting to provide fresh meat to their table as a supplemental staple. Some men farmed but also hunted simply for pleasure. However, the term *sportsman* in this study is primarily used to describe mostly antebellum enslavers and sportswriters, after the Civil War, middle- or upper-class urban dwellers who hunted for sport. The concept of the sportsman constantly evolved and blurred into other types of approaches to hunting. As we move deeper into the twentieth century, the concept of the sporting ethic continued to evolve, and more working-class men began to embrace the sporting mentality.

3. Theodore Catton, *Life, Leisure, and Hardship Along the Buffalo,* historic resource study of the Buffalo National River, National Park Service, Midwest Region, Omaha, NE, 2008, 51.

4. Timothy Flint, *Recollections of the Last Ten Years, Passed in Occasional Residences and Journeying in the Valley of the Mississippi* (1826; repr., New York: Da Capo, 1968), 253, 261. The lack of roads and the expanse of the swampy low ground along the river valleys served as considerable barriers to early settlement in Arkansas. This

difficulty hampered the populating of Arkansas until well after the Civil War. The low population and difficulty for visitors to reach Arkansas also played a key role in preserving the game and fish numbers for many years. The New Madrid earthquakes in 1811–1812 made thousands of eastern Arkansas acres "impassably muddy" and created much of the "Great Swamp." Most early accounts deal with the lowlands of Arkansas because travelers primarily used the waterways. Conevery Bolton Valencius, *The Lost History of the New Madrid Earthquakes* (Chicago: University of Chicago Press), 95, 104.

5. Flint, *Recollections*, 266, 271; "Game in Arkansas," *Forest and Stream*, September 14, 1882. The ague was another name for malaria. Linda Nash, *Inescapable Ecologies: A History of Environment, Disease, and Knowledge* (Berkeley: University of California Press, 2006), 49–82. Linda Nash explains that this nineteenth-century thinking is lost to most modern people.

6. Quoted in Conevery Bolton Valencius, *The Health of the Country: How American Settlers Understood Themselves and Their Land* (New York: Basic Books, 2002), 138. "Effluvia" is an unpleasant odor. Valencius also argues that the prevalence of sickness, along with the lack of adequate transportation, retarded the growth of the Arkansas delta area in the early territorial period.

7. James William Miller and Friedrich Gerstäcker, *In the Arkansas Backwoods: Tales and Sketches* (Columbia: University of Missouri Press, 1990), 144; Quoted in Valencius, *Health*, 19, 31, 146. Gerstäcker also cautioned of "poisonous vines" and that hunting horses had to wear long leggings or the thick brush would skin their legs to the bone.

8. Valencius, *Health*, 147–49.

9. Sallie W. Stockard, *The History of Lawrence, Jackson, Independence, and Stone Counties of the Third Judicial District of Arkansas* (Little Rock: Arkansas Democrat, 1904), 43, 48; Father Zenobius Membre, "Narrative of LaSalle's Voyage Down the Mississippi," in *Discovery and Exploration of the Mississippi Valley*, ed. John G. Shea (Albany, NY: Joseph McDonagh, 1903), 183; Jean-Baptiste Bénard de la Harpe, "Exploration of the Arkansas River by Benard de la Harpe, 1721–1722: Extracts from His Journal and Instructions," ed. and trans. Ralph A. Smith, *Arkansas Historical Quarterly* 10, no. 4 (Winter 1951): 353. Canebrakes stood so thick in some locations that even bison could not pass through them. For more information about the transition from canebrakes to agriculture, see Mart A. Stewart, "From King Cane to King Cotton: Razing Cane in the Old South," *Environmental History* 12, no. 1 (2007): 59–79.

10. Horace Kephart, "Lost in the Swamps-2," *Forest and Stream*, August 10, 1895, 112. *Chigre* or *chigger* is Arkansas vernacular.

11. Stockard, *History*, 54; M. F. Tateman, "Trailing a Turkey," *Shields Magazine* 1, no. 8 (November 1905), 367. "The Bowie-Knife," *New York Mirror*, July 14, 1838.

12. Kathleen Duval, *The Native Ground: Indians and Colonists in the Heart of a Continent* (Philadelphia: University of Pennsylvania Press, 2007), 8, 14, 15–17, 21, 39; Quoted in Morris S. Arnold, "The Significance of Arkansas's Colonial Experience," in

Cultural Encounters in the Early South: Indians and Europeans in Arkansas, ed. Jeannie Whayne (Fayetteville: University of Arkansas Press, 1995), 137.

13. Walter N. Vernon, *Methodism in Arkansas, 1816–1976* (Madison: University of Wisconsin Press, 1976), 11,13, 15. Missouri contained nearly 20,000 people in 1810 and 66,500 in 1820. Mississippi's population in 1810 was 31,000 and just over 75,000 in 1820. Louisiana's population in 1810 numbered 76,500, and 153,000 in 1820. United States Census, 1810 and 1820.

14. Brooks Blevins, *Arkansas/Arkansaw: How Bear Hunters, Hillbillies, and Good Ol' Boys Defined a State* (Fayetteville: University of Arkansas Press, 2009), 13.

15. Blevins, *Arkansas/Arkansaw,* 15.

16. Ted R. Worley and Russell W. Benedict, "Story of an Early Settlement in Central Arkansas," *Arkansas Historical Quarterly* 10, no. 2 (Summer 1951): 117–37.

17. Worley and Benedict, "Story," 117–37; Thomas Nuttall, *A Journal of Travels into the Arkansas Territory during the Year 1819* (Thomas H. Palmer, 1821), 119, 129.

18. Nuttall, *Journal of Travels,* 99–100.

19. Billy D. Higgins, "The Origins and the Fate of the Marion County Free Black Society," *Arkansas Historical Quarterly,* 54, no. 4 (Winter 1995): 427–43. Bull Shoals Lake covers much of the Little North Fork on the White River today. For more about free huntsman and their treatment in sporting literature, see Nicholas Proctor, *Bathed in Blood: Hunting and Mastery in the Old South* (Charlottesville: University of Virginia Press, 2002), 83.

20. Billy D. Higgins, *A Stranger and a Sojourner: Peter Caulder, Free Black Frontiersman in Antebellum Arkansas* (Fayetteville: University of Arkansas Press, 2005), 66–67, 137.

21. Higgins, *Stranger,* 137, 140.

22. Milton D. Rafferty, *Rude Pursuits and Rugged Peaks: School Craft's Ozark Journal, 1818–1819* (Fayetteville: University of Arkansas Press, 1996), 102–3. The larger vessels could not pass the Buffalo Shoals at the mouth of the Buffalo Fork because the water was too low. During the wetter times of the year, smaller boats could pass, and travelers could portage their pirogues and canoes if needed in the dryer periods; Higgins, *Stranger,* 141; "New Orleans Prices Current," *Arkansas Gazette* (Arkansas Post), June 18, 1822. Schoolcraft made these observations in 1819, the same year the Hall family settled in Marion County.

23. Rafferty, *Rude Pursuits,* 98.

24. Proctor, *Bathed,* 83.

25. J. Steele and James M. Campbell, eds. *Laws of Arkansas Territory* (Little Rock: J. Steel, 1835), 521.

26. For more information about the Native American, European, and American struggles in the Arkansas River valley for power, see DuVal, *Native Ground.*

27. Rafferty, *Rude Pursuits,* 64; William E. Bevens quoted in Mabel West, "Jacksonport, Arkansas: Its Rise and Decline," *Arkansas Historical Quarterly* 9, no. 4 (Winter 1950): 236.

28. John Shaw, "Shaw's Narrative," in *Collections of the State Historical Society of Wisconsin*, vol 2, ed. Lyman C. Draper (Madison: State Historical Society of Wisconsin, 1903), 199–202. After the disappointment of not reaching the Pacific Ocean and going bust at New Orleans, Shaw retreated to New Madrid, Missouri, just in time to experience the New Madrid earthquake in December 1811. The Embargo Act of 1807 closed United States ports to export in response to French and British molestation of neutral American ships during the Napoleonic Wars.

29. John Quincy Wolf and Charles Heinrich, "Journal of Charles Heinrich, 1849–1856," *Arkansas Historical Quarterly* 24, no. 3 (Autumn 1965): 246.

30. J. P. Young, *Standard History of Memphis, Tennessee* (Knoxville, TN: H. W. Crew, 1912), 66. Fort Assumption under the French. For more information about US factories in Arkansas, see Wayne Morris, "Traders and Factories on the Arkansas Frontier, 1805–1822," *Arkansas Historical Quarterly* 28, no. 1 (Spring 1969): 28–48.

31. Nuttall, *Journal of Travels*, 56. Memphis was started in 1819, not officially created until two years later. For more information, see James Roper, *The Founding of Memphis, 1818–1820* (Memphis: West Tennessee Historical Society, 1970); Isaac Rawlings was sent to establish the US government factory in 1814 and was still there in 1820.

32. Vivian Hansbrough, "The Crowleys of Crowley's Ridge," *Arkansas Historical Quarterly* 13, no. 1 (Spring 1954): 55. Depending upon where a fellow started, Cape Girardeau could take as long to reach as Memphis. If travelers were on the south end of the ridge, it would take longer because the ridge ran all the way up to just south of Cape Girardeau. Significant immigration into Arkansas did not start until after the War of 1812 when the government provided land grants to several veterans, and even then, the numbers remained low. Another boost came in 1819 when the Arkansas Territory was formed.

33. John Billingsley and Ted R. Worley, eds., "Letter from an Early Arkansas Settler," *Arkansas Historical Quarterly* 11, no. 4 (Winter 1952): 327–30.

34. "Soil and Climate of the American Territories: Translation of a Letter from an Enlightened French Immigrant," in *The Port-Folio*, vol. 4, no. 4 (Philadelphia: H. Maxwell, 1817).

35. *Arkansas Gazette*, December 4, 1819, January 15, 1820; George W. Featherstonhaugh, *Excursion through the Slave States: From Washington on the Potomac to the Frontier of Mexico* (London: J. Murray, 1844), 234; *Arkansas State Gazette*, July 24, 1839.

36. Joshua D. Rothman, *Flush Times and Fever Dreams: A Story of Capitalism and Slavery in the Age of Jackson* (Athens: University of Georgia Press, 2012), 2–3. Arkansas became a state in 1836. Nat Turner's rebellion occurred in Virginia in 1831. Enslavers began to develop a belief that northern abolitionists, politicians, and citizens conspired to outlaw slavery in the nation beginning in roughly 1830 and continuing through to the Civil War. Just across the Big Muddy from Arkansas, Mississippi experienced tremendous growth in settlers, the enslaved, cotton growth, and wealth. Between 1830 and 1836, 75,000 white settlers moved into Mississippi, doubling the

white population there. In 1830, the Arkansas Territory contained about 25,000 white and 4,500 black inhabitants. Ten years later, the state's white population numbered around 77,500, with nearly 20,000 enslaved people; United States Census, 1830, 1840.

37. *Arkansas Times and Advocate*, October 16, 1835, 1; *Randolph Recorder*, June 20, 1834.

38. Proctor, *Bathed*, 20–25; Dickson D. Bruce, *Violence and Culture in the Antebellum South* (Austin: Unversity of Texas Press, 1997), 200–203; Marks, *Southern Hunting*, 18, 24.

39. (Mrs. Isaac H.) Hilliard, diary, January 5, 1850, Isaac H. Hilliard Papers, Louisiana and Lower Mississippi Valley Collection, Louisiana State University, Baton Rouge, LA.

40. Ted Ownby, *Subduing Satan: Religion, Recreation, and Manhood in the Rural South, 1865–1920* (Chapel Hill: University of North Carolina Press,1990); Proctor, *Bathed*, 138–39, 162; Hayden Smith, "Knowledge of the Hunt: African American Guides in the South Carolina Lowcountry at the Turn of the Twentieth Century," in *Leisure, Plantations, and the Making of a New South*, ed. Julia Brock and Daniel Vivian (New York: Lexington Books, 2015), 134; Marks, *Southern Hunting*, 27. For the latest study about slavery in Arkansas, see Kelly Houston Jones, *A Weary Land: Slavery on the Ground in Arkansas* (Athens: University of Georgia Press, 2021).

41. Eugene D. Genovese, *Roll, Jordan, Roll: The World the Slaves Made* (New York: Pantheon Books, 1976), 486–87; Proctor, *Bathed*, 162–63; Smith, "Knowledge," 133; Interview with Dr. Simeon Bateman in Stockard, *History*, 48–51; Dan A. Rudd and Theo. Bond, *From Slavery to Wealth: the Life of Scott Bond; the Rewards of Honesty, Industry, Economy and Perseverance* (Madison, AR: Journal Printing, 1917), 258, 60–61. Benniah Bateman Sr. freed Jack in his 1849 will and an enslaved a woman named Liv. It is presumed that Liv was Jack's wife. Wills and Probate Court, Independence County, Arkansas, Wills 1833–1881, September 26, 1849, 115–20, Ancestry (website); Scott E. Giltner, *Hunting and Fishing in the New South: Black Labor and White Leisure after the Civil War* (Baltimore: Johns Hopkins University Press, 2008), 3; Ridge v. Featherston, 15 Ark. 159–60 (1854). During the Arkansas territorial period, the law proclaimed that "no slave or mulatto whatsoever shall keep or carry a gun, powder, shot, or club" and if caught, the weapons were seized and the enslaved person presented to the justice of the peace, and he/she could receive up to thirty-nine lashes as punishment. This seems to be the understood law even after Arkansas statehood. However, in 1846, the Arkansas state legislature altered the law so that enslaved people could carry guns with the written permission of their owner, presumably anywhere the enslaved traveled. Eight years later, several counties outlawed the carrying of firearms by enslaved people, "whether such weapon be the property of such slave or slaves, or not;" J. Steele and James M. Campbell, eds. *Laws of Arkansas Territory* (Little Rock: J. Steel, 1835), 521; Arkansas General Assembly, *Acts Passed at the Tenth Session of the General Assembly of the State of Arkansas* (Little Rock: Johnson and Yerkes, 1855), 135–36 Prior to 1819, laws concerning enslaved people and guns were held over from the

Spanish and French systems where the enslaved could carry weapons on their own plantations but only with written permission from the justice of the peace. For more see Orville W. Taylor, *Negro Slavery in Arkansas* (Durham, NC: Duke University Press, 1958), 29–30; Samuel H. Hempstead and E. H. English, *A Digest of the Statutes of Arkansas, Embracing All Laws of a General and Permanent Character in Force at the Close of the Session of the General Assembly of 1846: Together with Notes of the Decisions of the Supreme Court upon the Statutes* (Little Rock: Rearden and Garritt, 1848), 551. Any white man could ask for the gun permit from any free black person at any time, and if they could not produce one, the white man could confiscate and keep the firearm and the African American could face a fine as well; Arkansas General Assembly, *Acts Passed* (1855), 135–36; if the enslaved person was found guilty of the offense, their owner was to pay a fine of no less than twenty dollars and no more than one hundred dollars, and the individual who accosted the enslaved person could keep the seized weapon. Therefore, some Arkansas bondsmen not only had access to firearms but, in some cases, owned them. White owners had to oversee enslaved people using weapons, or the government had consequences for those caught breaking the law. The record is unclear why only these eleven counties narrowed the location where enslaved people could carry weapons under the 1846 gun law.

42. Proctor, *Bathed*, 126–27, 137; Marks, *Southern Hunting*, 28. Marks points out that the wealthy hunted during the day with lots of equipment and the enslaved hunted at night with very little.

43. John W. Blassingame, *The Slave Community: Plantation Life in the Antebellum South* (New York: Oxford University Press, 1979), 251–54; Genovese, *Roll*, 486–87; interview with Callie Washington, in George E. Lankford and Federal Writers' Project, *Bearing Witness: Memories of Arkansas Slavery: Narratives from the 1930s WPA Collection* (Fayetteville: University of Arkansas Press, 2003), 416; interview with John Bates, in Lankford and Federal Writers' Project, *Bearing Witness*, 316; Jones, *Weary Land*, 161–62; Proctor, *Bathed*, 126; Smith, "Knowledge," 133.

44. Interview with Henry Walker, in Federal Writers' Project, *Slave Narratives: A Folk History of Slavery in the United States from Interviews with Former Slaves*, typewritten records prepared by the Federal Writers' Project, 1936–1938, assembled by the Library of Congress project Work Projects Administration for the District of Columbia, vol. 2, *Arkansas Narratives*, part 7 (Washington, DC: 1941). Interview with Joe Bean, in Lankford and Federal Writers' Project, *Bearing Witness*, 382. Walker learned how to shoot from the Williams family while still in Nashville, Tennessee, before ending up in Hazen after the end of the Civil War. Williams brought young Walker to the Williams family farm in Arkansas near Little Rock.

45. Daniel Robinson Hundley, *Social Relations in Our Southern States* (New York: H. B. Price, 1860), 342–43. Hundley also claims that on large plantations across the South, enslaved people owned their own hunting dogs and hunted coon and possum with them. See also Proctor, *Bathed*, 120, 125, 137, 139, 149, 163–64.

46. Ruth Polk Patterson, *The Seed of Sally Good'n: A Black Family of Arkansas, 1833–1953* (Lexington: University Press of Kentucky, 1985), 26–28; Proctor, *Bathed*, 123, 141, 145, 162–64; Smith, "Knowledge," 134; Marks, *Southern Hunting*, 22–25; Rudd and Bond, *From Slavery to Wealth*, 258, 60–61; interview with Scott Bond, in Lankford and Federal Writers' Project, *Bearing Witness*, 74.

47. Interview with Jim Ricks, in Lankford and Federal Writers' Project, *Bearing Witness*, 49.

48. Interview with Doc Quinn, in Lankford and Federal Writers' Project, *Bearing Witness*, 242–43. Quinn explained that one of his enslaved companion's sole responsibilities was to care for the canines and drive predators from the fields.

49. Interview with Betty Brown, in Lankford and Federal Writers' Project, *Bearing Witness*, 129–31. Her mother was Mary-Ann Millan.

50. Quoted in John Soloman Otto and Ben Wayne Banks, "The Banks Family of Yell County, Arkansas: A 'Plain Folk' of the Highlands South," *Arkansas Historical Quarterly* 41, no. 2 (Summer 1982), 160; Marks, *Southern Hunting*, 28.

51. J. H. Atkinson and Charles McDermott, "A Memoir of Charles McDermott: A Pioneer of Southeastern Arkansas," *Arkansas Historical Quarterly* 12, no. 3 (Autumn 1953): 260.

52. William O. Ragsdale, *They Sought the Land: A Settlement in the Arkansas River Valley 1840–1870* (Fayetteville: University of Arkansas Press, 1997), 2.

53. Patricia A. Etter, ed., *An American Odyssey: The Autobiography of a 19th-Century Scotsman, Robert Brownlee, at the Request of His Children, Napa County, California, October 1892* (Fayetteville: University of Arkansas Press, 1986), 36; Cordelia Hambleton to Dear Brother, December 4, 1854, miscellaneous letter collection, J. N. Heiskell Papers, 1829–1973, Center for Arkansas History and Culture, University of Arkansas, Little Rock, AR.

54. Y. W. Etheridge, "Pioneers of Ashley County," *Arkansas Historical Quarterly* 16, no. 1 (Spring 1957): 64. Interview with Melvin Robinson in Stockard, *History*, 45; Stanley T. Baugh, "James and Julia Baugh," *Arkansas Historical Quarterly* 10, no. 2 (Summer 1951): 184, 196.

55. William P. Black to his mother, March 5, 1862, and S. S. Marrett to his wife, April 13, 1862, quoted in William L. Shea, "A Semi-Savage State: The Image of Arkansas in the Civil War," *Arkansas Historical Quarterly* 48, no. 4 (Winter 1989): 314; Sam Black, *A Soldier's Recollections of the Civil War*, printed by the *Minco Minstrel*, Micon, OK, 1912, 11.

56. Otto and Banks, "Banks Family," 158–59.

57. James Reed Eison and J. E. Lindsay, "A Letter from Dardanelle to Jonesville, South Carolina," *Arkansas Historical Quarterly* 28, no. 1 (Spring 1969): 74.

CHAPTER TWO

1. *St. Louis Globe-Democrat*, February 4, 1900, 47.

2. Arkansas had passed its first statewide game law in 1855, barring citizens from

hunting on Sunday. Arkansas General Assembly, *Acts Passed at the Tenth Session of the General Assembly of the State of Arkansas* (Little Rock: Johnson and Yerkes, 1855), 180. As elsewhere in the South, few authorities enforced the laws; Nicholas Proctor, *Bathed in Blood: Hunting and Mastery in the Old South* (Charlottesville: University of Virginia Press, 2002), 13. In 1875, Arkansas required "a $10 market hunting license for non-resident trappers, hunters, seiners, or netters of fish." Theodore S. Palmer, *Hunting Licenses: Their History, Objects, and Limitations* (Washington, DC: Government Printing Office, 1904), 55.

3. Sociologist Jan Dizard explains four main reasons modern outdoorsmen hunt: "to experience nature as a participant; to feel an intimate, sensuous connection to place; to take responsibility for one's food and to acknowledge our kinship with wildlife." Jan Dizard, "Why Hunt?" Center for Humans and Nature website, October 12, 2015. Men hunted, trapped, and fished for profit in the New World from the earliest days of European settlement. In this reference, "for profit" means hunting to sell meat primarily.

4. Proctor, *Bathed*, 150; Stuart A. Marks, *Southern Hunting in Black and White: Nature, History, and Ritual in a Carolina Community* (Princeton, NJ: Princeton, University Press, 1991), 6, 8–9.

5. "Sir Jennings Beckwith, of the Old School," *American Turf Register and Sporting Newspaper*, March 1834, 5.

6. "Sir Jennings Beckwith," 7; Julia Brock and Daniel Vivian, eds. *Leisure, Plantations, and the Making of a New South* (Washington, DC: Lexington Books, 2015), 152.

7. Friedrich Gerstäcker, *Wild Sports in the Far West* (Boston: Crosby, Nichols, 1859).

8. Ted R. Worley, "An Early Arkansas Sportsman: C. F. M. Noland," *Arkansas Historical Quarterly* 11, no. 1 (Spring 1952): 25–39.

9. Worley, "Early Arkansas Sportsman," 27; For information about how Noland shaped Arkansas's image, see C. Fred Williams, "The Bear State Image: Arkansas in the Nineteenth Century," *Arkansas Historical Quarterly* 39, no. 2 (Summer 1980): 99–111; and Brooks Blevins, *Arkansas/Arkansaw: How Bear Hunters, Hillbillies, and Good Ol' Boys Defined a State* (Fayetteville: University of Arkansas Press, 2009).

10. *The Arkansas Gazette* (Arkansas Post), October 14, 1823; "Jesse Bean Will," Wills and Probate Court, Independence County, Arkansas, 1841, Ancestry (website); *Arkansas Gazette* (Arkansas Post), August 8, 1832; "Abraham Ruddell Will," Wills and Probate Court, Independence County, Arkansas, 1840, Ancestry (website).

11. Sallie W. Stockard, *The History of Lawrence, Jackson, Independence, and Stone Counties of the Third Judicial District of Arkansas* (Little Rock: Arkansas Democrat, 1904), 54.

12. "Deer Hunting in Arkansas," *Spirit of the Times*, September 6, 1845.

13. Fire hunting can also use a fire light to cause deer eyes to reflect in the dark, allowing the hunter to shoot blinded animals. See Proctor, *Bathed*, 89, 93–98; Scott E. Giltner, *Hunting and Fishing in the New South: Black Labor and White Leisure after the Civil War* (Baltimore: Johns Hopkins University Press, 2008), 25–26. After the war,

men who followed the sporting ethic disapproved of fire hunting and other similar practices. Some records indicate that sportsmen and market hunters sometimes left the less desirable parts of animals in the field. Wasting meat is against the accepted ethical hunting and fishing practices of today.

14. "On the AR-KAN-SAW Prairies," *Forest and Stream*, November 15, 1882. "Byrne" was a regular contributor to the *Forest and Stream* newspaper.

15. Proctor, *Bathed*, 61; Marks, *Southern Hunting*, 4; "A Week's Hunt in Arkansas," *Turf, Field, and Farm*, January 5 and 12, 1872; "An Arkansas Deer Drive," *Forest and Stream*, October 6, 1887, 207.

16. Milton D. Rafferty, ed., *Rude Pursuits and Rugged Peaks: School Craft's Ozark Journal, 1818–1819* (Fayetteville: University of Arkansas Press, 1996), 63, 54–55.

17. Proctor, *Bathed*, 151; Marks, *Southern Hunting*, 4–5, 24. For more information about more modern sports and masculinity in the South, see Ted Ownby, "Manhood, Memory, and White Men's Sports in the Recent American South," *International Journal of the History of Sport* 15, no. 1 (1998): 103–18.

18. "A Hunt in Arkansas," *Forest and Stream*, May 9, 1889, 316.

19. George Engelmann, "The Hot Springs in Arkansas," in Jerome Jansma and Harriet H. Jansma, "George Engelmann in Arkansas Territory," *Arkansas Historical Quarterly* 50, no. 3 (Autumn 1991): 244–245; "Memphis Nimrods," *Memphis Daily Appeal*, January 4, 1884, 1. The italics are mine. For more information about early belief in the healing powers of springs in Arkansas, see Conevery Bolton Valencius, *The Health of the Country: How American Settlers Understood Themselves and Their Land* (New York: Basic Books, 2002), 152–58. Beginning after the Civil War, many councils and citizens claimed their towns and countryside could cure all sorts of ailments. See Gregg Mitman, "Geographies of Hope: Mining the Frontiers of Health in Denver and Beyond, 1870–1965," in Gregg Mitman, Michelle Murphy, and Christopher Sellers, *Landscapes of Exposure: Knowledge and Illness in Modern Environments* (Chicago: University of Chicago Press, 2004). See also Linda Nash, *Inescapable Ecologies: A History of Environment, Disease, and Knowledge* (Berkeley: University of California Press, 2006).

20. "Dardanelle Springs," *Weekly Arkansas Gazette*, April 28, 1841, 3.

21. "An Arkansas Outing," *Forest and Stream*, October 20, 1887, 247. This club is not to be mistaken for the Big Lake Shooting Club that was located near Big Lake in northeastern Arkansas. See also William Cronon, *Nature's Metropolis: Chicago and the Great West* (New York: W. W. Norton, 1991); and Florence Williams, *The Nature Fix: Why Nature Makes Us Happier, Healthier, and More Creative* (New York: W. W. Norton, 2017).

22. "For the St. Francis," *Forest and Stream*, November 17, 1900, 387. For more information about health and the outdoors, see Nash, *Inescapable Ecologies*; and Mitman, *Landscapes of Exposure*.

23. For more on the mention of class in women's preservation and conservation organizations, see Mrs. Louise Stephenson's comments in *Helena Weekly World*, April 30, 1902.

24. "Should Revise State Game Laws," *Daily Arkansas Gazette,* January 3, 1910, 6; "Hunting and Fishing," *Daily Arkansas Gazette,* August 14, 1910, 38; "Great Demand for Pheasants," *Arkansas Democrat,* September 25, 1910, 8.

25. Pump shotguns became popular in the later part of the nineteenth century. Although most Arkansas hunters continued to use a cheap single-shot shotgun, in the late 1880s, market hunters started buying the lever-action or pump Winchesters or Spencer shotguns because they could kill at a higher rate. More animals meant more money for the professional. Even some employers provided market hunters with access to pumps and ammunition. Sportsmen, usually with access to more funds, carried better firearms, like the 1905 semiautomatic Belgian-made Browning A5 or the Remington equivalent. A few even carried finely crafted double-barreled weapons, such as a Purdy or Holland & Holland, manufactured in England, or the Connecticut-made Parker. Nat Wetzel, "Cunning Wiles or the Hunters and the Hunted," *Texas Game and Fish* 11 (1953): 8–9, 29–30; Jay Vessels, "King of the Market Hunters," *Texas Game and Fish* 10, no. 12 (1952): 11–14; "Looking Back at Shotgun History," *American Rifleman* (online), May 23, 2016, accessed February 2, 2023; John M. Taylor, "The Spencer Pump: America's First Pump-Action Shotgun," *Outdoor Life* (online), January 20, 2022, accessed February 2, 2023; "$700 for a Gun," *Daily American,* January 4, 1885.

26. William Cronon, *Nature's Metropolis.*

27. "The World's Greatest Game Market," *Arkansas Conservationist* 1 (January, February, March 1925): 5, 15. According to a CPI calculator, one dollar in 1873 had a buying power of a little less than thirty dollars today.

28. "Approaching Nuptials," *Daily Arkansas Gazette,* April 6, 1873.

29. "Hunting in Arkansas," *Sporting Times,* January 2, 1869; *Daily Arkansas Gazette,* October 12, 1872, 4; George K. Holmes, *Wages of Farm Labor,* United States Department of Agriculture, Bureau of Statistics, Bulletin 99 (Washington, DC: Government Printing Office, 1912), 22. One dollar in 1868 had the same buying power as a little more than twenty dollars in 2023. CPI Calculator, Bureau of Labor and Statistics, accessed August 28, 2023. "Nimrod" refers to a biblical figure, the great-grandson of Noah, described in Genesis 10:8–12 as "the first on earth to be a mighty man. He was a mighty hunter before the Lord" (NIV). Nimrod became slang for a hunter in the nineteenth century.

30. "Shot Gun and Rifle," *Forest and Stream,* November 26, 1874, 246; "Shot Gun and Rifle," *Forest and Stream,* December 10, 1874, 281–82.

31. "Shot Gun and Rifle," *Forest and Stream,* December 10, 1874, 281–82; "On the AR-KAN-SAW Prairie," *Forest and Stream,* November 15, 1882.

32. Charley Hershey diary, 1883–1884, record number 188, Joplin Missouri Historical Society, 17–27. The Hershey brothers left Arkansas in 1884 after several highwaymen attempted to murder or attack them and steal their boats. Charley called these men "kuklux," but it is unclear whether they had any affiliation. Or $11,791.67 profit in today's CPI real price; Measuring Worth (website), accessed August 2, 2023.

33. "Memphis Nimrods," *Memphis Daily Appeal*, January 4, 1884, 1; "The Game Season," *St. Louis Globe-Democrat*, October 23, 1887, 12.

34. Proctor, *Bathed*, 30–31; Brock and Vivian, *Leisure*, 7, 23. For more on hunting plantations, see Giltner, *Hunting and Fishing*, 129–32.

35. Ted Ownby, *Subduing Satan: Religion, Recreation, and Manhood in the Rural South, 1865–1920* (Chapel Hill: University of North Carolina Press, 1990), 21; "Kentucky Hunting Club," *Shields Magazine*, July 1909, 38; "Big Wolf Drive Grows in Interest," *Daily Arkansas Gazette*, September 12, 1918, 4.

36. "Hunters Ready to Enter the Woods," *Daily Arkansas Gazette*, November 8, 1920, 10.

37. "Memphis Nimrods," *Memphis Daily Appeal*, January 4, 1884.

38. "Establishing Hunting Estates," *St. Louis Globe-Democrat*, June 18, 1893, 31.

39. "St. Louis Hunting and Fishing Clubs," *St. Louis Post-Dispatch*, December 23, 1900.

40. Saikku, *This Delta, This Land: An Environmental History of the Yazoo-Mississippi Floodplain* (Athens: University of Georgia Press), 31; "Memphis Nimrods," *Memphis Daily Appeal*, January 4, 1884, 1; "The Hunting Clubs of Memphis," *Memphis Avalanche*, September 2, 1888, 2; *St. Louis Globe-Democrat*, December 11, 1898, 43. Wheatley was a real estate developer and an Irish Setter breeder. His dogs competed in many field trials across the country in the 1870s and 1880s. Guido was Wheatley's pen name when he wrote to *Forest and Stream*. *Memphis Evening Herald*, November 14, 1877; "The Hunting Clubs of Memphis," *Memphis Avalanche*, September 2, 1888, 2.

41. Brock and Vivian, *Leisure*, 7.

42. "Memphis Nimrods," *Memphis Daily Appeal*, January 4, 1884, 1. William Bard Edrington and F. P. Poston were Memphis lawyers. Edrington, along with Tate brother Robert, was known as the sportsmen who first brought Chesapeake Bay retrievers to Memphis. *Memphis Daily Appeal*, October 12, 1890. Arthur Wheatley owned several as well. *Memphis Daily Avalanche*, November 27, 1887; *St. Louis Globe-Democrat*, December 11, 1898, 43. The Osceola Club later became known as the Memphis Club because of all the members from Memphis.

43. "Memphis Nimrods," *Memphis Daily Appeal*, January 4, 1884, 1. Dr. Mitchell was instrumental in fighting the yellow fever epidemic in Memphis during the outbreaks in the 1870s. William Speer, *Sketches of Prominent Tennesseans* (1888; repr., Baltimore: Genealogical Publishing, 2010), 556; Duane Huddleston, Sammie Rose, and Pat Wood, *Steamboats and Ferries on the White River* (Fayetteville: University of Arkansas Press, 1998), 76.

44. "Arkansas and the South," *Forest and Stream*, August 26, 1899, 165.

45. "The Hunting Clubs of Memphis," *Memphis Avalanche*, September 2, 1888, 2. Wapanocca Lake was a shallow lake about a mile and a half wide and three and a half miles long with marshy land surrounding it. Phil Gwynn and his wife are buried on the old grounds of the club.

46. "The Hunting Clubs of Memphis," *Memphis Avalanche,* September 2, 1888, 2; "Wapanocca of Tennessee," *Forest and Stream,* September 24, 1898, 245–46.

47. "Memphis Nimrods," *Memphis Daily Appeal,* January 4, 1884, 1.

48. *The Sportsman's Paradise on the Buffalo Island Route, Arkansas* (St. Louis: Woodward and Tiernan, 1897), 12, 15; "St. Louis Hunting and Fishing Clubs," *St. Louis Post-Dispatch,* December 23, 1900. The Brooklyn Club (1899) and the Loda Hunting and Fishing Club (1900) at Knobel, Arkansas, were two other St. Louis–filled clubs.

49. Lynn Morrow, "St. Louisans at the Knobel: Sport Tourism and Conservation in Arkansas," Southeast Missouri State University Press website, accessed August 12, 2021; *Sportsman's Paradise,* 12.

50. "Shooting and Fishing," *St. Louis Globe-Democrat,* September 12, 1897, 40. When both Missouri and Arkansas introduced hunting and fishing seasons, the members of Moark could hunt or fish the side of the island where the game or fish they pursued was in season in that state. Isaac Taylor was the chief architect of the buildings for the St. Louis World's Fair in 1904. *St. Louis Republic,* September 22, 1901; *Kansas City Times,* May 10, 1891; *Sun Herald* (Biloxi, MO), April 27, 1940; *Plattsmouth Journal,* April 9, 1903. Some other St. Louis men also owned the Brookland Hunting and Fishing Club, located eight miles from the Iron Mountain railroad depot at Powell. "On the Spreads of the St. Francis," *St. Louis Globe-Democrat,* October 8, 1899, 33.

51. Will Wildwood, *The Sportsman's Directory and Year Book* (Milwaukee: Pond and Goldeye, 1892), 66, 86; *Daily Arkansas Gazette,* June 2, 1886. W. W. Wright, George W. Watson, James Kennedy, Peter Benz, E. S. Rockwood, and Charles Garrett were members of the Ouachita Rod and Gun Club. *Arkansas Democrat,* May 30, 1890; "Arkansas Fox Hunters," *Forest and Stream,* August 19, 1899, 145; *Arkansas Democrat,* July 10, 1900, December 16, 1892; March 18, 1916; February 20, 1900.

52. Ledger, Point Gun Club Records, 1892–1911, Center for Arkansas History and Culture, University of Arkansas, Little Rock, AR; *Fort Smith Times,* March 26, 1903; "Bryan to visit Van Buren," *Daily Arkansas Gazette,* April 3, 1909; *Fort Smith Times,* January 8, 1904.

53. "Arkansas Club," *Forest and Stream,* October 23, 1897, 327.

54. J. M. Rose, "Grassy Lake Hunting Club," *Forest and Stream,* June 30, 1894, 555; J. M. Rose, "The Grassy Lake Club," *Forest and Stream,* April 10, 1897, 287.

55. "The Christian County Hunting Club," *Forest and Stream,* July 15, 1899, 45–46; "Kentucky Hunting Club," *Shields Magazine,* July 1909, 38.

56. *Fayetteville Weekly Democrat,* August 28, 1871, 2; *Osceola Times,* March 25, 1876, 3; *Pine Bluff Daily Graphic,* August 22, 1897; John Gaskins, *Life and Adventures of John Gaskins in the Early History of Northwest Arkansas* (Eureka Springs: Printed and for sale by the author, 1893), 2.

57. George P. Rawick, ed., *The American Slave: A Composite Autobiography* (Westport, CT: Greenwood, 1972–1979), vol. 10, pt. 5, 28; vol. 8, pt. 2, 231; vol. 9, pt. 3, 140.

58. Interview with Joe Golden, in Federal Writers' Project, *Slave Narratives*, vol. 2, part 3, 48–50.

59. Interview with Warren McKinney, in Federal Writers' Project, *Slave Narratives*, vol. 2, part 5; Interview with George Braddox, in George E. Lankford and Federal Writers' Project, *Bearing Witness: Memories of Arkansas Slavery; Narratives from the 1930s WPA Collections* (Fayetteville: University of Arkansas Press, 2003), 305. The narrative states that Braddox hunted with ex-Confederate Colonel Alexander Filip Yopp and his dogs. Henry Walker also fed Yopp's dogs and attended the colonel on foxhunts. Ruth Adams Yopp was the widow of A. F. Yopp, a woman forty-seven years his junior at his death in 1929. See "Alexander Yopp," Arkansas Department of Vital Records; Little Rock, AR; Marriage Certificates, Ancestry (website); 1923, Microfilm Reel: #3, Arkansas State Archives.

60. *Southern Watchmen* (Athens, GA), February 12, 1868; *Memphis Avalanche*, January 7, 1868.

61. *Daily Arkansas Gazette*, December 5, 1911, 4; "Uses Shotgun for Well Rope," *Arkansas Democrat*, November 17, 1922, 8; Scott Giltner, *Hunting and Fishing*, 5.

62. Giltner, *Hunting and Fishing*, chaps. 3, 4; Hayden Smith, "Knowledge of the Hunt: African American Guides in the South Carolina Lowcountry at the Turn of the Twentieth Century," in Brock and Vivian, *Leisure*, 139. These camp workers certainly had other employment because working for the clubs, with rare exceptions, did not pay enough or last long enough during the year to provide a living wage.

63. Ted Ownby, *Subduing Satan*, 26; Giltner, *Hunting and Fishing*, 2.

64. Giltner, *Hunting and Fishing*, 88–92; Smith, "Knowledge," 139.

65. Donna Goldstein, "The Life and Legacy of Henry Norwood," 2015, vertical file, South Sebastian County Historical Society, Greenwood, AR.

66. "Officials Return Minus B'ar Meat," *Daily Arkansas Gazette*, November 10, 1912, 5.

67. "Big Wolf Drive Grows in Interest," *Daily Arkansas Gazette*, September 12, 1918, 4; Smith, "Knowledge," 139–40.

68. Smith, "Knowledge," 140–43; "Republican Convention," *Memphis Daily Appeal*, August 18, 1876; "Tried for his Life," *Memphis Daily Appeal*, February 1, 1884. Not much information exists about Gwynn. After the Civil War, he was active in the Memphis Republican party, serving on a committee to send delegates to the state Republican convention in 1876. The racism or lack thereof of Nash Buckingham remains an area of contention among outdoor writers and readers. See Jim Casada, "Nash Buckingham, Southern Gentleman," Outdoor Witers Association of American website, accessed March 29, 2023; "Forums," Classic Fly Rod Forum (website), accessed March 29, 2023.

CHAPTER THREE

1. For more about the evolution of Arkansas and its reputation, see Brooks Blevins, *Arkansas/Arkansaw: How Bear Hunters, Hillbillies, and Good Ol' Boys Defined a State* (Fayetteville: University of Arkansas Press, 2009). According to Tara Kelly, hunting

narratives in sporting publications also positively impacted the late nineteenth-century middle class's attitude change toward conservation laws from opposition to support. Big game hunting narratives that appeared after 1880 and continued until the 1920s were the first to capture the reader's attention nationally. New narratives from the new Gilded Age sportsmen replaced the big game stories in most sporting publications as time passed. "Manly display and the Second Industrial Revolution, hunting and writing, the fair stalk and the railway trip were thus all linked from the beginning," Kelly concludes. Robert L. Kelly, *The Hunter Elite: Manly Sport, Hunting Narratives, and American Conservation, 1800–1925* (Lawrence: University of Kansas Press, 2018), 11, 40–41. For the growth in leisure activities, see Dickson D. Bruce, *Violence and Culture in the Antebellum South* (Austin: University of Texas Press), 198; Joel Shrock, *The Gilded Age* (Westport, CT: Greenwood, 2004), chap. 7.

2. (Mrs. Isaac H.) Hilliard, diary, April 12, 1850, Isaac H. Hilliard Family Papers, Louisiana and Lower Mississippi Valley Collection, Lousiana State University Libraries, Baton Rouge, LA; "Shot Gun and Rifle," *Forest and Stream*, December 10, 1874, 281–82; "Hunting in Arkansas," *Sporting Times*, January 2, 1869, 5. Although Arkansas contained myriads of wildlife, it also contained (as seen in chapter 1) a host of environmental challenges, such as heat, mosquitoes, swamps and lowlands, and few roads. As the knowledge of these conditions spread out from Arkansas, the combination of much game and rough country created a tempting lure to outdoorsmen who wanted to challenge themselves. See Conevery Bolton Valencius, *The Health of the Country: How American Settlers Understood Themselves and Their Land* (New York: Basic Books, 2002), 146.

3. "Memphis Nimrods," *Memphis Daily Appeal*, January 4, 1884, 1; "Shot Gun and Rifle, *Forest and Stream*, December 10, 1874, 281–82; "On the AR-KAN-SAW Prairie," *Forest and Stream*, November 15, 1882. A snipe is a wading shorebird with a long slender bill and brownish coloring.

4. *Daily Arkansas Gazette*, October 12, 1872, 4; "Hunting in Arkansas," *Sporting Times*, January 2, 1869, 5; "A Week's Hunting in Arkansas," *Turf, Field, and Farm*, January 12, 1872, 181–82.

5. "Answers to Correspondents," *Forest and Stream*, December 28, 1876; "Grouse Hunting in Arkansas," *Rod and Gun*, September 11, 1875, 355.

6. *Daily Arkansas Gazette*, July 8, 1877, 2.

7. "Return of a Sporting Party," *St. Louis-Globe*, October 30, 1875, 8. J. G. Prather was a steamboat captain on the Ohio and lower Mississippi. Captain George W. Ford was assistant harbor and wharf commissioner in St. Louis. Daniel Able was the secretary of St. Louis's O'Fallon Park. Joe Rickey was a popular silver lobbyist in Jefferson City, Missouri, and later, the gin rickey was named after him.

8. Kelly, *Hunter Elite*, 11, 42–45. See also Karl Jacoby, *Crimes against Nature: Squatters, Poachers, Thieves, and the Hidden History of Conservation* (Berkeley: University of California Press, 2001); and John F. Reiger, *American Sportsmen and the Origins of*

Conservation (Corvallis: Oregon University Press, 2001). "Arkansas Trapper," *Forest and Stream*, January 25, 1882, 508. Hunters have retold their experiences around campfires for millennia and, later, in print to relive their experiences and teach listeners lessons. Bruce, *Violence*, 210–11.

9. *Forest and Stream*, December 14, 1882, 388.

10. Elliot Gorn and Warren Goldstein, *A Brief History of American Sports* (New York: Hill and Wang, 1993), 67; W. David Sloan, and Lisa Mullikin Parcell, eds., *American Journalism: History, Principles, Practices* (Jefferson, NC: McFarland, 2002), 199; "Col. S. D. Bruce," *New York Times*, January 23, 1898.

11. Larry K. Menna and Thomas L. Altherr, *Sports in North America: The Origins of Modern \Sports, 1820–1840* (Ann Arbor: University of Michigan Press, 1995), 35–36, 49.

12. "American Turf Register and Sporting Newspaper," *Arkansas Gazette*, October 5, 1831, 4.

13. Thomas Bangs Thorpe, "The Big Bear of Arkansas," *Spirit of the Times*, March 27, 1841; S. D. Barnes, "In the Bruin's Stronghold," *Forest and Stream*, September 25, 1890, 186. For more information on Thomas Bangs Thorpe, see Milton Rickels, *Thomas Bangs Thorpe: Humorist of the Old Southwest* (Baton Rouge: Louisiana State University Press, 1962); see also the section on southwestern humorists in Blevins, *Arkansas/Arkansaw*, 24–28.

14. "A Grand Arkansas Hunt," *Spirit of the Times*, February 26, 1842.

15. "A Week's Hunt in Arkansas," *Turf, Field, and Farm*, January 5 and 12, 1872. After the Civil War, General Wade Hampton III, born in South Carolina, became widely known as a sportsman who hunted bear near Lake Washington on his plantation in Mississippi. He reportedly killed many of his bear with a knife. Theodore Roosevelt may have come to hunt bear in Mississippi because of Wade Hampton. Arkansas farmer Opps in Helena guided Kentucky bear hunters annually in the 1880s and 1890s.

16. From Our Correspondent, "Tally-Ho!," *Forest and Stream*, November 14, 1874, 230.

17. "Answers to Correspondents," *Forest and Stream*, September 10, 1874, 82; Little Rock, Arkansas, city directory, 1890; *Daily Arkansas Gazette*, October 19, 1883, 3.

18. "Forest, Field and Stream," *Time*, June 16, 1930. *Field and Stream* absorbed *Forest and Stream* in 1930 and kept the name, *Field and Stream*. For more information on Grinnell, see Carolyn Merchant, *Spare the Birds! George Bird Grinnell and the First Audubon Society* (New Haven, CT: Yale University Press, 2016). Ohio hunters in "A Wet Deer Hunt in Arkansas," *Forest and Stream*, April 25, 1896, 333. Massachusetts hunters in *Forest and Stream*, May 23, 1896, 418.

19. *S & D Reflector* 7, no. 1 (March 1970), 5–7. The *CO* sank in 1897 while towing another ship, the *Oakland*.

20. M. La Rue Harrison, "Railroad Letter," *Fayetteville Weekly Democrat*, December 4, 1869, 2. The P. & G. E. R. W. Co. never built a rail line in northeastern Arkansas but did build in northwestern Arkansas several years later.

21. "For the St. Francis," *Forest and Stream*, November 17, 1900, 387.

22. "St. Louis Siftings," *Forest and Stream*, November 8, 1883, 286.

23. George Franklin Cram, Cram's township and railroad map of Arkansas, Chicago, 1895, Library of Congress online catalog.

24. "St. Louis Siftings," *Forest and Stream*, November 8, 1883, 286.

25. William Harris, ed. *The Sportsman's Guide to the Hunting and Shooting Grounds of the United States and Canada*, (New York: Angler's Publishing, 1888), xiv; Arista C. Shewey, *Shewey's Guide and Map to the Happy Hunting Grounds of Missouri and Arkansas: The Paradise of the Hunter and Fisherman* (St. Louis: Shewey, 1892); "Hunting and Fishing Notes," *St. Louis Globe-Democrat*, November 4, 1894, 10. Easton rode bicycles in the late 1890s for hundreds of miles, once completing a trip from St. Louis to San Francisco in 1894. "St. Louis Cyclists," *St. Louis Post-Dispatch*, September 29, 1894; "Hunting and Fishing Notes," *Forest and Stream*, December 2, 1894, 11.

26. *St. Louis Republican*, November 14, 1900, 9; *Rich Hill Tribune*, October 10, 1901, 3.

27. Harris, *Sportsman's Guide*.

28. Shewey, *Shewey's Guide*, 10, 14.

29. *The Sportsman's Paradise on the Buffalo Island Route, Arkansas* (St. Louis: Woodward and Tiernan, 1897), 12.

30. Charles S. Nichols, *With Rod and Gun in Arkansas* (Chicago: Chicago & Eastern Railroad, 1900); *St. Louis Republican*, November 3, 1900, 7; *Forest and Stream*, March 17, 1894, 202.

31. *Feathers and Fins on the Frisco*, issued by the passenger department, St. Louis & San Francisco Railroad, St. Louis, 1900.

32. *Where to Hunt American Game* (Lowell: United States Cartridge Company, 1898), 22–24.

33. *Where to Hunt American Game* (Lowell, MA: United States Cartridge Company, 1898), 22–24; Nichols, *With Rod and Gun*, 15.

34. "A Water Haul," *Forest and Stream*, January 17, 1889, 515.

35. "A Ducking Outfit," *St. Louis Globe-Democrat*, October 6, 1889, 32.

CHAPTER FOUR

1. Early sportsmen were more apt to overkill and judge success with the number of animals and birds taken during a hunt. As the sporting ethic evolved, true success during the pursuit became less about the number of animals harvested but the way that the man conducted himself in prosecuting a fair chase. For example, during a hunt through the Dakotas, Theodore Roosevelt, one of the most famous sportsmen in American history, harvested over 170 animals in forty-seven days. Daniel Herman, *Hunting in the American Imagination* (Washington, DC: Smithsonian Institute Press, 2001), 244–45.

2. Herman, *Hunting*, 246–47. For more information about class conflict and sport

hunting, see Herman, *Hunting*, 246–69. Resident Arkansas market hunters also sometimes shipped game and fish to out-of-state markets. However, these men resided in Arkansas and their fellow citizens tolerated their activity until the growth of the conservation movement in the late nineteenth and early twentieth centuries. See the Big Lake Records, Bureau of Biological Survey, United States Fish and Wildlife Service, record group 22, National Archives, College Park, MD.

3. Arkansas General Assembly, *Acts, Resolutions, and Memorials of the General Assembly of the State of Arkansas* (Little Rock: William E. Woodruff, Jr., 1875), 176–77, 250; "Sportsmen Would License Hunters," *Daily Arkansas Gazette*, May 12, 1910, 9. A seine is a net that hangs vertically in the water. It has floats at the upper edge to hold one end at the surface and sinkers at the lower to hold the other end down. Fishermen use it to encircle the fish. Daniel Herman argues that when conservationists asked the government to protect wildlife, they were protecting them for the wealthy and middle class. "They made the government the engine with which to abolish market and subsistence hunting," he concludes. Herman, *Hunting*, 11–12.

4. William H. Deaderick, "A History of Arkansas Ornithology," *American Midland Naturalist* 26, no. 1 (July 1941): 210. Dr. William Heiskell Deaderick practiced medicine in Marianna, Arkansas, and Hot Springs. His foraminifera collection is preserved at the Smithsonian. On a different note, Arkansas passed an amendment to the criminal law in 1855 for "Sabbath breaking" setting a penalty for "hunting or shooting" on Sundays. Evidently, fishing was a more pious activity. "An Act to Amend the Criminal Law of this State under the head of Sabbath Breaking,"in Arkansas General Assembly, *Acts Passed at the Tenth Session of the General Assembly of the State of Arkansas* (Little Rock: Johnson and Yerkes, 1855), 180; Theodore S. Palmer, *Hunting Licenses: Their History, Objects, and Limitations* (Washington, DC: Government Printing Office, 1904), 14.

5. Palmer, *Hunting Licenses*, 9–12. For a listing of national game laws, see Charles B. Reynolds, *The Game Laws in Brief: Laws of the United States and Canada Relating to Game and Fish Seasons; For the Guidance of Sportsmen and Anglers* (New York: Forest and Stream, 1894), also 1896, 1904, 1905, 1911.

6. "Nonresidents Shooting in Poinsett County, Arkansas," *American Field*, July 2, 1887, 3–4.

7. "The Game Law," *Arkansas Democrat*, January 27, 1887, 8; For more on the continued argument to repeal the law, see *Arkansas Democrat*, February 5, 1887, 4.

8. *American Field*, January 15, 1887, 49; "Nonresidents Shooting in Poinsett County, Arkansas," *American Field*, July 2, 1887, 3–4. *American Field*, January 15, 1887, 1.

9. *Southern Standard*, January 22, 1887, 7; "Nonresidents Shooting in Poinsett County, Arkansas," *American Field*, July 2, 1887, 3–4.

10. "The Proposed Maine License," *Forest and Stream*, January 18, 1902.

11. *Arkansas Democrat*, January 21, 1879. Furbush was an African American carpetbagger and photographer who had moved to Phillips County, Arkansas, in 1870. He served in the Arkansas legislature as a Republican and was part of a state political

fusion system that combined Republican political power with Democrat economic power to reduce violence and foster cooperation. For more information about Furbush, see Blake Wintory, "William Hines Furbush: An African American, Carpetbagger, Republican, Fusionist and Democrat," *Arkansas Historical Quarterly* 63, no. 2 (Summer 2004): 107–65.

12. *Arkansas Democrat*, January 21, 1879.

13. *Arkansas Democrat*, January 21, 1879.

14. *Arkansas Democrat*, January 21, 1879; February 12, 1879.

15. "A Game Law," *Daily Arkansas Gazette*, January 21, 1879. For more about hunting and fishing as a common right, see Steven Hahn, "Hunting, Fishing, and Foraging: Commons Rights and Class Relations in the Postbellum South," *Radical History Review*, no. 26 (1982): 37–64; Scott E. Giltner, *Hunting and Fishing in the New South: Black Labor and White Leisure after the Civil War* (Baltimore: Johns Hopkins University Press, 2008), 8–9, 15–17, 46–47, 170; Gary Kulick, "Dams, Fish, and Farmers: Defense of Public Rights in eighteenth century Rhode Island," in *The Countryside in the Age of Capitalist Transformation: Essays in the Social History of Rural America*, ed. Steven Hahn and Jonathan Prude (Chapel Hill: University of North Carolina Press, 1985), 25–50; William Elliott, *Carolina Sport by Land and Water: Including Incidents of Devil-Fishing, Wildcat, Deer, and Bear Hunting, etc.* (1846; repr., Columbia: University of South Carolina Press, 1994).

16. *Arkansas Democrat*, March 5, 1879, 1.

17. *Arkansas Democrat*, March 7, 1879; March 16, 1879; and March 27, 1879.

18. *Arkansas Democrat*, February 19, 1879, 1.

19. *Southern Standard*, February 22, 1879.

20. *Southern Standard*, February 22, 1879.

21. *Arkansas Democrat*, April 26, 1879.

22. "A Week's Hunting in Arkansas," *Turf, Field, and Farm*, January 12, 1872; *Southern Standard*, February 22, 1879; Giltner, *Hunting and Fishing*, 5.

23. William E. Spencer, "Stop the Sale of Game," *Forest and Stream*, March 17, 1894, 225.

24. "Some American Game Birds," *Forest and Stream*, October 17, 1896, 304.

25. Giltner, *Hunting and Fishing*, 8.

26. Arkansas General Assembly, *Acts, Resolutions, and Memorials of the General Assembly of the State of Arkansas* (Little Rock: William E. Woodruff, Jr., 1879), 60–61.

27. Arkansas General Assembly, *Acts, Resolutions, and Memorials* (1879), 60–61; *Acts and Resolutions of the General Assembly of the State of Arkansas* (Little Rock: Gazette Printing, 1887), 10. Several studies cover how fishing conservation laws often lead to the passage of game laws nationally. See Gary Kulik, "Dams, Fish, and Farmers: Defense of Public Rights in Rhode Island," in Hahn and Prude, *Countryside*, 25–50; Harry Watson, "The Common Rights of Mankind: Subsistence, Shad, and Commerce in the Early Republican South," *Journal of American History* 83, no. 1 (June 1996):

13–43. Watson studies the fight over fishing rights in the South Atlantic coast starting in the 1750s. This fight also included argumentation about the common rights of man. In other words, who owned the fish and had access to them. By extension, the argument soon included wild animals as well. Some citizens argued that all people should have uninhibited access to wild game and fish, while others claimed that the state or federal government had jurisdiction over them. See also Samuel P. Haynes, *The Aquatic Frontier: Oysters and Aquaculture in the Progressive Era* (Amherst: University of Massachusetts Press, 2019); Erik Reardon, *Managing the River Commons: Fishing and New England's Rural Economy* (Amherst: University of Massachusetts Press, 2021); Mark Kurlansky, *Cod: A Biography of the Fish that Changed the World* (New York: Walker, 1997); and Tom Downey, "Riparian Rights and Manufacturing in Antebellum South Carolina: William Gregg and the Origins of the 'Industrial Mind,'" *Journal of Southern History* 65, no. 2 (February 1999): 77–108.

28. "Rod and Gun in Arkansas," *Forest and Stream*, August 2, 1883, 183.

29. "St. Louis Siftings," *Forest and Stream*, November 8, 1883, 286; "On the Spreads of the St. Francis," *Forest and Stream*, October 8, 1899, 33.

30. "Ducks By the Barrelful," *Forest and Stream*, December 27, 1888, 457.

31. "Diminution of the Ducks," *Forest and Stream*, January 17, 1889, 516. Club members were primarily nonresidents.

32. *Acts and Resolutions of the General Assembly of the State of Arkansas* (Little Rock: Gazette Printing, 1889), 61, 84–85.

33. Arkansas General Assembly, *Acts and Resolutions of the General Assembly of the State of Arkansas* (Little Rock: Gazette Printing, 1889), 61, 84–85.

34. Leonidas P. Sandels and Joseph M. Hill, eds., *A Digest of the Statutes of Arkansas Embracing All Laws of a General Nature in Force at the Close of the Session of the General Assembly of One Thousand Eight Hundred and Ninety-Three* (Columbia, MO: E. W. Stephens, 1894), 844–48.

35. Sandels and Hill, *Digest*, 844–48. For more on the illegal transportation of game for sale, see Theodore S. Palmer and H. W. Olds, *Laws Regulating The Transportation and Sale of Game*, United States Department of Agriculture Bulletin No. 14 (Washington, DC: Government Printing Office, 1900); *Southern Standard*, February 26, 1887, 2; The Convention between the United States and Great Britain (on behalf of Canada) for the Protection of Migratory Birds, or the Migratory Bird Treaty, was signed into law in August 1916; Philip Garone, *The Fall and Rise of the Wetlands of California's Great Central Valley* (Berkeley: University of California Press, 2020), 133–35.

36. "The Sale of Game," *Forest and Stream*, November 6, 1890, 312; *Forest and Stream*, March 17, 1894, 204; The sunk or sunken lands refers to land that sank during the New Madrid earthquake of 1812, for more information, see Nancy Hendricks, "Sunk Lands," *Encyclopedia of Arkansas* (online), accessed January 1, 2022.

37. Sandels and Hill, *Digest*, 846; Charles B. Reynolds, *Game Laws* (1896), 6; *Forest and Stream*, November 17, 1900, 387.

38. Sandels and Hill, *Digest,* 846; Reynolds, *Game Laws* (1896), 6; *Forest and Stream,* November 17, 1900, 387.

39. Prairie chickens were so low that they were among the first birds to receive protection from the Arkansas legislature. Sandels and Hill, *Digest,* 847; Paul R. Litzke, "Arkansas and the South," *Forest and Stream,* September 23, 1899, 248.

40. "Answers to Correspondents," *Forest and Stream,* December 28, 1876.

41. *Forest and Stream,* March 17, 1894, 203.

42. "Save the Birds," *Arkansas Democrat,* January 20, 1879, 4; "Game Bag and Gun," *Forest and Stream,* March 28, 1878, 164.

43. Sandels and Hill, *Digest,* 847.

44. "Chicago and the West," *Forest and Stream,* October 1, 1898, 266; "The Game Law," *Arkansas Democrat,* January 27, 1887, 8.

45. Theodore S. Palmer, *Digest of Game Laws* (Washington, DC: Government Printing Office, 1901), 46–48.

46. "Arkansas Note," *Forest and Stream,* January 31, 1883, 470; "Duck Shooting at Wapanocca," *Forest and Stream,* November 24, 1894, 445; "Shooting and Fishing," *Forest and Stream,* September 12, 1897, 40. One positive note is that the party that harvested 430 ducks in one day only kept thirty and donated the other 400 to the mental institution and the orphans and widows of Memphis. *St. Louis Post-Dispatch,* December 8, 1895, 38.

47. William Harris, ed., *The Sportsman's Guide to the Hunting and Shooting Grounds of the United States and Canada* (New York: Angler's Publishing, 1888), 8.

48. J. M. Rose, "Grassy Lake Hunting Club," *Forest and Stream,* June 20, 1894, 555.

49. J. M. Richardson, "Decimation of Game," *Daily Arkansas Gazette,* February 14, 1895, 4. Richardson was a wealthy planter in Devall's Bluff, a Confederate captain, and owned a mansion called "The Cedars" there. In 1897, one report called Richardson the "largest cotton planter in the world." *Daily Arkansas Gazette,* December 31, 1895; *Southern Standard,* November 19, 1897, 1.

50. Paul R. Litzke, "Arkansas and the South," *Forest and Stream,* September 23, 1899, 248.

51. Edmund Hough, "The Prairie Chicken in Arkansas," *Forest and Stream,* April 22, 1899.

52. *Arkansas Democrat,* August 21, 1891; *Daily Arkansas Gazette,* August 21, 1891.

53. George Bird Grinnell and Charles Sheldon, eds., *Hunting and Conservation* (New York: Arno and the *New York Times,* 1925), 201–2; Zachary Michael Jack, ed., *The Green Roosevelt: Theodore Roosevelt in Appreciation of Wilderness, Wildlife, and Wild Places* (New York: Cambria, 2010), 359.

54. Tara K. Kelly, *The Hunter Elite: Manly Sport, Hunting Narratives, and American Conservation, 1880–1925* (Lawrence: University of Kansas Press, 2018), 11. See also Karl Jacoby, *Crimes Against Nature: Squatters, Poachers, Thieves, and the Hidden History of Conservation* (Berkeley: University of California Press, 2001); and John F. Reiger,

American Sportsmen and the Origins of Conservation (Corvallis: Oregon University Press, 2001). Kelly explains that the common thread between the wealthy, elite hunters and the middle-class hunters and conservationists was the concept of "self-controlled manliness that remained part of American cultural life well into the 1920s."

55. *Arkansas Democrat*, August 21, 1891; and *Daily Arkansas Gazette*, August 21, 1891.

56. *Arkansas Democrat* (Little Rock), August 21, 1891; and *Daily Arkansas Gazette*, August 21, 1891.

57. *Arkansas Democrat*, August 21, 1891.

58. *Daily Arkansas Gazette*, June 7, 1892.

59. *Daily Arkansas Gazette*, July 16, 1892.

60. *Pine Bluff Daily Graphic*, September 9, 1897, 1. James Turner Lloyd was a Pine Bluff grocer who started a sporting goods business, the J. T. Lloyd Company, in 1899. In 1908, a Lloyd Sporting Goods store opened at 601 Main Street in Little Rock where they sold "explosives, arms, fishing tackle, athletic goods, baseball goods, football goods, and all other goods." The company later became involved in the automobile business. *Daily Arkansas Gazette*, July 26, 1908; *Pine Bluff Daily Graphic*, September 10, 1897.

61. *Arkansas Democrat*, July 27, 1898. John M. Rose was an attorney and the son of Uriah M. Rose, founder of the famed Rose Law Firm in Little Rock. Ancestry (website). Paul R. Litzke was the ASSA secretary in 1898. By 1900, he was a representative and writer for *Forest and Stream* magazine and in 1901, he worked for the Peters Cartridge Company. *Arkansas Democrat*, February 20, 1900; *Daily Arkansas Gazette*, January 22, 1902.

62. *Arkansas Democrat*, July 27, 1898. Elected at the 1898 meeting were President W. R. Duley, Little Rock, Vice President J. K. Thibault, Pulaski County, Treasurer E. T. Reaves, Little Rock, and Secretary Paul R. Litzke.

63. *Helena Weekly World*, October 19, 1898, 4. The ringneck was called the Denny pheasant in Oregon and named after United States consul general O. N. Denny, who had imported the Chinese pheasant to America. A few years later, Baxter County sportsmen in north Arkansas contacted the Bureau of Fisheries in Washington to obtain rainbow trout. They wanted to stock the North Fork River with the game fish. They received a shipment in the summer of 1907. *Baxter Bulletin*, February 22, 1907, 1.

64. *Helena Weekly World*, December 21, 1898.

65. Mrs. Louise M. Stephenson was instrumental in getting the Arkansas legislature to enact "Bird Day" in the public schools in 1897 and a law outlawing the killing of nongame birds. National Science Club, *Proceedings of the National Science Club* (Washington, DC: Judd and Detweiler, 1897). Stephenson was a member of the American Ornithologists' Union Committee on the Protection of North American Birds. She was a reporter for the Biological Survey from 1894–1916. She was later called the "Mother of Arkansas Ornithology" in Deaderick, "History of Arkansas Ornithology," 211.

66. Paul R. Litzke, "Arkansas and the South," *Forest and Stream*, January 15, 1900, 28.

CHAPTER FIVE

1. Lawmakers intended these two licensing laws to cause nonresidents to purchase permits from the local county clerk to hunt in their county. The 1875 law was the first of its kind in the nation and was supposedly aimed specifically at market hunters, but few local officials enforced either law.

2. Theodore S. Palmer and H. W. Olds, *Digest of Game Laws for 1901* (Washington, DC: Government Printing Office, 1901), 47; "The Proposed Maine License," *Forest and Stream*, January 18, 1902. Oregon was the other state. The law read, "All nonresident trappers, hunters, seiners, or netters of fish who may follow trapping, hunting, seining, or netting of fish [as employment] in the State." Washington, Oregon, Montana, Wyoming, North and South Dakota, Nebraska, Minnesota, Iowa, Wisconsin, Illinois, Michigan, Indiana, Florida, South Carolina, West Virginia, Delaware, Pennsylvania, Maryland, and part of Virginia required licenses.

3. *Helena Weekly World*, April 30, 1902. The Arkansas Federation of Women's Clubs was founded in 1897. Clubs that joined the federation included those focused on philanthropy and the general improvement of the state. Julianne H. Sallee, "General Federation of Women's Clubs of Arkansas," *Encyclopedia of Arkansas* (online), accessed January 20, 2022.

4. *Helena Weekly World*, April 30, 1902.

5. *Helena Weekly World*, April 30, 1902.

6. *Helena Weekly World*, April 30, 1902. The nature study movement grew in popularity during the Progressive Era. Many national programs appeared to provide children with an appreciation for nature through direct contact. This new approach became a part of Progressive public education reform. Kevin Armitage argues that this movement laid the foundation for modern American environmentalism. The United States Department of Agriculture's Bureau of Economic Ornithology argued the benefit of birds to agriculture in many pamphlets and bulletins, specifically targeting farmers, starting in 1889. For the only complete study of the nature study movement, see Kevin Armitage, *The Nature Study Movement: The Forgotten Popularizer of America's Conservation Ethic* (Lawrence: University Press of Kansas, 2009). For a listing of forty-five USDA Farmers' Bulletins published from 1889 to 1913, see Mabel Colcord, Ina L. Hawes, and Angelina Carabelli, compilers, *Check List of Publications on Entomology*, United States Department of Agriculture Library, vol. 20 (January 1930), 152–53.

7. *Pine Bluff Daily Graphic*, February 5, 1903.

8. *Osceola Times*, November 15, 1902, 6; *Fort Smith Times*, May 26, 1901, 4.

9. "Chicago's Trade in Wild Game," *American Sportsman*, December 27, 1873, 204; *Pine Bluff Daily Graphic*, February 5, 1903; "Arkansas State Sportsmen's Association," Jonesboro, Arkansas, January 16, 1903, in Joseph Taylor Robinson Papers,1900–1954, Special Collections Department, University of Arkansas Libraries, Fayetteville, AR. In 1903, the ASSA sent a press release to every Arkansas politician and press outlet. In the release, they recommended that Arkansas legislators make changes to the law.

The ASSA wanted alterations to the open seasons on deer, turkeys, and quail and a five-year ban on prairie chicken hunting. They requested annual limits of ten turkeys and four deer and daily limits of twenty-five quail, fifty ducks, and twenty-five fish. The organization wanted fishing limited to hook and line only. Ending the practice of intrastate shipping of game and fish.

10. Theodore S. Palmer, *Hunting Licenses: Their History, Objects, and Limitations* (Washington, DC: Government Printing Office, 1904), 14, 19, 64. Within a few years, legislators in Louisiana, Arkansas, and Missouri had passed legislation to ban market hunters. Davis's biography claimed that Davis was "a splendid shot, passionately loved hunting quail and always kept a number of splendidly trained pointer and setters." For more on Jefferson "Jeff" Davis and his hunting, see Charles Jacobson, *The Life Story of Jeff Davis: The Stormy Petrel of Arkansas Politics* (Little Rock: Parke-Harper, 1925), 19–20, 121; Raymond Arsenault, *The Wild Ass of the Ozarks: Jeff Davis and the Social Bases of Southern Politics* (Knoxville: University of Tennessee Press, 1988), 198; Claude Van Dyne to My Dear Friend, December 17, 1908, and A. J. Ronaldson to Hon. Jeff Davis, January 1, 1910, Jeff Davis Papers, 1849–1986, Special Collections Department, University of Arkansas Libraries, Fayetteville, AR.

11. *Arkansas Democrat*, February 6, 1903, 4.

12. This group was upset over the DeRossitt proposal; "Memphis is Wrought Up," *Arkansas Democrat*, February 6, 1903, 4.

13. *Monticellonian*, February 12, 1903, 1; "Memphis is Wrought Up," *Arkansas Democrat*, February 6, 1903, 4. Frank William DeRossitt (born October 1, 1859, in Kentucky) moved to Arkansas in 1879. In 1905, he owned and farmed seven hundred acres of land. His farm included a lake considered one of the best in the state for fishing. He was a St. Francis County representative 1897–1905 and Arkansas state senator 1906–1910; "Frank William DeRossitt," Find a Grave (website), accessed February 1, 2022; House of Representatives composite photo of the Thirty-Fourth General Assembly of the State of Arkansas, 1903, black-and-white photograph, Arkansas Digital Archives, accessed February 3, 2022.

14. *Arkansas Democrat*, February 13, 1903, 4.

15. *Arkansas Democrat*, February 27, 1903, 4.

16. *Southern Standard*, April 16, 1903, 1; *Monticellonian*, April 16, 1903, 1; Arkansas Senate, *Journal for the Senate of Arkansas, Thirty-Fourth Regular Session* (Little Rock: Arkansas Democrat, 1903), 91, 94, 106, 121, 122, 124, 134.

17. *Daily Arkansas Gazette*, March 5, 1903; Arkansas Senate, *Journal* (1903), 151, 178, 179, 197. The Fifteenth District was composed of the counties of Ashley and Chicot, Mississippi Delta counties.

18. Arkansas Secretary of State, *Arkansas: Biennial Report of the Secretary of the State for the Years 1903–1904* (Little Rock, Tunnah and Pittard, 1905), 185; the Fourteenth District was composed of Lee and Phillips counties.

19. *Daily Arkansas Gazette*, March 5, 1903, 4; The Twenty-Third District was

composed of Fulton, Izard, and Baxter counties. Joseph H. Acklen was a lawyer, state game warden of Tennessee, leader of the Big Lake Hunting Club, as well as a member of numerous other hunting and fishing clubs. Wappanocca was a hunting club across the Arkansas River from Memphis.

20. *Daily Arkansas Gazette,* March 5, 1903, 4; the Twenty-Second District was composed of Crittenden and St. Francis counties.

21. *Daily Arkansas Gazette,* March 5, 1903, 4.

22. *Daily Arkansas Gazette,* March 5, 1903, 4.

23. *Daily Arkansas Gazette,* March 11, 1903, 3.

24. *Daily Arkansas Gazette,* March 11, 1903, 3; *Arkansas Democrat,* March 12, 1903, 6; *Daily Arkansas Gazette,* March 11, 1903, 3. The Thirty-Third District was composed of Scott and Polk counties.

25. *Arkansas Democrat,* March 12, 1903, 6; *Daily Arkansas Gazette,* March 11, 1903, 3.

26. *Daily Arkansas Gazette,* March 11, 1903, 3; Arkansas Senate, *Journal* (1903), 91. The Seventeenth District was composed of Drew and Desha counties. The Twelfth District was composed of Lonoke and Prairie counties.

27. *Daily Arkansas Gazette,* March 13, 1903, 3; Arkansas Senate, *Journal* (1903).

28. *Daily Arkansas Gazette,* March 13, 1903, 3. James T. Lloyd was a charter member of the ASSA. He was the 1897 president of the organization. Another fishing bill, Senate Bill No. 19, that Killough had written came up the following week. After two representatives said they wanted their counties exempted, the bill was indefinitely postponed. *Arkansas Democrat,* March 18, 1903, 2.

29. *Arkansas Democrat,* March 8, 1903, 1; *Gentry Journal-Advance,* May 1, 1903; *Daily Arkansas Gazette,* July 4, 1903, 4. Frank W. DeRossitt was from Forrest City where he owned a seven-hundred-acre farm where he grew corn and cotton. The farm contained a large fish-filled lake that was famous for large catches. He was first elected to the Arkansas legislature in 1897 and again in 1901 and 1903, *Forrest City Times at Souvenir,* 1905, 55.

30. *Southern Standard,* April 30, 1903, 1; *Pine Bluff Daily Graphic,* February 22, 1903, 2; and April 27, 1903, 1; *Arkansas Democrat,* April 25, 1903, 1; *Osceola Times,* December 19, 1903, 6; Arkansas Senate, *Journal* (1903), 337.

31. Grover Cleveland, *Fishing and Shooting Sketches* (New York: Outing, 1906), 54; Baird Callicott, introduction to *Companion to a Sand County Almanac: Interpretive and Critical Essays* (Madison: University of Wisconsin Press, 1987). Thomas R. Dunlap argues that the early conservation movement, which evolved into the modern environmental movement, was essentially a religious movement where nature was the cathedral, the scripture, and the divine location. The early conservationists asked, "What purpose do humans have in the world, and what must they do to fulfill it?" Men such as Ralph Waldo Emerson (romanticism) found divine meaning in the wilderness. This idea, according to Dunlap, started as a reaction to the Enlightenment and its emphasis on scientific reason and secularism. Thomas R. Dunlap, *Faith in*

Nature: Environmentalism as Religious Quest (Seattle: University of Washington Press, 2004), 13. Daniel Herman claims modern hunters "enter the wild to become part of it, not to prove their superiority over other forms of life." This return to nature allows hunters to regain sanity (and saintliness) in a world divorced of nature." Daniel Herman, *Hunting in the American Imagination* (Washington, DC: Smithsonian Institution Press, 2001), 278–81.

32. *Forest and Stream*, May 2, 1903, 348.

33. *Forest and Stream*, May 2, 1903, 1.

34. *Daily Arkansas Gazette*, May 14, 1903, 2.

35. *Monticellonian*, May 7, 1903, 2.

36. *Daily Arkansas Gazette*, May 14, 1903, 2; *Baxter Bulletin*, June 12, 1903, 1; *Arkansas Democrat*, April 25, 1903, 1.

37. *Daily Arkansas Gazette*, May 14, 1903, 2; *Baxter Bulletin*, June 12, 1903, 1; *Arkansas Democrat*, April 25, 1903, 1.

38. *Forest and Stream*, May 9, 1903, 1. Joseph Acklen was the "nonresident" that the magazine mentioned.

39. *Forest and Stream*, May 16, 1903, 386. Warren is in the southeastern part of the state in Drew County near Lake Village and Monticello.

40. *Pine Bluff Daily Graphic*, May 24, 1903, 4.

41. Gen. 1:25–27 (NIV); *Southern Standard*, July 30, 1903, 1.

42. *Forest and Stream*, June 27, 1903.

43. *Fort Smith Times*, June 30, 1903, 1. Senator James P. Clarke, a lawyer from Helena, was an Arkansas representative (1886), an Arkansas senator (1888), governor of Arkansas (1895), and a United States senator (1903–1916); "James Paul Clarke," Old Statehouse Museum, Arkansas Heritage (website), accessed January 1, 2019.

44. *Daily Arkansas Gazette*, June 27, 1903. Captain William Barton Mallory was involved in brick manufacturing, insurance, cotton buying, and wholesale grocery. Charles Patrick and Joseph Mooney, eds., *The Mid-South and Its Builders: Being the Story of the Development and a Forecast of the Future of the Richest Agricultural Region in the World* (Memphis, TN: Mid-South Biographic and Historical Association, 1920), 584–85. Interestingly, the Wapanocca Hunting Club became the Wapanooca National Wildlife Refuge in 1961; "Wapannoca National Wildlife Refuge," State Parks (website), accessed May 5, 2022; Several Arkansas state senators accused the Wapanocca Club of attempting to influence game legislation, including writing some of the legislation for some Arkansas politicians.

45. *Forest and Stream*, July 25, 1903; *Daily Arkansas Gazette*, June 27, 1903; *Daily Arkansas Gazette*, December 4, 1904, 6.

46. *Daily Arkansas Gazette*, July 4, 1903, 4; *Pine Bluff Daily Graphic*, July 26, 1903, 2; *Daily Arkansas Gazette*, July 24, 1903, 2; *Arkansas Democrat*, December 15, 1903, 8; T. S. Palmer, *Hunting Licenses*, 47, 55, 64; Charles B. Reynolds, *The Game Laws in Brief: Laws of the United States and Canada Relating to Game and Fish Seasons; For the*

Guidance of Sportsment and Anglers (New York: Forest and Stream, 1894), 6. For clarification, the original nonresident law passed in 1875 and required a ten-dollar license to trap or hunt game, sein, trap, or net fish. In 1897, the fee rose to twenty-five dollars. In 1903, nonresidents were excluded from hunting or fishing (except in MS County) until the law was declared unconstitutional.

47. *Pine Bluff Daily Graphic*, November 5, 1903, 1; and *Daily Arkansas Gazette*, December 4, 1904, 6.

48. *Daily Arkansas Gazette*, July 4, 1903, 4; and *Osceola Times*, December 10, 1904, 4. The italicized portion is the necessary addition. The argument continued surrounding the Federal Migratory Bird laws and the Migratory Bird Treaty of 1916. Not many studies have covered this aspect of conservation in the United States, however, readers can consult Kurkpatrick Dorsey, *The Dawn of Conservation Diplomacy: U.S.-Canadian Wildlife Protection Treaties in the Progressive Era* (Seattle: University of Washington Press, 2009); United States Department of Agriculture, *The Migratory Bird Treaty: The Decision of the Supreme Court of the United States Sustaining the Constitutionality of the Migratory Bird Treaty and Act of Congress to Carry It into Effect*, Department Circular 102, Washington, DC, 1920; Helen Ossa, *They Saved Our Birds: The Battle Won and the War to Win* (New York: Hippocrene Books: 1973).

49. *Osceola Times*, December 19, 1903, 7.

50. *Daily Arkansas Gazette*, November 3, 1904, 6.

51. *Daily Arkansas Gazette*, November 8, 1904, 4; United States Census, Statistics on Agriculture, 1900; Sam B. Hilliard, *Hog Meat and Hoecake: Food Supply in the Old South, 1840–1860*, 71–73; Christopher Morris, *Big Muddy: An Environmental History of the Mississippi and Its Peoples from Hernando De Soto to Hurricane Katrina* (Oxford: Oxford University Press, 2012), 136; Mikko Saikku, *This Delta, This Land: An Environmental History of the Yazoo-Mississippi Floodplain* (Athens, University of Georgia Press, 2005), 230–31.

52. *Monticellionian*, May 7, 1903, 1. Besides the Whitley and DeRossitt bills: Representative Scrimshire's turkey bill, barring the catching of wild turkey by means of baiting or dead falls. Arkansas Senate *Journal* (1903), 138, 149, 161. Representative Hale's law against killing deer, turkey, or fowl for the purpose of selling in Mississippi County, Representative Campbell's Bill Number 197 requiring nonresidents to have a license to hunt passed, but the DeRossitt law passed after it and therefore made it null and void. Arkansas Senate, *Journal* (1903), 183, 269. Senator Logan's bill banning fishing except with a hook and line in Madison, Carroll, Washington, and Benton counties. Arkansas Senate, *Journal* (1903), 231, 247, 254, 257, 295, 299, 300, 332, 336. Representative Stockard's bill for a bounty of hawks and foxes. Arkansas Senate, *Journal* (1903), 268, 272, 296.

53. *Pine Bluff Daily Graphic*, December 11, 1904, 1; *Daily Arkansas Gazette*, December 13, 1904, 2. For more information about William D. Jones, see Bar Association of Arkansas, *Report of the Proceedings of the Bar Association of Arkansas*, Little Rock, 1932,

195. For more information about W. B. Sorrells, see Dallas Tabor Herndon, *Centennial History of Arkansas*, (Chicago: S. J. Clarke, 1922), 1027. For more information on the 1887 Interstate Commerce Act, see Interstate Commerce Act, 1887, Milestone Documents, National Archives (online), accessed May 10, 2022. A. J. Stewart was constable of Vaugine Township in 1900 and was elected Pine Bluff chief of police in 1905 and resigned in 1909 under accusations of misconduct from city attorney W. B. Sorrells; *Pine Bluff Daily Graphic*, March 15, 1900, 1; April 11, 1905, 1; November 12, 1906, 1; and September 7, 1909, 1; *Daily Arkansas Gazette*, April 6, 1906, 2; *Pine Bluff Daily Graphic*, May 23, 1905, 1.

54. *Daily Arkansas Gazette*, December 17, 1904, 2; *Pine Bluff Daily Graphic*, December 16, 1904, 1; *Southern Standard*, December 22, 1904, 1.

55. *Daily Arkansas Gazette*, January 11, 1905, 3; *Arkansas Democrat*, January 14, 1905, 2; *Daily Arkansas Gazette*, January 19, 1905, 3; *Nashville News*, January 21, 1905, 3; *Monticellonian*, January 26, 1905, 1; *Southern Standard*, February 9, 1905, 1; *Daily Arkansas Gazette*, February 11, 1905, 3; *Nashville News*, February 11, 1905, 3; *Arkansas Democrat*, February 16, 1905, 3; *Monticellonian*, February 16, 1905, 1; *Daily Arkansas Gazette*, February 16, 1905, 6; *Arkansas Democrat*, February 19, 1905, 7; *Daily Arkansas Gazette*, March 8, 1905, 3; and April 4, 6, and 8,1905; *Arkansas Democrat*, May 4, 1905, 2; Arkansas Senate, *Journal* (1903), 45, 61, 71, 90, 91, 94, 106, 121, 122, 124, 132 ,134, 138, 149 ,151, 161, 178, 179, 183, 190, 197, 206, 231, 239, 252, 253, 247, 254, 257, 269, 280, 282, 283, 295 ,299, 300, 327, 332, 336; *Mountain Echo*, July 7, 1905, 2; *Baxter Bulletin*, July 21, 1905, 4.

56. "Shall We Allow Automatic Guns," *Recreation* 19 (November 1903), 397–98; *Recreation* 21 (July 1904), 39, 46, 60, xxxii, xlix; *Recreation* 21 (August 1904), 109, 111, 112, 126; *Recreation* 22 (January 1905), 47–48. This semiautomatic banning bill appeared before several state legislative houses in 1905. Shields also published communications between the magazine and the Winchester Arms company over the editors condemning their semiautomatic shotgun. *Recreation* 20 (January 1904), 54–60. For more information about Winchester shotguns, see Dennis Addler, *Winchester Shotguns* (New York: Simon and Schuster, 2015).

57. Theodore S. Palmer, "Federal Game Protection: A Five Year Retrospect," Reprint from the *Yearbook of Department of Agriculture for 1905* (Washington: Government Printing Office, 1905), 546–55.

58. Palmer, "Federal Game Protection," 546–55.

59. John Castle, *Supplement to Kirby's Digest of the Statues of Arkansas* (Indianapolis: Bobbs-Merrill, 1911), 290–93; Arkansas General Assembly, *Public and Private Acts and Joint and Concurrent Resolutions and Memorials of the General Assembly of the State of Arkansas* (Little Rock: Democrat Print, 1907), 912–13.

60. Arkansas General Assembly, *Public and Private Acts* (1907), 186–87, 520–21, 617–19, 675–76, 812–13, 838–39, 860.

61. "Morning Session," *Daily Arkansas Gazette*, February 11, 1909, 3. By 1900,

Attorney A. G. Little was working and living in Blytheville. He served as the Blytheville mayor before becoming a member of the House. Later, he became the lawyer for the Mississippi County market hunters. "The New Official Family of Mississippi County," *Osceola Times*, November 5, 1908, 5. For more information on Donaghey, see Cal Ledbetter, *Carpenter from Conway: George Washington Donaghey as Governor of Arkansas, 1909–1913* (Fayetteville: University of Arkansas Press, 1993); and George W. Donaghey, *Autobiography of George W. Donaghey* (Benton, AR: L. B. White, 1939).

62. Arkansas General Assembly, *Public and Private Acts and Joint and Concurrent Resolutions and Memorials of the General Assembly of the State of Arkansas*, (Little Rock: Democrat Print, 1909), 1131–32.

63. Arkansas Senate, *Journal of the Senate of Arkansas, Thirty-Sixth Regular Session* (Hot Springs: Sentinel-Records, 1907), 31.

64. Arkansas Senate, *Journal of the Senate of Arkansas, Thirty-Seventh Session* (Little Rock: Tunnah and Pittard, 1909), 256; *Baxter Bulletin*, April 30, 1909, 4. Daily bag limits in the bill included quail and partridge at twenty-five; goose and brant at ten; ducks, plover, and snipe at twenty-five; and one dear per day while in season. It also barred killing pheasants, grouse, or prairie chickens until 1914 and banned the hunting of deer with dogs and hunting at night.

65. Arkansas Senate, *Public and Private Acts* (1909), 190–91, 257–58, 584–85, 758–60, 931–32, 1004–5, 1051–52, 1114–15, 1121–25.

66. Arkansas Senate, *Public and Private Acts* (1909), 190–91, 257–58, 584–85, 758–60, 931–32, 1004–5, 1051–52, 1114–15, 1121–25.

CHAPTER SIX

1. *Daily Arkansas Gazette*, January 16, 1906, 7; *Daily Arkansas Gazette*, October 21, 1906, 16.

2. *Fort Smith Times*, July 9, 1906, 1; *Arkansas Democrat*, October 10, 1906, 2.

3. *Daily Arkansas Gazette*, October 21, 1906, 16; "After Game Law Violators," *Green Forest Tribune*, September 6, 1912; "Five Violators of the State Game Laws Get Fines Stayed," *Arkansas Democrat*, October 15, 1921; "Other Cases," *Daily Arkansas Gazette*, May 18, 1921.

4. *Arkansas Democrat*, January 20, 1909, 7; *Daily Arkansas Gazette*, January 21, 1909, 3; *Fort Smith Times*, January 21, 1909, 1. James T. M. Holt was a Missionary Baptist farmer from Bingen who later served in the Arkansas Senate. *Southern Standard*, December 26, 1912, 7; "Hon. J. T. M. Holt," *Daily Arkansas Gazette*, April 18, 1899.

5. "Many Want to Be Warden," *Arkansas Democrat*, January 14, 1903, 3.

6. "Game Wardens Are Needed," *Daily Arkansas Gazette*, January 23, 1909.

7. "Game Wardens Are Needed," *Daily Arkansas Gazette*, January 23, 1909; Protection of Game," *Arkansas Democrat*, October 27, 1909, 6; The idea of all Arkansawyers joining and working together was one of Visart's standard speeches that he used several times. Visart is not differentiating between races but groups or classes. In this case, the

"shiftless and roving class" references criminals, mainly thieves who carry guns, and claim they are hunting, but are scouting for opportunities to commit crimes. However, for information about how some states used game laws to control African-American access to game and fish, see Steven Hahn, *A Nation under Our Feet: Black Political Struggles in the Rural South from Slavery to the Great Migration* (Cambridge: Harvard University Press, 2003); and Scott E. Giltner, *Hunting and Fishing in the New South: Black Labor and White Leisure after the Civil War* (Baltimore: Johns Hopkins University Press, 2008).

8. "Was Burning Day at Warren," *Daily Arkansas Gazette*, May 18, 1910, 3.

9. *Daily Arkansas Gazette*, March 1, 1909, 3.

10. "Our Game Laws and Other States," *Daily Arkansas Gazette*, March 16, 1909, 1, 9. Arkansas not being progressive is an argument that many conservationists used in their states.

11. "Want a Game Warden," *Arkansas Democrat*, August 20, 1909, 1.

12. "Names a Warden," *Arkansas Democrat*, September 24, 1909, 10.

13. "Want State Game Warden," *Arkansas Democrat*, August 20, 1909, 1.

14. "Protection of Fish and Game," *Arkansas Democrat*, September 19, 1909, 15. For more information about Tip T. Omohundro, see *Arkansas Democrat*, January 24, 1901; *Pine Bluff Daily Graphic*, December 1, 1902; *Fort Smith Times*, December 20, 1907.

15. "Protection of Fish and Game," *Arkansas Democrat*, September 19, 1909, 15. They met with the governor on Wednesday, September 15, 1909.

16. "Protection of Fish and Game," *Arkansas Democrat*, September 19, 1909, 15.

17. "Sportsmen's Association," *Arkansas Democrat*, September 19, 1903, 15.

18. "Non-residents Prohibited," *Arkansas Democrat*, September 19, 1909, 15.

19. "Preparing Game Law Digest," *Arkansas Democrat*, September 24, 1909, 3. For more about Oswald C. Ludwig, see Mary L. Kwas, *A Pictorial History of Arkansas's Old State House* (Fayetteville: University of Arkansas Press, 2010), 137.

20. "Sympathy for the Orphans," *Arkansas Democrat*, May 29, 1910, 6.

21. Fay Hempstead, *Historical Review of Arkansas: Its Commerce, Industry, and Modern Affairs* (Chicago: Lewis, 1911); *Daily Arkansas Gazette*, March 4, 1886, 3; "Earnest Vivian Visart," United States Department of Agriculture, *The Official Records of the United States Department of Agriculture* 5, no. 1 (Washington, DC: Government Printing Office, 1926); "The Crumbling of a Chinese Wall," *Recreation* 47, no. 1 (June 1912), 270.It is rumored that the family left after the Belgian Revolution in the 1830s. Another story claims that Maria Phillipe Julian Joseph Visart fled after his youngest son, Hippolyte Visart de Bocarmé, murdered his brother-in-law, Gustave Fougnies. The murder occurred in the 1850s, however.

22. "Visart/Tucker Bond for Marriage," Carroll County Arkansas, December 11, 1895, Ancestry (website), accessed February 9, 2022; "Earnest V. Visart," Cedar, Carroll, Arkansas, United States Census Bureau, *Twelfth Census of the United States*, 1900; "In Circuit Court," *Fort Smith Times*, April 21, 1901, 8; October 4, 1901, 1; "State Notes," *Fort Smith Times*, December 12, 1901, 4.

23. "Flora Lola Tucker Visart," Find a Grave (website), accessed February 9, 2022; *Fort Smith Times*, December 4, 1904, 6, as well as October 20, 1905, 1.

24. "Arrest Few More," *Fort Smith Times*, December 18, 1905, 1.

25. *Fort Smith Times*, June 6, 1906, 4; "Fort Smith and Van Buren," *Arkansas Democrat*, June 28, 1906, 1.

26. "Corporations," *Arkansas Democrat*, September 15, 1907, 10; "Argenta Wholesale Cigar and Tobacco Company," *Daily Arkansas Gazette*, March 1, 1908, 7.

27. "Incorporated Tomorrow," *Arkansas Democrat*, March 16, 1908, 9; "Sporting News," *Fort Smith Times*, March 16, 1908, 4; "Visart to Leave," *Daily Arkansas Gazette*, May 1, 1908, 10; "Conway—New Wholesale Grocery," *Daily Arkansas Gazette*, October 16, 1908, 3.

28. "New Fishing Club Planned," *Arkansas Democrat*, August 20, 1908, 9.

29. "New Fishing Club Planned," *Arkansas Democrat*, August 20, 1908, 9.

30. "Game Warden Appoints Deputy Here," *Pine Bluff Daily Graphic*, September 24, 1909, 1; "Names a Game Warden," *Arkansas Democrat*, September 24, 1909, 10.

31. "Cleaning out the River," *Arkansas Democrat*, October 31, 1909, 14. For more examples, see "For Using Illegal Nets," *Arkansas Democrat*, March 27, 1910, 13; "Game Wardens Busy," *Arkansas Democrat*, April 10, 1910, 7; *Daily Arkansas Gazette*, April 10, 1910, 19; "Argenta Notes," *Arkansas* Democrat, May 14, 1910, 9; "Four Were Arrested," *Arkansas Democrat*, May 27, 1910, 10; "Fish Nets Burned," *Arkansas Democrat*, May 28, 1910, 7; *Arkansas Democrat*, July 16, 1910, 10. For more about fishing rights and legislation, see Steven Hahn, "Hunting, Fishing, and Foraging: Commons Rights and Class Relations in the Postbellum South," *Radical History Review*, no. 26 (1982): 37–64; Tom Downey, "Riparian Rights and Manufacturing in Antebellum South Carolina: William Gregg and the Origins of the 'Industrial Mind,'" *Journal of Southern History* 65, no. 2 (February 1999): 77–108; Harry Watson, "The Common Rights of Mankind: Subsistence, Shad, and Commerce in the Early Republican South," *Journal of American History* 83, no. 1 (June 1996): 13–43.

32. "First Arrest for Game Violations," *Arkansas Democrat*, September 28, 1909, 7; "For Violation of Game Law," *Arkansas Democrat*, October 1, 1909, 4.

33. "First Arrest for Game Violations," *Arkansas Democrat*, September 28, 1909, 7; "For Violation of Game Law," *Arkansas Democrat*, October 1, 1909, 4. For more about G. T. Cazort, see "Would Protect Fishing Grounds," *Fort Smith Times*, November 14, 1904, 4; "Game Warden to Be Given Contest," *Arkansas Democrat*, October 13, 1909, 2.

34. "Enforcing Game Laws," *Arkansas Democrat*, October 3, 1909, 12.

35. "Turned Game Loose," *Arkansas Democrat*, October 6, 1909, 12; "No Birds in Cages," *Arkansas Democrat*, November 22, 1909, 4; "Brief Outline of State Game Laws," *Daily Arkansas Gazette*, November 29, 1909, 5; "Will Go After Sunday Hunters," *Arkansas Democrat*, January 6, 1910, 9; "Arkansas Sunday Laws," *Liberty: A Magazine of Religious Freedom* 16, no. 1 (1919): 117. Arkansas had a state law forbidding Sunday labor, horse racing, gambling, hunting, or baseball since 1885.

36. "For Killing Quail," *Arkansas Democrat,* October 14, 1909, 10; "State Game Warden Visart," *Arkansas Democrat,* October 24, 1909, 3.

37. "Is Game Warden Legally Appointed?," *Arkansas Democrat,* October 25, 1909, 3.

38. "Net 900 Feet Long," *Arkansas Democrat,* January 25, 1910, 2; "Game Warden Visart Sued," *Daily Arkansas Gazette,* January 25, 1910, 8; "Visart Faces Litigation," *Daily Arkansas Gazette,* June 7, 1910, 7; "Visart Discharged," *Arkansas Democrat,* June 21, 1910, 3; "Ike Sangston, Who Was," *Arkansas Democrat,* June 22, 1910, 3; "Visart Wins Damage Suit," *Daily Arkansas Gazette,* June 22, 1910, 10. C. E. Sangston sued Visart in June 1910. Newspapers reported that "several other suits [are] pending" against the warden.

39. "Visart Receives Letters," *Arkansas Democrat,* March 22, 1910, 4.

40. "Cleaning Out River," *Arkansas Democrat,* October 31, 1909, 14; "Visart Visits Morrilton," *Daily Arkansas Gazette,* April 24, 1910, 6. See also "Visart Meets Success," *Arkansas Democrat,* April 24, 1910, 9; "Visart at Cotton Plant," *Daily Arkansas Gazette,* April 28, 1910, 2; "State Game Warden E. V. Visart," *Arkansas Democrat,* May 20, 1910, 3; "Sportsmen of Garland County," *Arkansas Democrat,* August 6, 1910.

41. "Found Large Wolf in Maumelle Bottoms," *Arkansas Democrat,* January 31, 1910, 1; W. F. Bancroft, "National Gathering of Game Wardens and Commissioners," in South Dakota Department of Game and Fish, *Annual Report* (Sioux Falls: Mark D. Scott, 1910), 13–14.

42. "Visart Meets Success," *Arkansas Democrat,* April 24, 1910, 9; "State Game Warden Visart," *Arkansas Democrat,* November 7, 1909, 3.

43. "Trip Was a Failure," *Arkansas Democrat,* November 9, 1909, 4.

44. "Must Be a Resident," *Star Progress* (Berryville), November 12, 1909, 4.

45. "Protection of Game," *Arkansas Democrat,* October 27, 1909, 6.

46. "Visart Investigates Exportation of Game," *Daily Arkansas Gazette,* December 17, 1909, 1; "Arkansas Game in Memphis Markets," *Daily Arkansas Gazette,* December 20, 1909, 6; "Should Revise Game Laws," *Daily Arkansas Gazette,* January 3, 1910, 6; "Watching Game Fish Shipments," *Daily Arkansas Gazette,* June 5, 1910, 7; "To Ask Highest Fine," *Arkansas Democrat,* June 5, 1910, 10. Visart returned to Memphis during the summer of 1910 to investigate illegal fish shipments from Arkansas to the city.

47. "Will Go After Sunday Hunters," *Arkansas Democrat,* January 6, 1910, 9.

48. "Visart Investigates Exportation of Game," *Daily Arkansas Gazette,* December 17, 1909, 1; "After Pot Hunters," *Arkansas Democrat,* December 17, 1909, 7; *Arkansas Democrat,* December 20, 1909, 6; "Game Shipments, *Arkansas Democrat,* December 20, 1909, 6; "Should Revise Game Laws," *Daily Arkansas Gazette,* January 3, 1910, 6; "To Ask Highest Fine," *Arkansas Democrat,* June 5, 1910, 10. Visart called the Mississippi County exemption the worst game law in Arkansas. This exemption allowed all other market hunters to catch the game and fish anywhere in the state, then ship it to Mississippi County, then ship it out of the state from there.

49. "After Exporters of Game," *Batesville Guard,* December 24, 1909, 6; "Lacey Act,

Regulating Interstate Commerce in Wild Animals," Act of March 4, 1909, Sec. 241–35, Stat. 1137, in United States Department of Agriculture, Bureau of Biological Survey, *Migratory-Bird Treaty-Act Regulations and Text of Federal Laws Relating to Game and Birds*, 1930. It was originally passed in 1900 and was the first federal law protecting wildlife.

50. "E. O. Visart, State Game Warden," *Arkansas Democrat*, December 31, 1909, 3; "Confiscates Nine Barrels of Ducks," *Daily Arkansas Gazette*, January 1, 1910, 8; "Unfortunates Given a Treat," *Arkansas Democrat*, January 11, 1910, 10. He also sent confiscated fish to the School for the Blind. For more, see "State Game Warden E. V.," *Arkansas Democrat*, March 7, 1910, 3; "Blind Eat Fish Taken by Poachers," *Daily Arkansas Gazette*, March 9, 1910, 7.

51. "State Game Warden Visart," *Arkansas Democrat*, January 10, 1910, 3; "Express Company Fined," *Arkansas Democrat*, February 10, 1901, 4; "Fined for Shipping Game out of State," *Daily Arkansas Gazette*, February 10, 1910, 7. Harrisburg is in Poinsett County. The other case was in Rector, in Clay County.

52. "Game Warden Gets a Tip," *Daily Arkansas Gazette*, February 2, 1910, 14.

53. "Many Game Law Violators Punished," *Daily Arkansas Gazette*, December 10, 1909, 7; "Visart Gets Convictions," *Daily Arkansas Gazette*, December 8, 1909, 11.

54. "Fish and Game Illegally Taken," *Daily Arkansas Gazette*, November 27, 1909, 10.

55. "Fish and Game Illegally Taken," 10; "Visart Gets Convictions," *Daily Arkansas Gazette*, December 8, 1909, 11; "Visart in St. Francis," *Daily Arkansas Gazette*, December 16, 1909, 5. Hatchie Coon is an area near Marked Tree, Arkansas.

56. "E. V. Visart, State Game Warden," *Arkansas Democrat*, April 8, 1910, 3; Game Wardens Burn Launch Load of Nets," *Daily Arkansas Gazette*, April 19, 1910, 2; "Visart Meets Success," *Arkansas Democrat*, April 24, 1910, 9; "Burns Illegal Fish Nets," *Daily Arkansas Gazette*, May 2, 1910, 2; "Illegal Seines Burned," *Arkansas Democrat*, May 2, 1910, 6. Southerners had a considerable history with organized groups chasing criminals or conducting vigilante actions like lynchings. There are many studies on this subject, but the latest for Arkansas is Guy Lancaster, ed. *Bullets and Fire: Lynching and Authority in Arkansas, 1840–1950* (Fayetteville: University of Arkansas Press, 2018).

57. "C. E. Visart, State Game," *Arkansas Democrat*, May 28, 1910, 3; "Hunting and Fishing," *Daily Arkansas Gazette*, September 4, 1910, 46; "Bag Limits for Our Game Laws," *Arkansas Democrat*, September 8, 1910, 4. Warren deputy game warden R. D. Garrison later arrested four men for dynamiting fish in April 1910. Malvern deputy game warden J. C. Ballew arrested two men dynamiting fish in the Ouachita River in May 1910. Cross County deputy game warden J. W. Luellen arrested John Sifford for dynamiting. Newport deputy game warden George M. Johnson arrested T. F. Benton, Joseph Snead, and W. E. Brantley for dynamiting in September 1910. In 1921, Camp Pike soldier Byron Carr accidentally killed himself when the dynamite fuse he used for fishing burned too quickly, exploding in his hands. See "Investigating Dynamiting," *Daily Arkansas Gazette*, June 2, 1921, 2.

58. "Brief Outline of State Game Laws," *Daily Arkansas Gazette*, November 29, 1909, 5; "Arkansas Game Law," *Osceola Times*, December 2, 1909, 17; "The General State Game Laws," *Southern Standard*, December 2, 1909, 4; "State Game Laws," *Baxter Bulletin*, December 17, 1909, 8; "Brief Outline of State Game Laws," *Pine Bluff Daily Graphic*, December 17, 1909, 3; "Visart Meets Success," *Arkansas Democrat*, April 24, 1910, 9; "Illegal Seines Burned," *Arkansas Democrat*, May 2, 1910, 6. Many of the inquiries about game laws came specifically from ladies worried about keeping songbirds. They wanted to know if Visart was coming to arrest them. (See appendix for 1909 outline of Arkansas game laws.)

59. "Under Direction of State," *Arkansas Democrat*, June 2, 1910, 3; "To Protect Fish and Game," *Batesville Guard*, June 10, 1910, 7.

60. "Sportsmen Uphold Visart," *Daily Arkansas Gazette*, November 16, 1909, 8; "No Birds in Cage," *Arkansas Democrat*, November 22, 1909, 4; "Will Go After Sunday Hunters," *Arkansas Democrat*, January 6, 1910, 9. Convictions included W. W. Webb (ten-dollar fine) for selling dear meat and Ed Hicks, Grover Hicks, O. E. Hicks, and Tom Lingo (five-dollar fine) for illegal fishing and sales.

61. "Will Go After Sunday Hunters," *Arkansas Democrat*, January 6, 1910, 9; "Visart Investigates Exportation of Game," *Daily Arkansas Gazette*, December 17, 1909, 1.

62. "Comiskey and Ban B. Escape Arrest in South," *St. Louis Post-Dispatch*, February 3, 1910, 18; *Arkansas Democrat*, February 3, 1910, 2; "More Convictions," *Arkansas Democrat*, February 4, 1910, 7.

63. "Will Go After Sunday Hunters," *Arkansas Democrat*, January 6, 1910, 9; "Visart Investigates Exportation of Game," *Daily Arkansas Gazette*, December 17, 1909, 1; "Visart Talks of Game Laws," *Arkansas Democrat*, August 14, 1910, 6; "Hunting and Fishing," *Daily Arkansas Gazette*, August 14, 1910, 38.

64. "Should Revise Game Laws," *Daily Arkansas Gazette*, January 3, 1910, 6; "Will Go After Sunday Hunters," *Arkansas Democrat*, January 6, 1910, 9.

65. "To Sportsmen," *Daily Arkansas Gazette*, February 27, 1910, 8; March 20, 1910, 11; April 3, 1910, 8; April 17, 1910, 44.

66. "Seeks Rainbow Trout," *Arkansas Democrat*, March 16, 1910, 2; "V. E. Visart, State Game," *Arkansas Democrat*, March 27, 1910, 3; "Hopes for Many Pheasants," *Daily Arkansas Gazette*, August 7, 1910. Visart made a statement in the summer of 1910 that he was interested in building a state "propagating farm" for deer. Trout are not native to Arkansas. For more about Arkansas trout, see "Trout," Arkansas Game and Fish Commission website, accessed April 25, 2022; Keith Sutton, "How Trout Came to Arkansas," *Arkansas Democrat-Gazette* (online), accessed April 25, 2022; "Salmonids," *Encyclopedia of Arkansas* (online), accessed April 25, 2022.

67. "Mammoth Springs National Fish Hatchery," United States Fish and Wildlife Service website, accessed February 21, 2022; "A Shipment of Black Bass," *Arkansas Democrat*, August 6, 1910; "Hunting and Fishing, *Daily Arkansas Gazette*, August 14, 1910, 38; "Hunting and Fishing," *Daily Arkansas Gazette*, September 4, 1910, 46.

68. "Gets Neimeyers Pond," *Arkansas Democrat*, July 24, 1910, 2; "Stocking Lakes with Fish," *Daily Arkansas Gazette*, August 7, 1910, 35. Visart's attempts are a forerunner of the Arkansas Game and Fish owned and managed lakes.

69. "State Game Warden E. V.," *Arkansas Democrat*, May 29, 1910, 3; "To Protect Fish and Game," *Batesville Guard*, June 10, 1910, 7. The introduction of nonnative species did not begin in Arkansas. Countries around the globe imported and exported nonnative fish and wildlife, which started as the Europeans pushed to colonize the world during the 1500s. In the United States, during the late nineteenth century, conservationists brought in birds and fish, such as the Chinese pheasant and the brown trout, to replace the decreasing native species. During the early 1900s, the American Game Protective and Propagation Association made it part of their mission to distribute new species into the United States. Thomas R. Dunlap, "Remaking the Land: The Acclimation Movement and Aglo Idea of Nature," *Journal of World History* 8, no. 2 (Fall 1997), 307–10. For more about pheasant acclimation, see George Laycock, *The Alien Animals: The Story of Imported Wildlife* (Garden City, NY: 1970), 19–37. For more on biological exchange, see Alfred Crosby, *The Columbian Exchange* (Westport, CT: Greenwood, 1973); Peter M. Minard, *All Things Harmless, Useful, and Ornamental: Environmental Transformation through Species Acclimatization, from Colonial Australia to the World* (Chapel Hill: University of North Carolina Press, 2019); *The American Game Protective and Propagation Association: The Story of its Origin, Purposes, History and Policies*, New York, 1915; James B. Trefethen, *An American Crusade for Wildlife* (New York: Winchester Press; Boone and Crockett Club, 1975). For more about game farming and propagation, see Drew Swanson, "Growing Wild: Visions of Wildlife Management as Agricultural Science in American Forests and Fields," *Agricultural History* 97, no. 4 (2023): 177–214.

70. "A New Industry," *Arkansas Democrat*, June 14, 1910, 2; "State Game Warden E. V. Visart," *Arkansas Democrat*, September 27, 1910, 3. Visart displayed some of the pheasants at his home on 1821 Park Avenue in Little Rock so interested parties could view the birds.

71. "Value of Ring-Neck Pheasant," *Daily Arkansas Gazette*, June 19, 1910, 13; "Samples of Pheasant's Eggs," *Arkansas Democrat*, June 20, 1910, 3.

72. "Applying for Pheasants Eggs," *Arkansas Democrat*, June 23, 1910, 4; "To Stock State with Partridges," *Arkansas Democrat*, June 29, 1910, 2; "Great Demand for Pheasants," *Arkansas Democrat*, September 25, 1910, 8.

73. "Game Warden Checkmated," *Daily Arkansas Gazette*, April 22, 1910, 1; "V. E. Visart," *Arkansas Democrat*, April 28, 1910, 3; "Visart To Probe Gillette Trouble," *Daily Arkansas Gazette*, May 5, 1910, 5; "Warden's Case Continued," *Arkansas Democrat*, April 26, 1910, 9; "Game Warden Case Again Postponed," *Arkansas Democrat*, April 30, 1910, 3. For more about challenges, see "Mob at Amity Forces Warden to Leave Town," *Nashville News*, May 26, 1923, 1; "Use of Seines," *Nashville News*, June 20, 1923, 4; "Fish Net Case Ended," *Nashville News*, August 11, 1923, 2.

74. "Sportsmen Thankful," *Arkansas Democrat*, May 17, 1910, 2; "Bag Limits for Our Game Laws," *Arkansas Democrat*, September 8, 1910, 4. A Remington 25–35 was a Model 8. It was Remington's answer to the famous Winchester 25–35 and 30–30. They were manufactured from 1907 to 1915. Dave Campbell, "Remington Model 8: A Look Back," *American Rifleman* (online), accessed August 28, 2023.

CHAPTER SEVEN

1. "Hunter's License," *Arkansas Democrat*, May 12, 1910, 4; The Arkansas Fish and Game Protective Association began in 1912. United States Fish and Wildlife Service, *Officials and Organizations Concerned with Wildlife Protection* (Washington, DC: Government Printing Office, 1914), 11.

2. "Sportsmen Would License Hunters," *Daily Arkansas Gazette*, May 12, 1910, 9; "Hunter's License," *Arkansas Democrat*, May 12, 1910, 4; *Ideal Hunting and Fishing Grounds*, pamphlet issued by H. C. Townsend, general passenger agent, Missouri Pacific Railway Co., Iron Mountain route, 1900, 85; "State Game Warden Visart," *Arkansas Democrat*, August 14, 1910.

3. "Enforcing Arkansas Game Laws," *Forest and Stream*, February 8, 1913, 176.

4. "Enforcing Arkansas Game Laws," *Forest and Stream*, February 8, 1913, 176; Christopher Morris, *The Big Muddy: An Environmental History of the Mississippi and Its Peoples from Hernando De Soto to Hurricane Katrina* (Oxford: Oxford University Press, 2012), 119–21. *Forest and Stream* contended that some Arkansas counties lost 75 percent of their deer population in the Mississippi River flood of 1912.

5. E. A. McIlhenny to E. V. Visart, January 28, 1913, Edward Avery McIlhenny Papers, 1911–1947, Louisiana and Lower Mississippi Valley Collections, Louisiana State University Libraries, Baton Rouge, LA; "Arkansas Great State for Game," *Daily Arkansas Gazette*, February 3, 1913, 6; Shane K. Bernard, *Tabasco: An Illustrated History* (Avery Island, LA: McIlhenny, 2007), 132; John Glenn, Lillian Brandt, and F. Emerson Andrews, *Russell Sage Foundation 1907–1946* (New York: Russell Sage Foundation, 1948). See also Robert L. Kirby, *Kirby Smith's Confederacy: The Trans-Mississippi South, 1863–1865* (Tuscaloosa: University of Alabama Press, 1972), 68–70.

6. "Arkansas Great State for Game," *Daily Arkansas Gazette*, February 3, 1913, 6.

7. E. V. Visart to Mr. E. A. McIlhenny, January 6, 1912, McIlhenny Papers. When Arkansas senator Jeff Davis died in office, Governor Robinson wanted to replace him. Later, when Davis won the senate seat, he resigned from the governorship. With no lieutenant governor, Little Rock was thrown even deeper into turmoil when a disagreement broke out over who would fill the vacant office. After much wrangling, Attorney General Hal L. Norwood ruled that Futrell was the rightful holder of the office. Later, an Arkansas Supreme Court case affirmed the decision. For more about Jeff Davis, see Raymond Arsenault, *The Wild Ass of the Ozarks: Jeff Davis and the Social Bases of Southern Politics* (Knoxville: University of Tennessee Press, 1988). For more about the political maelstrom, see Timothy P. Donovan, Willard B. Gatewood Jr., and Jeannie

M. Whayne, eds. *The Governors of Arkansas: Essays in Political Biography* (Fayetteville: University of Arkansas Press, 1995).

8. E. A. McIlhenny to E. V. Visart, January 6, 1913, McIlhenny Papers. He wrote three letters on the same day to Visart. One covering the newspaper article, one concerning the "Arkansas Conservation Law," and the other asking for the new governor's inauguration day and the date the legislature prepared to meet. Robinson announced his intention to run for Davis's old office on January 8. On March 10, 1913, he resigned as governor to take the position. See Stuart Towns, "Joseph T. Robinson and Arkansas Politics, 1912–1913," *Arkansas Historical Quarterly* 24, no. 4 (Winter 1965): 291–307.

9. E. V. Visart to E. A. McIlhenny, January 9, 1913, mislabeled 1912, typed letter 1, McIlhenny Papers; *Batesville Guard*, January 10, 1913, 15. Joseph "Joe" T. Robinson was a life-long outdoorsman who owned many quail hunting dogs and often fished with political and law colleagues. Joe T. Robinson to H. N. Thomason, April 15, 1903; A. B. Grace to Joe T. Robinson, June 6, 1908; Joe T. Robinson to A. B. Grace, June 8, 1908; Judge James Gould to Hon. Joe T. Robinson, June 10, 1908; Joe T. Robinson to Judge James Gould, June 11, 1908; J. E. Jones to Hon. Joe T. Robinson, February 20, 1909; Joe T. Robinson to Hon. J. E. Jones, February 24, 1909; F. E. Brown to Hon. Joe T. Robinson, June 26, 1909; Clement S. Ucker to Hon. Joseph G. Robinson, March 23, 1911, Joseph T. Robinson Papers, 1900–1954, Special Collections Department, University of Arkansas Libraries, Fayetteville, AR.

10. E. V. Visart to E. A. McIlhenny, January 9, 1913, handwritten letter 2, McIlhenny Papers; E. A. McIlhenny to E. V. Visart, January 13, 1913, McIlhenny Papers. In May 1909, while an Arkansas United States representative, Robinson heard conservationist and the head United States forester Gifford Pinchot speak at the meeting of the Southern Clubs at the Arlington Hotel in Washington, DC. In November 1912, the National Conservation Exposition had asked Governor Robinson to serve as chairman of the Arkansas Exposition board to gather state participants for the exposition in Knoxville, Tennessee. Don Carlos Ellis, United States Forest Service secretary and vice president of the exposition requested that Robinson form a board of A. M. Van Auken of Blytheville, Sid B. Redding of Little Rock, Thomas C. McRae of Prescott, and Mrs. C. H. Burr of Conway. Ellis claimed that everyone he suggested was a nationally known conservationist. E. L. Givens to Hon. Joseph T. Robinson, May 29, 1909; and Don Carlos Ellis to Joe T. Robinson, November 13, 1912, Robinson Papers.

11. "Makes a Plea for Game Protection," *Daily Arkansas Gazette*, January 10, 1913, 11; "An Appeal for Wild Life to the People of Arkansas," *Arkansas Democrat*, January 10, 1913, 9; E. V. Visart to E. A. McIlhenny, January 10, 1913, McIlhenny Papers.

12. *Batesville Guard*, January 10, 1913, 4.

13. E. V. Visart to E. A. McIlhenny, January 14, 1913, McIlhenny Papers; Towns, "Robinson," 300–307.

14. S. D. Barnes, "The Crumbling of a Chinese Wall," *Recreation* 47, no. 1 (June 1912): 269–270; Charlie Daniels, *Historical Report of the State of Arkansas* (Fayetteville:

University of Arkansas Press, 2008), 164–69. A. G. Little resigned in 1911 and Clyde Robinson took his place.

15. Copy of meeting minutes, Phillips County Rod and Gun Club, January 20, 1913, McIlhenny Papers; Copy of the meeting minutes, Porter Lake Hunting Club, January 20, 1913, McIlhenny Papers; E. V. Visart to E. A. McIlhenny, January 23, 1913, McIlhenny Papers.

16. E. V. Visart to E. A. McIlhenny, January 23, 1913, McIlhenny Papers.

17. Arkansas House of Representatives, *Journal of the House of Representatives of Arkansas, Thirty-Ninth Regular Session* (Little Rock: Southern Press, 1913), 726–27; E. V. Visart to E. A. McIlhenny, January 23, 1913, McIlhenny Papers. The Arkansas Legislature's shift in thinking toward wildlife from simply a food source to a natural resource fits into the American Progressive movement that was occurring during that time.

18. "Arkansas Great State for Game," *Daily Arkansas Gazette*, February 3, 1913; E. A. McIlhenny to E. V. Visart, January 28, 1913, letter 2, McIlhenny Papers; "Arkansas Great State for Game," *Daily Arkansas Gazette*, February 3, 1913, 6; "Offers Arkansas Free Game Refuge," *Daily Arkansas Gazette*, February 4, 1913, 10.

19. "Would Protect Game," *Commercial Appeal*, January 15, 1913, 11.

20. E. V. Visart to E. A. McIlhenny, January 28, 1913, McIlhenny Papers; E. A. McIlhenny to Jno. Dymond, Jr., January 29, 1913, McIlhenny Papers; "Another Game Bill in the Senate Now," *Batesville Guard*, February 13, 1913.

21. Ten years later, in 1923, Arkansas Game and Fish commissioner Lee Miles wrote the Federal authorities at the United States Biological Survey, a department that Visart worked for by that time, to stop Visart from appearing before the Arkansas state legislature to lobby for new game laws. "I am sure his appearance will react unfavorably," said Visart's boss, George A. Lawyer. This occurrence illustrates that some of Visart's enemies remained. See Lawyer to Biological Survey, October 1, 1923, telegram, box 385, Misc.-State Law, General Correspondence-State Laws-Arkansas, Bureau of Biological Survey, United States Fish and Wildlife Service, record group 22, National Archives, College Park, MD.

22. E. V. Visart to E. A. McIlhenny, February 19, 1913, McIlhenny Papers.

23. Visart to McIlhenny, February 19, 1913.

24. E. V. Visart to E. A. McIlhenny, February 20, 1913, wire messages 1–4, McIlhenny Papers, E. V. Visart to E. A. McIlhenny, February 21, 1913, wire message, McIlhenny Papers.

25. "Want M'Illhenny To Aid in Fight," *Daily Arkansas Gazette*, February 22, 1913; E. V. Visart to E. A. McIlhenny, February 22, 1913, McIlhenny Papers.

26. E. V. Visart to E. A. McIlhenny, February 25, 1913, McIlhenny Papers.

27. E. A. McIlhenny to E. V. Visart, January 13, 1913, McIlhenny Papers; Arkansas House of Representatives, *Journal* (1913), 726–27.

28. E. V. Visart to E. A. McIlhenny, March 4, 1913, March 17, 1913, and March 30, 1913, McIlhenny Papers.

29. E. A. Visart to E. A. McIlhenny, March 4, 1913, McIlhenny Papers.

30. E. A. Visart to E. A. McIlhenny, March 4, 1913, McIlhenny Papers; E. A. Visart to E. A. McIlhenny, March 19, 1913, McIlhenny Papers.

31. Arkansas Senate, *Journal of the Senate of Arkansas, Thirty-Ninth Regular Session* (Little Rock: Southern Press, 1913), 264; Arkansas House of Representatives, *Journal* (1913), 394, 729–30. State representative Andrew V. Smith of Bradley County also introduced House Joint Resolution No. 7, an amendment to the Arkansas State Constitution, that passed both houses. The title read, "For the regulation of fishing and hunting and the protection and propagation of fish and game," but the bill included no funding or enforcement measures. Therefore, supportive legislature members overcame some resistance to pass a generalized conservation item, but not enough to obtain specific laws. It served as a conciliatory measure.

32. E. V. Visart to E. A. McIlhenny, March 14, 1913, McIlhenny Papers.

33. E. A. McIlhenny to E. V. Visart, March 27, 1913, McIlhenny Papers.

34. "The Governor Keeps in Touch," *Arkansas Democrat*, April 3, 1913; "Daily Flood Bulletin," *Arkansas Democrat*, April 12, 1913, 1; "Thousands of Dollars of Damage Done Here by Rains," *Hot Springs New Era*, April 9, 1913, 1. In 1912, spring floods raised the water level so high it had driven people to higher ground across the Mississippi River in Tennessee. They had no possessions or food. It is understandable that if families found themselves caught without food, they would poach wildlife to survive. See Mayor E. H. Crump to Senator Davis, April 8, 1912, Jeff Davis Papers, 1849–1986, Special Collections Department, University of Arkansas Libraries, Fayetteville, AR. Futrell became governor temporarily after the resignation of Robinson in March 1913.

35. E. V. Visart to E. A. McIlhenny, April 5, 1913, McIlhenny Papers; E. V. Visart to E. A. McIlhenny, April 12, 1913, McIlhenny Papers.

36. E. V. Visart to E. A. McIlhenny, April 4, 1913, McIlhenny Papers.

37. E. V. Visart to E. A. McIlhenny, April 11, 1913, and April 12, 1913, McIlhenny Papers; "Plans to Protect Game During Flood," *Daily Arkansas Gazette*, April 4, 1913, 10; "Protector of Game a Paragould Visitor," *Unknown*, April 8, 1913, McIlhenny Papers.

38. "Plans to Protect Game," *Daily Arkansas Gazette*, April 10, 1913; "Party Will Aid Marooned Game," *Daily Arkansas Gazette*, April 13, 1913, 26.

39. E. V. Visart to E. A. McIlhenny, April 22, 1913, April 11, 1913, and April 12, 1913, McIlhenny Papers; "To Protect Wild Deer," *Daily Arkansas Gazette*, April 3, 1913, 9.

40. E. V. Visart to E. A. McIlhenny, April 11, 1913, McIllhenny Papers.

41. E. A. McIlhenny to E. V. Visart, April 21, 1913, McIlhenny Papers.

42. "Patrol System is Saving Game," *Daily Arkansas Gazette*, April 21, 1913.

43. "First Check for the State Guard," *Arkansas Democrat*, April 26, 1913; E. V. Visart to E. A. McIlhenny, April 31, 1913, McIlhenny Papers.

44. Arkansas Fish and Game Protective Association membership card, undated, McIlhenny Papers; E. V. Visart to E. A. McIlhenny, April 27, 1913, April 31, 1913, May 1, 1913, McIlhenny Papers.

45. E. V. Visart to E. A. McIlhenny, July 26, 1913, McIlhenny Papers; E. V. Visart to Hon. Joe T. Robinson, April 20, 1917, Robinson Papers. In 1917, Visart was still fighting to get the federal government to set aside some of the National Forest in Arkansas as a national game preserve. He asked Arkansas senator Joseph T. Robinson for assistance. For a definition of the Weeks-McClean Act, see "Weeks-McLean Act Law and Legal Definition," USLegal (website), accessed March 31, 2022.

46. E. V. Visart to E. A. McIlhenny, July 26, 1913, McIlhenny Papers.

47. "Visart Named to Enforce Laws," *Daily Arkansas Gazette,* October 12, 1913, 31; "State News," *Batesville Daily Guard,* October 13, 1913, 1, and October 15, 1913, 2; *Arkansas Democrat,* October 15, 1913, 1; "Visart Commissioned," *Daily Arkansas Gazette,* October 15, 1913, 9; *Jonesboro Daily Tribune,* October 17, 1913, 2; "E. V. Visart to E. A. McIlhenny, March 15, 1915, McIlhenny Papers.

48. "County Game Laws Are held Invalid," *Daily Arkansas Gazette,* November 18, 1913, 1; E. V. Visart to E. A. McIlhenny, November 14, 1913, McIlhenny Papers.

49. Lewis v. State, 110 Ark. 204, 161 S. W. 154 (1913); "County Game Laws Are held Invalid," *Daily Arkansas Gazette,* November 18, 1913, 1; "92 Bills Launched in House in Week," *Daily Arkansas Gazette,* January 17, 1917, 3. In January 1917, Representative Mosely of Cleveland County attempted to introduce a bill that prohibited nonresidents from hunting in his county. Representative Belote of Searcy County reminded him that the Supreme Court had ruled such laws unconstitutional because Arkansas game belonged to all Arkansas citizens as a whole.

50. "Recommendations of Game and Fish Commissioners," *Forest and Stream,* September 14, 1912, 336.

51. "The New Tell-at-a-Glance Game Law System," *Forest and Stream,* February 15, 1913, 199; Thomas S. Palmer, circular for the Bureau of Biological Survey, United States Department of Agriculture, August 22, 1914.

52. "Game Cannot Be Shipped from the State," *Daily Arkansas Gazette,* January 27, 1914, 3; E. V. Visart to E. A. McIlhenny, January 27, 1914, McIlhenny Papers.

53. "Morning Session," *Daily Arkansas Gazette,* February 11, 1909, 3. Back in February 1909, Mississippi County representative Anthony George Little had introduced a petition from the inhabitants of his district, asking for exemptions from the Whitley laws for four of the county's townships to allow shipping game out of the state. Therefore, anyone who wanted to ship game or fish from Arkansas sent it to Mississippi County first and then out of the state, a terrible loophole; "Morning Session," *Daily Arkansas Gazette,* February 11, 1909, 3; Arkansas General Assembly, *Public and Private Acts and Joint and Concurrent Resolutions and Memorials of the General Assembly of the State of Arkansas* (Little Rock: Democrat Print, 1909), 1131–32.

54. J. N. Adams, Jake River, and J. K. McMasters v. Jonesboro, Lake City and Eastern Railroad, No. 242, Ark. 2nd Circuit, pleas before Honorable W. J. Driver, Judge, January 29, 1914, William H. Bowen School of Law Library, University of Arkansas, Little Rock; Jonesboro, Lake City and Eastern Railroad v. J. N. Adams et al., 117 Ark.

54, 174 S.W. 527 (1915), abstract and briefs, case 3306, December 3, 1914, William H. Bowen School of Law Library, University of Arkansas, Little Rock; *Arkansas Democrat*, February 15, 1915, 1; *Daily Arkansas Gazette*, February 16, 1915, 8.

55. "Game Commission is Planned by Futrell," *Arkansas Democrat*, January 15, 1915, 12; See also "New Game Law is Recommended By Senate Members, *Arkansas Democrat*, January 28, 1915, 2; E. V. Visart to E. A. McIlhenny, February 8, 1915, McIlhenny Papers; "From Fort Smith Times Record," *Arkansas Democrat*, April 3, 1914, 14; Christopher P. Yingling, "The Arkansas Game and Fish Commission," senior thesis, SM350, Spring 1985, Torreyson Library Archives, University of Central Arkansas, Conway, AR, 1; "The New Game Law," *Star Progress*, April 2, 1915, 1; "Statewide Game Bill is Passed in Lower House," *Arkansas Democrat*, February 27, 1915, 4; *Star Progress*, April 2, 1915, 1.

56. E. V. Visart to E. A. McIlhenny, February 8, 1915, McIlhenny Papers; Green Benton to E. A. Visart, February 2, 1915, McIlhenny Papers.

57. Arkanasas Senate, *Journal of the Senate of Arkansas, Fortieth Regular Session* (Little Rock: Southern Press, 1915), 127; Arkansas House of Representatives, *Journal of the House of Representatives of Arkansas, Fortieth Regular Session* (Little Rock: Southern Press, 1915), 553; E. V. Visart to E. A. McIlhenny, February 26, 1915, McIlhenny Papers; E. A. McIlhenny to E. A. Visart, March 6, 1913, McIlhenny Papers. On March 11, 1915, Arkansas governor George W. Hays signed Act 124, creating the Arkansas Game and Fish Commission. The governor was a sportsman and supported stricter game regulations. Hays penned a letter to Arizona governor W. P. Hunt: "I think that the time has arrived when we should begin to conserve our wild animal life else it will soon be entirely exhausted." See George W. Hays to George W. P. Hunt, November 29, 1915, "William T. Hornady Wildlife Conservation Scrapbook," William T. Hornaday Papers, 1888–1937, Wildlife Conservation Society Archives, vol. 7, New York, 103.

58. *Southern Standard*, March 4, 1915, 6; "From Fort Smith Times Record," *Arkansas Democrat*, April 3, 1914, 14; Arkansas Game and Fish Commission, *First Annual Report* (Fort Smith: Calvert-McBride, June 15, 1916), 5.

59. "Review of 1915 Game Legislation," *Forest and Stream*, October 1915, 634; Arkansas House of Representatives, *Journal* (1915). For a synopsis of the first statewide game act, see appendix Z. For a complete listing of the Futrell Act, see "New Game Law Given in Full," *Batesville Daily Guard*, March 24, 1915, 1.

60. "Game Bill Author Says Funds Usable," *Daily Arkansas Gazette*, April 4, 1915, 16.

61. E. Pickins to Mr. Govner, February 24, 1915, box 1, George W. Hays Papers, 1913–1917, Arkansas State Archives, Little Rock, AR. Pickins wrote from Arkadelphia; J. S. Gibson to Gov Hayze, March 9, 1915, box 1, Hays Papers. Black Rock is located in northeastern Arkansas, near Powhatan.

62. W. J. Johnson to Mr Govner Hays, March 15, 1915, box 1, Hays Papers. Waco was located near Manila in Mississippi County, not far from Big Lake.

63. "Governor Names Game Commission," *Southern Standard*, March 25, 1915,

6; Arkansas Game and Fish Commission, *First Annual Report*, 5; *The Field Dog Stud Book* (Chicago: American Field Publishing, 1922), 119, 254 ; Fay Hempstead, *Historical Review of Arkansas: Its Commerce, Industry, and Modern Affairs* (Chicago: Lewis, 1911), 990; American Poland-China Record Association, *American Poland-China Record*, vol. 73 (Columbia, MO: E. W. Stephens, 1920), 472; *The Field Dog Stud Book* (Chicago: American Field Publishing, 1904), 147; *The Field Dog Stud Book* (Chicago: American Field Publishing, 1919), 58; United Mine Workers v. Coronado Coal Co., 259 U.S. 344 (1922), transcript of record, Supreme Court of the United States, October term 1921, vol. 1; "Calendar of Coming Meetings," *Good Roads* 14, no. 6 (1917): 78; Dallas Tabor Herndon, *Centennial History of Arkansas* (Chicago: S. J. Clark, 1922), 1120; "Game Commission Chooses Secretary," *Daily Arkansas Gazette*, March 25, 1915, 10. Judge Miles became the vice president of the International Association for the Protection of Migratory Birds in 1920.

64. "Game Commission Begins Business," *Pine Bluff Daily Graphic*, March 25, 1915, 5; "New Game Law to Be Enforced," *Monticellonian*, October 7, 1915, 1; "The New Game Law," *Southern Standard*, March 25, 1915, 5; "After Years of Defeat, Far-Seeing Sportsmen Protect Game and Fish," *Arkansas Democrat*, November 28, 1917, 4.

65. "One State Game Warden Selected," *Daily Arkansas Gazette*, March 28, 1915, 22; "Game Commission," *Nashville News*, March 27, 1915, 1. Arkansas Game and Fish Commission, *First Annual Report*, 3. The AGFC praised Futrell as the main advocate for the bill that created the commission.

66. "Game Commission Begins Business," *Pine Bluff Daily Graphic*, March 25, 1915, 5.

67. "No Appropriations to Pay Salaries," *Daily Arkansas Gazette*, March 30, 1915, 8; *Batesville Daily Guard*, March 30, 1915, 1; *Monticellonian*, April 1, 1915, 4.

68. "Game Bill Author Says Funds Usable," *Daily Arkansas Gazette*, April 4, 1915, 16.

69. *Southern Standard*, April 15, 1915, 6; "May Digest Game Laws," *Daily Arkansas Gazette*, April 16, 1915, 8; "The State Fish and Game Commission," *Arkansas Democrat*, April 17, 1915, 13; "Game Commission Asks Opinion on Use License Fee," *Arkansas Democrat*, April 16, 1915, 1; "Opinion Asked on License Fee," *Monticellonian*, April 22, 1915, 1; *Southern Standard*, May 20, 1915, 2.

70. *Southern Standard*, May 20, 1915, 2.

71. Arkansas Game and Fish Commission, *First Annual Report*, photograph caption, inside cover; "One State Game Warden Selected," *Daily Arkansas Gazette*, March 28, 1915, 22; "Speculate as to Chief," *Daily Arkansas Gazette*, April 8, 1915, 2; "New Texarkana Officers," *Daily Arkansas Gazette*, April 13, 1917, 2.

72. "Kills Self by Accident," *Daily Arkansas Gazette*, April 27, 1916, 2; "Greene County Roster Filled," *Arkansas Democrat*, January 3, 1921, 6; "Liquor Seized: Arrests Made," *Arkansas Democrat*, March 21, 1922; "Herschel Neely," United States Census, 1910, Ancestry (website), accessed March 2, 2022; "Joseph A. Galvin," United States Census, 1910, Ancestry (website), accessed March 2, 2022; "George B. Rison," Arkansas Death Certificates, 1929, Ancestry (website), accessed March 3, 2022.

73. Arkansas Game and Fish Commission, *First Annual Report,* photograph caption, inside cover; "One State Game Warden Selected," *Daily Arkansas Gazette,* March 28, 1915, 22; City of Texarkana v. Hudgins Produce Co., 112 Ark. 17, 163–64 S.W. 737 (1914); "Clide J. March," United States Census, 1910, Ancestry (website), accessed March 3, 2022; "C. John March," United States Census, 1920, Ancestry (website), accessed March 3, 2022; "John C. Marsh," *Commercial Appeal,* December 31, 1936; "Counterfeiting Ring Charged," *Daily Arkansas Gazette,* February 8, 1915, 2; "Washington E. Clibourne," United States Census, 1920, Ancestry (website), accessed March 3, 2022; "James Fernandez," U.S., City Directories, 1910, Ancestry (website), accessed March 3, 2022; "Refuse to Remove Fort Smith Chief," *Daily Arkansas Gazette,* October 21, 1917, 22; "Car Thieves Arrested," *Daily Arkansas Gazette,* April 24, 1915, 2; *Southern Standard,* June 17, 1915, 6; "Mrs. Ada Clibourne," *Daily Arkansas Gazette,* November 8, 1918, 14. Wash Clibourne left the commission in 1919 and built houses in Conway. Fernandez went back to his old job as police chief by 1917. Rison quit the commission after five years but came back the next year. He worked for the AGFC until he died in 1929.

74. "Jim Fernandez a Deputy Game Warden," *Daily Herald,* May 27, 1915, 5; "Game Wardens Appointed," *Arkansas Democrat,* April 5, 1919, 4; "Game Warden Resigns," *Daily Arkansas Gazette,* April 11, 1920, 10. Since the Arkansas Game and Fish have retained few of their records from those early years, it is difficult to ascertain whether applicants simply applied for these positions or had recommendations from the community. We do know that Calvert played a key role in hiring Fernandez, both of the men resided in Fort Smith.

75. "Discuss Market Fishermen," *Daily Arkansas Gazette,* June 13, 1915, 6; Lee Miles to E. A. McIlhenny, June 5, 1915, McIlhenny Papers.

76. *Monticellonian,* June 3, 1915, 5; "Must Heed New Game Law," *Daily Arkansas Gazette,* June 7, 1915, 3; "Deputy Warden Tours South Arkansas," *Arkansas Democrat,* June 18, 1915, 6; "Fishermen are Discharged," *Pine Bluff Daily Graphic,* July 9, 1915, 7; Arkansas Game and Fish Commission, *First Annual Report,* 3; "Negro Fisherman Taken; 4 Hoop Nets Destroyed," *Daily Arkansas Gazette,* July 21, 1915, 12.

77. "New Game Law Is Recommended by Senate Members," *Arkansas Gazette,* January 28, 1915, 2; "New Game Law Given in Full," *Batesville Daily Guard,* March 24, 1915, 1; St. Louis, MO Indiana and Arkansas Lumber And Mfg Co. to Governor Hays, March 5, 1915; A.C. Lange, General Superintendent to Governor Hays, nd; Snoden, et al. to Governor Hays, nd; J. A. Riechman to Governor Hays, March 4, 1915; Anderson Tully Co. to Governor Hays, nd, Box 1, Folder 25, Hays Papers.

78. Arkansas Game and Fish Commission, *First Annual Report,* 14.

79. "Attack Warden's Boats," *Monticellonian,* June 29, 1916, 1.

80. Arkansas Game and Fish Commission, *First Annual Report,* 14.

81. "Prosecuting Violators of Fish and Game Law," *Fayetteville Daily Democrat,* July 27, 1916, 1; "New Game Law Given in Full," *Batesville Daily Guard,* March 24, 1915, 1.

82. "Games Fines Only $83.78," *Daily Arkansas Gazette,* November 13, 1915, 9;

"Hunters Fined $1 Each," *Daily Arkansas Gazette,* November 28, 1915; "Nonresidents Blamed," *Daily Arkansas Gazette,* January 15, 1916, 8.

83. "Games Fines Only $83.78," *Daily Arkansas Gazette,* November 13, 1915, 9; "Lack of License Blanks," *Daily Arkansas Gazette,* January 4, 1916, 9; Arkansas Game and Fish Commission, *First Annual Report,* 34.

84. "Game Wardens to Be Dismissed; Suit in Court," *Arkansas Democrat,* January 14, 1916, 12. "Mandamus—A (writ of) mandamus is an order from a court to an inferior government official ordering the government official to properly fulfill their official duties or correct an abuse of discretion," Cornell Law School website, Legal Information Institute, accessed March 22, 2022. Several of the game wardens had to seek other employment. For example, former warden Wash E. Clibourne ran for Faulkner County Sheriff. "Faulkner County Politics," *Daily Arkansas Gazette,* January 25, 1916, 7.

85. "Nonresidents Blamed," *Daily Arkansas Gazette,* January 15, 1916, 8; *Batesville Daily Guard,* January 18, 1916, 5; For the case against the Big Lake sportsmen, see State v. Stokes, 117 Ark. 192 (1915).

86. "Game Wardens to Be Dismissed; Suit in Court," *Arkansas Democrat,* January 14, 1916, 12. The other two funds collected licensing fees into the state treasury and drew on the money through warrant draws as the AGFC did. *Batesville Daily Guard,* January 18, 1916, 5.

87. "State to Protect Game," *Daily Arkansas Gazette,* January 30, 1916, 8; "Ouachita Spreading Out," *Daily Arkansas Gazette,* January 30, 1916, 8; "1,000 Deer in Arkansas," *Hot Springs New Era,* February 19, 1916, 1; "Lowlands are Flooded," *Daily Arkansas Gazette,* January 30, 1916, 8; "Wild Game Marooned," *Daily Arkansas Gazette,* February 3, 1916, 7.

88. Arkansas Game and Fish Commission, *First Annual Report,* 10–13. Scott Giltner argues that proponents of game conservation laws used the African American hunter and fisherman to convince other white voters to support their cause. They blamed Black folks for the decrease in game and fish. Therefore, he argues, game and fish legislation sometimes targeted African Americans specifically. In Arkansas, during the political debates and or in newspapers or speeches surrounding the passage of conservation laws, there is only passing mention of African Americans. Scott E. Giltner, *Hunting and Fishing in the New South: Black Labor and White Leisure after the Civil War* (Baltimore: Johns Hopkins University Press, 2008), chap. 5.

89. "Explains Work of Game Commission," *Daily Arkansas Gazette,* November 12, 1922, 12. Miles represents individual racism when he placed the caption and the photograph in the commission report.

90. "Wild Game Marooned," *Daily Arkansas Gazette,* February 3, 1916, 7; "Wild Game Marooned," *Monticellonian,* February 17, 1916, 8; Arkansas Game and Fish Commission, *First Annual Report,* 10; "Demurrer Sustained," *Monticellonian,* February 10, 1916, 1. There is no information on how the wardens paid the locals to feed and look after the animals.

91. "Wardens Report Game Protected in Flood," *Arkansas Democrat*, March 4, 1916, 16; "Wild Game Cared For," *Daily Arkansas Gazette*, March 4, 1916, 8; Arkansas Game and Fish Commission, *First Annual Report*, 6; "500 Deer Hunters Get State Permit," *Arkansas Democrat*, November 11, 1916, 3.

92. "Seek Game Fund Test," *Daily Arkansas Gazette*, May 6, 1916, 8; "Game Warden Wants $13," *Daily Arkansas Gazette*, May 7, 1916, 14; *Arkansas Democrat*, May 8, 1916, 6.

93. "Game Fund Suit Is Taken Up on Appeal," *Arkansas Democrat*, June 26, 1916, 6; "Game Fund Case," *Pine Bluff Daily Graphic*, June 27, 1916, 8; "Specific Act of Assembly Needed to Validate Funds," *Arkansas Democrat*, July 3, 1916, 1; "Game Law Upheld," *Monticellonian*, July 13, 1916, 1.

94. "Game and Fish Wardens May Have to Quit Job," *Batesville Daily Guard*, July 5, 1916, 1.

95. "Proclamations Issued," *Daily Arkansas Gazette*, July 9, 1916, 8; *Monticellonian*, July 20, 1916, 1.

96. Arkansas Game and Fish Commission, *First Annual Report*, 24, 30; "500 Deer Hunters Get State Permit," *Arkansas Democrat*, November 11, 1916, 13.

97. "For Violation of the game law," *Arkansas Democrat*, November 24, 1916, 10.

98. Arkansas Game and Fish Commission, *First Annual Report*, 28.

99. Arkansas Game and Fish Commission, *First Annual Report*, 32–33.

100. Arkansas Game and Fish Commission, *First Annual Report*, 33–34.

101. "After Years of Defeat, Far-Seeing Sportsmen Protect Game and Fish," *Arkansas Democrat*, November 28, 1917, 4.

CHAPTER EIGHT

1. "Governor's Message," in Arkansas House of Representatives, *Journal of the House of Representatives of Arkansas, Forty-First Regular Session* (Little Rock: Southern Printing, 1917), 13, 23. Law required the AGFC to provide a yearly report to the governor. In its first annual statement, the AGFC claimed that Arkansas conservation had progressed the previous year. They further sought wanted (1) a five-year ban on prairie chicken hunting, (2) a squirrel season, (3) prohibition on squirrel sales, (4) reduction of duck bag limits from twenty-five to fifteen, (5) a bag limit on geese, (6) a ban on killing does, bag limits on bass, (7) licensing for each net or seine, (8) allow the AGFC to ship fish for scientific and stocking purposes, (9) enable the sale of fish and game raised on private lands and waters, (10) permit the AGFC to create bird hatcheries, (11) a gun license, (12) make dynamiting fish a felony, and (13) create a fur-bearing season. Arkansas Game and Fish Commission, *First Annual Report* (Fort Smith: Calvert-McBride, June 15, 1916), 32–33.

2. "Governor's Message," in Arkansas House of Representatives, *Journal* (1917), 49–91, 61; J. R. Gammeter to Gov. Charles Brough, September 15, 1920, Charles Hillman Brough Papers, 1895–1935, Special Collections Department, University of Arkansas Libraries, Fayetteville, AR. Governor Brough occasionally took part in outdoor activities

like golf, hunting, and fishing. He sometimes took visiting nonresident friends on duck and turkey hunts.

3. Lon Slaughter and E. D. Wall, "Reminiscences of the Late Lon Slaughter as Told to Mrs. E. D. Wall, September 3, 1920," *Arkansas Historical Quarterly* 8, no. 2 (Summer, 1949), 167–69; Arkansas Senate, *Journal of the Senate of Arkansas, Forty-First Regular Session* (Little Rock: Southern Printing, 1917), 334, 346, 525; "Proposed Game Law Would Tax Dealers," *Daily Arkansas Gazette*, February 19, 1917, 5; "Some Provisions of the New Game Law," *Osceola Times*, February 23, 1917, 7; Arkansas Game and Fish Commission, *Second Annual Report* (Fort Smith: Calvert-McBride, April 1, 1918), 8; "In the Open with the Gun and Rod," *Arkansas Democrat* April 14, 1917, 4.

4. "Some Provisions of the New Game Law," *Osceola Times*, February 23, 1917, 7; "Enumerates Benefits of Game Commission," *Pine Bluff Daily Graphic*, February 25, 1917, 2; Arkansas Senate, *Journal* (1917), 488, 525; "Some Provisions of the New Game Law," *Osceola Times*, February 23, 1917, 7.

5. Arkansas Senate, *Journal* (1917), 461, 491, 492, 716, 774.

6. "Stormy Fight on Game Fund," *Daily Arkansas Gazette*, March 7, 1917, 11; Dallas Tabor Herndon, *Centennial History of Arkansas* (Chicago: S. J. Clark, 1922), 694–95; Charlie Daniels, *Historical Report of the State of Arkansas* (Fayetteville: University of Arkansas Press, 2008), 172.

7. "Stormy Fight on Game Fund," *Daily Arkansas Gazette*, March 7, 1917, 11; *Daily Arkansas Gazette*, March 7, 1917, 11.

8. *Daily Arkansas Gazette*, March 7, 1917, 11–12.

9. Arkansas Senate, *Journal* (1917), 461, 491, 492, 716, 774; *Daily Arkansas Gazette*, March 7, 1917, 13–14; "Game Protection Is Given Big Boost By House After Battle," *Arkansas Democrat*, March 7, 1917, 16.

10. "Game Wardens Get Boat," *Daily Arkansas Gazette*, March 27, 1917, 3; "Migratory Birds Winter in Southeastern Arkansas," *Daily Arkansas Gazette*, July 4, 1920, 32. The *Mary B* had no motor and the wardens used a gasoline-powered boat, called the *Green Beetle*, to move it up and down the rivers. In 1920, the Henry Plant family cared for the boat, keeping it prepared for the wardens and protecting it from thieves or vandals.

11. "Game Commission to Police State," *Arkansas Democrat*, April 4, 1917, 3; "In the Open with the Gun and Rod," *Arkansas Democrat*, April 21, 1917, 4.

12. *Arkansas Democrat*, April 14, 1917, 4.

13. "In the Open with the Gun and Rod," *Arkansas Democrat*, June 2, 1917, 4; "Confiscate Illegal Seines," *Daily Arkansas Gazette*, March 4, 1917, 12; "Game Warden Busy," *Daily Arkansas Gazette*, June 7, 1917, 4; Arkansas Game and Fish Commission, *Second Annual Report*, 7; For more about dynamiting, see "Investigate report of Dynamiting Stream," *Arkansas Democrat*, June 11, 1917, 2; "Dynamite the Best Bait," *Batesville Daily Guard*, July 14, 1917, 1; "Enforcing Game Laws," *Daily Arkansas Gazette*, July 16, 1917, 5; "D. G. Moore Game Warden in This Section," *Baxter Bulletin*, July 27, 1917, 2.

14. "Net 900 Feet Long," *Arkansas Democrat*, January 25, 1910, 2; "Game Warden

Visart Sued," *Daily Arkansas Gazette*, January 25, 1910, 8; "Visart Faces Litigation," *Daily Arkansas Gazette*, June 7, 1910, 7; "Visart Discharged," *Arkansas Democrat*, June 21, 1910, 3; "Ike Sangston, Who Was," *Arkansas Democrat*, June 22, 1910, 3; "Visart Wins Damage Suit," *Daily Arkansas Gazette*, June 22, 1910, 10; "Dutch Jones Held: Destroyed A Seine," *Daily Arkansas Gazette*, June 10, 1917, 9; "W. A. "Dutch" Jones is released on Bond," *Daily Arkansas Gazette*, June 11, 1917, 8.

15. "Dutch Jones Held: Destroyed A Seine," *Daily Arkansas Gazette*, June 10, 1917, 9; "W. A. 'Dutch' Jones is released on Bond," *Daily Arkansas Gazette*, June 11, 1917, 8.

16. "Dutch Jones is Fined for Trespass," *Arkansas Democrat*, June 21, 1917, 6; "Dutch Jones Fined $100," *Daily Arkansas Gazette*, June 22, 1917, 3.

17. Arkansas Game and Fish Commission, *Second Annual Report*, 1. Little had served as Miller County clerk in Texarkana.

18. "Now Open Season for Deer, Bear and Hunting Lies," *Daily Arkansas Gazette*, November 12, 1917, 7; "Hunters Getting Ready for Sport," *Arkansas Democrat*, September 18, 1917, 6.

19. "Little Predicts Big Season for Hunters," *Arkansas Democrat*, September 29, 1917, 4. The Biological Survey also printed annual game law booklets that contained state and federal regulations. Appearing in newspapers and magazines, advertisements explained that anyone wishing a free copy of the booklet should write to the Washington bureau. See "Game Law Booklet Free to All Huntsmen," *Daily Arkansas Gazette*, August 27, 1920, 6.

20. "The Food Proclamation," *Arkansas Democrat*, November 7, 1917, 7; "Hotels Cannot Serve Protected Game Birds," *Arkansas Democrat*, November 7, 1917, 13.

21. Arkansas Game and Fish Commission, *Second Annual Report*, 7, 10.

22. Arkansas Game and Fish Commission, *Second Annual Report*, 12.

23. Arkansas Senate, *Journal of the Senate of Arkansas, Forty-Secondnd Regular Session* (Little Rock: Southern Printing, 1919), 91; Arkansas Game and Fish Commission, *Second Annual Report*, 14–15.

24. Herbert Hoover quoted in Arkansas Game and Fish Commission, *Second Annual Report*, 20; Arkansas Senate, *Journal* (1919), 92.

25. Arkansas Game and Fish Commission, *Second Annual Report*, 29–30; Arkansas Senate, *Journal* (1919), 91–92; J. E. Mons to Hon. Jeff Davis, April 10, 1912, and May 10, 1912, Jeff Davis Papers, 1849–1986, Special Collections Department, University of Arkansas Libraries, Fayetteville, AR. The federal government established the Mammoth Springs Federal Fish Hatchery in 1903. For more on Mammoth Springs, see "Mammoth Springs Federal Fish Hatchery," United States Fish and Wildlife Service website.

26. Arkansas Game and Fish Commission, *Second Annual Report*, 23.

27. Arkansas Game and Fish Commission, *Second Annual Report*, back cover.

28. Arkansas Game and Fish Commission, *Second Annual Report*, 23.

29. "Deer Plentiful Over Arkansas," *Arkansas Democrat*, November 22, 1918, 4; Arkansas Game and Fish Commission, *Second Annual Report*, 23, 32.

30. "Is for Running Game," *Daily Arkansas Gazette*, January 26, 1919, 4; "Act 276, March 17, 1919," *Osceola Times*, June 27, 1919, 5.

31. Arkansas Game and Fish Commission, *Second Annual Report*, 31–32. For more on game hogs, see *Recreation* 21, 1905; William T. Hornady, *Our Vanishing Wildlife: Its Extermination and Preservation* (New York: C. Scribner's Son, 1913), 244–46; William T. Hornady, "Will You Have a Gameless Continent?," *Independent*, May 4, 1911, 946–49.

32. Arkansas Game and Fish Commission, *Second Annual Report*, 31–32.

33. "Baldy Vinson," *Arkansas Democrat*, March 28, 1917, 5.

34. "Proposed Changes," *Star Progress*, February 7, 1919, 4; "Changes Game Law," *Daily Arkansas Gazette*, January 21, 1919, 3; Arkansas Senate, *Journal* (1919), 490.

35. "Fight on Game Bill," *Daily Arkansas Gazette*, January 28, 1919, 3; "Attack Game Bill," *Arkansas Democrat*, January 28, 1919, 4: "Mr. Dunaway Presents One of His Typical Presents," *Daily Arkansas Gazette*, January 28, 1919, 1; Arkansas Senate, *Journal* (1919), 266.

36. "House Votes, 57 to 21, to Abolish the Game Commission," *Arkansas Democrat*, February 13, 1919, 1; *Monticellonian*, February 21, 1919, 1.

37. "Choate Bill Reaches Senate," *Arkansas Democrat*, February 18, 1919, 7; "Committee Reports," February 20, 1919, 3; "Kill Choate Bill," *Arkansas Democrat*, February 25, 1919, 1.

38. "Fish Commission Loses," *Daily Arkansas Gazette*, February 28, 1919, 12.

39. "Fish Commission Loses," *Daily Arkansas Gazette*, February 28, 1919, 12; "Refuse Appropriations from Fish-Game Fund," *Fayetteville Daily Democrat*, February 28, 1919, 1.

40. "New Fish and Game Bill," *Daily Arkansas Gazette*, March 7, 1919, 15; *Monticellonian*, March 14, 1919, 1; "Game Wardens Are Appointed," *Arkansas Democrat*, April 5, 1919, 4; "Appoint 8 Game Wardens," *Daily Arkansas Gazette*, April 5, 1919, 10. The commission appointed Wash E. Clibourne, Med Donaldson, George B. Rison, Jr, Leo Harrel, A. G. Stedman, and Arthur Thomas.

41. "New Fish and Game Bill," *Daily Arkansas Gazette*, March 7, 1919, 15; *Monticellonian*, March 14, 1919, 1; "Act 276, March 17, 1919," *Osceola Times*, June 27, 1919, 5; "Act No. 276," *Southern Standard*, June 19, 1919, 2. If fishermen only used hoop nets, they only paid $25, and if only trotlines, they paid $10 annually. For Act 276, see appendix B.

42. "'Axers' Control for A Time," *Daily Arkansas Gazette*, March 11, 1919, 3; "Reversal Complete," *Arkansas Democrat*, March 11, 1919, 1.

43. "Dynamiting Fish Charge," *Daily Arkansas Gazette*, May 10, 1919, 13; "A. L. Roberts and J. D. Roberts," *Arkansas Democrat*, May 10, 1919, 8.

44. "After Game Law Violators," *Mountain Echo*, November 6, 1919, 3; Arkansas Game and Fish Commission, *Third Bi-Ennial Report* (Fort Smith: McBride-Calvert Publishing, 1920), 8.

45. "Protect Fish and Game," *Daily Arkansas Gazette*, May 8, 1919, 10.

46. "Enforcing Game Law," *Daily Arkansas Gazette*, April 7, 1919, 4.

47. "Will Organize for Protection of Game," *Daily Arkansas Gazette*, July 25, 1920, 29.

48. "Water Drive Deer from Lowlands," *Arkansas Democrat*, April 5, 1920, 4; "Game Wardens Sent to Watch the River," *Little Rock Daily News*, April 9, 1920, 8; "Water Drives Game Out," *Daily Arkansas Gazette*, April 11, 1920, 10 and May 1, 1920, 6; "Game Warden," *Nashville News*, July 28, 1920, 1.

49. Dick Brundidge to Mr. E. W. Nelson, January 26, 1920; D. G. Beauchamp to Chief, Bureau of Biological Survey, January 30, 1920; Lee Miles to Dr. E. W. Nelson, February 12, 1920, box 382, General Correspondence-Treaty Act-Arkansas, Bureau of Biological Survey, United States Fish and Wildlife Service, record group 22, National Archives, College Park, MD.

50. "Bad Luck to Have Game Wardens as His Guests," *Arkansas Democrat*, February 12, 1919, 3; "Waives Examination," *Daily Arkansas Gazette*, February 18, 1919, 7; "Held Under Game Law," *Daily Arkansas Gazette*, March 6, 1920, 11.

51. "Bureau Compiles More Statistics," *Daily Arkansas Gazette*, September 6, 1920, 3.

52. "A Problem to Face," *Batesville Daily Guard*, October 1, 1920, 1.

53. Arkansas Game and Fish Commission, *Third Bi-Ennial Report*, 8–9, 10, 17.

54. Arkansas Game and Fish Commission, *Third Bi-Ennial Report*, 19.

55. "Pot Hunters are Busy," *Arkansas Democrat*, November 26, 1920, 17; "Lee County to Ask Special Game Law," *Arkansas Democrat*, December 13, 1920, 3; Arkansas Game and Fish Commission, *Third Bi-Ennial Report*, 9.

CHAPTER NINE

1. Sam B. Hilliard, *Hog Meat and Hoecake: Food Supply in the Old South, 1840–1860* (Carbondale: Southern Illinois University Press, 1972), 77.

2. Arkansas Senate, *Journal of the Senate of Arkansas, Forty-Third Regular Session* (Little Rock: Southern Printing, 1921), 52–53; "Brother Don't Quit," *Arkansas Conservationist* 1, no. 4 (1925): 12.

3. "Arkansas Senate Passes a Bill to Abolish Pen Body," *Pine Bluff Daily Graphic*, January 29, 1921, 1; "McRae For Governor," *Southern Standard*, August 5, 1920, 3; "Says Arkansas Ring Politics are Shattered," *Hot Springs New Era*, August 20, 1920, 5. For more about Governor McRae, see W. David Baird, "Thomas Chipman McRae, 1921–1925," in Timothy P. Donovan, Willard B. Gatewood Jr., and Jeannie M. Whayne, eds., *The Governors of Arkansas* (Fayetteville: University of Arkansas Press, 1995), 157–64.

4. "Would Abolish State Game and Fish Commission," *Arkansas Democrat*, January 29, 1921, 1; "The Amende Honorable," *Hot Springs New Era*, February 18, 1921, 1.

5. Arkansas Senate, *Journal* (1921), 407–8; Arkansas General Assembly, *General Acts of the Forty-Third General Assembly of Arkansas* (Little Rock: Democrat Printing and Lithographing, 1921), 669; "State Spent 6 Millions During 1921," *Arkansas Democrat*, January 3, 1922, 1; Arkansas Game and Fish Commission, *Fourth Bi-Ennial Report* (Fort Smith: McBride-Calvert, 1922), 20–21. In 1921, the AGFC spent $26,015.87

for the calendar year. Between July 1, 1920, and June 30, 1921, the AGFC transferred $35,370.60 to the state treasury. Between July 1, 1921, and June 30, 1922, the AGFC transferred $45,027.90 to the state treasury. The AGFC spent $31,245.23 during the period July 1, 1921, and June 30, 1922.

6. Chief U.S. Game Warden to Hon. Lee Miles, April 7, 1921, box 433, Depredations to Migratory Bird Laws-Game Protection Misc. Correspondence 1919–1925, Bureau of Biological Survey, United States Fish and Wildlife Service, record group 22, National Archives, College Park, MD.

7. Arkansas Senate, *Journal* (1921), 407–8; Arkansas General Assebly, *General Acts* (1921), 209–10.

8. Arkansas Senate, *Journal* (1921), 471–72. Many Arkansawyers, including Joe T. Robinson, felt that federal forest rangers "abuse[d] their authority." Some citizens even argued that the federal government should abolish all national forests located in Arkansas. Joe T. Robinson to Hon. David Y. Thomas, July 7, 1911, Joseph Taylor Robinson Papers, 1900–1954, Special Collections Department, University of Arkansas Libraries, Fayetteville, AR.

9. Lee Miles to Henry C. Wallace, August 23, 1921, box 382, General Correspondence-Treaty Act-Arkansas, records of the Bureau of Biological Survey; Ray P. Holland, "The Public Shooting Ground—Refuge Bill," *Bulletin of the American Game Protective Association* 10 (July 1921): 1–5. Sportsmen Indiana senator Harry S. New had submitted S. B. 1452 and Kansas representative Dan R. Anthony Jr. introduced H. B. 5823 in spring 1921. Both bills were almost identical. Anthony's bill died in the House within a few months. When similar bills appeared in 1926, H. B. 7479 and S. B. 2607, the AGFC produced an educational sheet for distribution to the Arkansas public and publication in newspapers. See "Not Propaganda but Facts," Arkansas Game and Fish Commission, press release, 1926, General Correspondence-Treaty Act, box 382, records of the Bureau of Biological Survey.

10. United States House of Representatives, *Migratory Bird Refuges and Public Shooting Grounds: Hearings before the Committee on Agriculture, House of Representatives, February 16 and 17, 1922* (Washington, DC: Government Printing Office, 1922), 57–58; "Heard at Washington," *Bulletin of the American Game Protective Association*, April 2022, 15; "Sport For All," *New York Times*, March 19, 1922, 38; "Proposed Federal License to Hunt Birds," *Orlando Sentinel*, February 14, 1922, 3; "Public Shooting Grounds $5,000,000 Annually," *Bureau County Tribune*, February 24, 1922, 7; "Federal Hunting License Measure Probable," *Pittsburg Post-Gazette*, March 26, 1922, 29; "Looks Like Nothing Doing in Public Shooting Grounds," *Chicago Tribune*, February 19, 1923. The bills called for a one-dollar licensing fee to hunt migratory birds. Some reports claimed that the bills could affect five million hunters nationally. Opponents argued that it meant federal control of the game and fish, something they thought states should control. Although the New's bill passed the Senate, it failed in the House in February 1923 by a vote of 154 to 135.

11. "$30,000 More is Needed for Game Preserve," *Arkansas Democrat,* January 31, 1921, 2. Where the AGFC kept this money is unclear, or even if they had any.

12. "Organized Effort," *Arkansas Conservationist* 1, no. 2 (July/August/September 1924), 3–4; *Star Progress,* November 13, 1925, 2; "Big County Drive by Isaac Walton League," *Baxter Bulletin,* November 13, 1925, 1; "Arkansas Division of Izaak Walton League Is Founded November, 20th," *Arkansas Conservationist* 1, no. 4 (1925): 13. Although the Audubon Society was not established in Arkansas until much later, the state counted one hundred members of the Bird Lovers Society by 1900. This organization worked for the protection of birds. See "Bird Savers," *Arkansas Democrat,* July 7, 1900, 6.

13. "Arkansas Division of Izaak Walton League Is Founded November, 20th," *Arkansas Conservationist* 1, no. 4 (1925): 11. For more information on the Izaak Walton League, see the Izaak Walton League of America website, accessed May 6, 2020.

14. "Oppose Abolition of Game Commission," *Arkansas Democrat,* May 13, 1920, 1; "Want a Fish Hatchery," *Daily Arkansas Gazette,* July 21, 1920, 2; "To Enforce Game Laws," *Daily Arkansas Gazette,* September 22, 1921, 3; "Streams in Crawford County Stocked," *Arkansas Democrat,* November 13, 1920, 2; "Plan Sportsmen's Club For Jackson County," *Arkansas Democrat,* March 10, 1922, 3; *Daily Arkansas Gazette,* March 27, 1920, 4; "Name All the Birds," *Daily Arkansas Gazette,* January 22, 1922, 20. Several county groups provided information to the AGFC about illegal activity in their area. See "Game Wardens Finds and Destroys Nets," *Arkansas Democrat,* August 30, 1921, 6; "Game Warden War on Unlawful Seins," *Little Rock Daily News,* November 23, 1921, 5; "Game Laws Violated," *Daily Arkansas Gazette,* December 8, 1921, 6; *Arkansas Democrat,* January 8, 1922, 2. By 1923, the Arkadelphia Sportsmen's Club became the Clark County Game and Fish Protective Association. See "$25.00 Reward," *Southern Standard,* July 19, 1923, 6.

15. "Scout-O-Grams," *Star Progress,* April 17, 1925, 1.

16. "Sportsmen of State Elect Newport Man," *Arkansas Democrat,* July 30, 1920, 4; "Game Laws Violated," *Daily Arkansas Gazette,* December 8, 1921, 6.

17. *Arkansas Democrat,* March 5, 1921, 4; "Owens Bill Passed in the Senate," *Batesville Daily Guard,* March 5, 1921, 1; Arkansas Senate, *Journal* (1921), 976; "2nd Cigarette Bill," *Nashville News,* March 9, 1921, 1. Senate Bill No. 376 attempted to amend multiple sections of the 1917 and 1919 acts that created the commission. It passed the Senate but failed in the House. For more information about this bill, see Arkansas Senate, *Journal* (1921), 822.

18. "Newport Man Appointed By the Governor," *Batesville Daily Guard,* April 8, 1921, 1; "Vol Walker Renamed on Game Commission," *Fayetteville Daily Democrat,* April 8, 1921, 3; "The Game and Fish Will," *Arkansas Democrat,* April 14, 1921, 9; "Three Elected Game Wardens," *Arkansas Democrat,* April 16, 1921, 8; "Game Wardens Named," *Daily Arkansas Gazette,* April 17, 1921, 13; Arkansas Game and Fish Commission, *Fourth Bi-Ennial Report,* 20–21.

19. "Investigates Dynamiting," *Daily Arkansas Gazette,* June 2, 1921, 2; "Streams

Are Dynamited," *Daily Arkansas Gazette,* July 23, 1921, 2; "After Violators of Fishing Laws," *Little Rock Daily News,* August 4, 1921, 8; "Reward for Dynamiters," *Daily Arkansas Gazette,* July 27, 1921, 2; *Daily Arkansas Gazette,* September 28, 1921, 6.

20. "Replenish Our Game Supply," *Daily Arkansas Gazette,* July 4, 1921, 4; Arkansas Game and Fish Commission, *Fourth Bi-Ennial,* 8–10. When shooters fired high-powered rifles into the water, it presented the same stunning effect as dynamite, but in a smaller, more confined space. Neosho, Missouri, had a rainbow trout hatchery. Mississippi had a catfish hatchery. The United States government had a black bass fishery.

21. Lee Miles, "Additional Protection Needed for Wild Life, Which Still Abounds in Arkansas Forests," *Arkansas Democrat,* November 6, 1921, 11. By this time, Miles continued to serve on the Arkansas Game and Fish Commission but also was vice president of the International Association of Game, Fish, and Conservation and a member of the shooting regulation advisory board to the United States Biological Survey.

22. "Open Seasons for Game in 1922," *Forest and Stream,* September 1922, 404–5.

23. "Arkansas Sheriffs Plan More 'Dry' Enforcement," *Arkansas Democrat,* December 30, 1921, 1; "Sheriffs in Most Successful Session," *Little Rock Daily News,* December 31, 1921, 3.

24. "To Save Game in Flooded Sections," *Daily Arkansas Gazette,* May 31, 1922, 1.

25. "Game Laws Violated," *Daily Arkansas Gazette,* April 14, 1922, 7.

26. Arkansas Game and Fish Commission, *Fourth Bi-Ennial,* 7; "Did You Know," *Arkansas Conservationist* 1, no. 4 (1925), back cover; "Helena," *Arkansas Democrat,* June 19, 1922, 2; "Fish Are Dynamited in White River," *Arkansas Democrat,* June 23, 1922, 2; "Thousands of Fish Dead in Barber's Lake," *Arkansas Democrat,* July 20, 1922, 2; "$100 Reward is Offered," *Daily Arkansas Gazette,* July 22, 1922, 2.

27. Lee Miles, "Explains Work of Game Commission," *Daily Arkansas Gazette,* November 12, 1922, 12; Arkansas Games and Fish Commission, *Fourth Bi-Ennial Report,* 11. In 1922, the AGFC also created posters that included all of the current regulations. They posted them in businesses and law offices across the state.

28. J. N. Heiskell, "Valuable Work for Arkansas," *Daily Arkansas Gazette,* November 15, 1922, 6.

29. "Enforcing the Game Laws," *Arkansas Democrat,* November 27, 1922, 4; "Game Hogs Still Working," *Daily Arkansas Gazette,* November 22, 1922, 6.

30. "To Fight Game Hogs," *Daily Arkansas Gazette,* November 28, 1922, 2; "After Game Hogs," *Baxter Bulletin,* December 1, 1922, 1; "Farmers Protect Quail," *Daily Arkansas Gazette,* November 28, 1922, 2. For more about public conservation education, see Kevin C. Armitage, *The Nature Study Movement: The Forgotten Popularizer of America's Conservation Ethic* (Lawrence: University Press of Kansas, 2009).

31. "Many Nonresident Hunters Are Here," *Daily Arkansas Gazette,* November 26, 1922, 4; "RPH to George," December 28, 1922, box 385, General Correspondence-State Laws-Arkansas, records of the Bureau of Biological Survey; "Special Delivery to My dear Ray," December 30, 1922, box 385, General Correspondence-State Laws-Arkansas,

records of the Bureau of Biological Survey; Lee Miles to Mr. Geo. A. Lawyer, April 7, 1923, box 433, Depredations to Migratory Bird Laws-Game Protection Misc. Correspondence 1919–1925, records of the Bureau of Biological Survey. Ray P. Holland worked as a sportswriter, editor of *Field and Stream*, a United States game warden, and the editor of the AFGPA's Bulletin. As a US game warden, he became involved in the case *Missouri v. Holland* in 1919 over the constitutionality of the Migratory Bird Treaty of 1916 and the Migratory Bird Treaty Act of 1918. During on trip to Arkansas, Holland contracted typhoid fever and nearly died. Holland's papers are located at Wesleyan University in Connecticut.

32. Arkansas Game and Fish Commission, *Fourth Bi-Ennial Report*, 10–11.

33. Arkansas Game and Fish Commission, *Fourth Bi-Ennial Report*, 17.

34. *Arkansas Conservationist* 1, no. 4 (1925): 5.

35. "State Game Wardens Appeals Dismissal," *Hot Springs New Era*, January 5, 1923, 8; "Hunters Obtain Permits," *Daily Arkansas Gazette*, January 7, 1923, 9; "Fine License Violators," *Hot Springs New Era*, January 9, 1923, 8. Judge Vernal Ledgerwood assisted Hot Springs mayor Leo McLaughlin to operate a political machine in Garland County for almost twenty years (1927–1947). Many citizens considered Ledgerwood an expert fisherman.

36. "Paradise For Fishermen and Hunters in Arkansas," *Daily Arkansas Gazette*, January 18, 1923, 30, 38; "Better Enforcement of Laws and More Effective Work of Wardens Responsible, Commission Secretary Declares," *Arkansas Democrat*, February 11, 1923, 4.

37. Lee Miles to Geo. A. Lawyer, February 26, 1923, box 359, Game Protection-Regulations Arkansas, records of the Bureau of Biological Survey.

38. GAL to Hon. Lee Miles, March 3, 1923; Lee Miles to Geo. A. Lawyer, March 12, 1923, box 359, Game Protection-Regulations Arkansas, records of the Bureau of Biological Survey; "Conserving Our Game," *Arkansas Conservationist* 1, no. 4 (1925): 6, 15.

39. Lee Miles to Dr. E. W. Nelson, August 9, 1923; Chief of Bureau, Memorandum for Mr. Henderson, September 8, 1923, box 359, Game Protection-Regulations Arkansas, records of the Bureau of Biological Survey; GAL to Lee Miles, April 23, 1923, box 433, Depredations to Migratory Bird Laws-Game Protection Misc. Correspondence 1919–1925, records of the Bureau of Biological Survey.

40. Jno. W. Allen to W. C. Henderson, May 25, 1923, box 385, Misc. State Laws-General Correspondence (May 1922-Oct 1928) State Laws- Arkansas, records of the Bureau of Biological Survey.

41. "Ring-Neck Pheasant," *Arkansas Conservationist* 1, no. 2 (1924), 4; "To Raise Wild Turkey," *Baxter Bulletin*, October 2, 1915, 1.

42. *Hot Springs New Era*, March 10, 1923, 1.

43. "Fish," *Baxter Bulletin*, June 20, 1924, 2; "Fish and Game Conservation," *Baxter Bulletin*, December 12, 1924, 2; *Baxter Bulletin*, December 12, 1924, 4; *Mountain Echo*, November 27, 1924, 1; "Marion County Will Help," *Baxter Bulletin*, December

5, 1924, 1; Dell Brown, "The Basses of Arkansas," *Arkansas Conservationist* 1, no. 2 (1924): 10, 13.

44. "Fish and Game Conservation," *Baxter Bulletin*, December 12, 1924, 2.

45. "Fish and Game Conservation," *Baxter Bulletin*, December 12, 1924, 2; *Baxter Bulletin*, December 12, 1924, 4.

46. "What Pennsylvania Has Done, Arkansas Can Do," *Baxter Bulletin*, March 6, 1925, 1; *Arkansas Conservationist* 1, no. 4 (1925): 7, 15.

47. "Work on Fish Pond Started," *Baxter Bulletin*, May 15, 1915, 1; "To Wage War on Gar," *Baxter Bulletin*, September 25, 1915, 1.

48. *Arkansas Conservationist* 1, no. 2 (1924): 5.

49. "Brother Don't Quit," *Arkansas Conservationist* 1, no. 4 (1925): 13.

50. *The Deer* 1, no. 1 (April/May/June 1924); *Arkansas Conservationist* 1, no. 2, (1924); Arkansas Game and Fish Commission, *Arkansas Wildlife: A History* (Fayetteville: University of Arkansas Press, 1998), 131; "Arkansas," *American Game*, October 1925, 18. The last issue of the *Arkansas Conservationist* was published in 1931.

51. *Arkansas Conservationist* 1, no. 2, (1924): 11–12.

52. *Arkansas Conservationist* 1, no. 2, (1924): 13.

53. "McRae's Swan Song," *Osceola Times*, January 16, 1923, 2; Arkansas Senate, *Journal of the Senate of Arkansas, Forty-Fifth Regular Session* (Little Rock: Southern Printing, 1925), 39.

54. *Arkansas Conservationist* 1, no. 4 (1925): 5, 6.

55. Arkansas Senate, *Journal* (1925), 53–55. Terral had served as deputy state superintendent of public instruction from 1912 to 1916.

56. Arkansas Senate, *Journal* (1925), 53–55; Terral also had Ku Klux Klan support during his 1924 gubernatorial run. Although not his idea, during his term as governor, Terral oversaw the opening of the first state park on Petit Jean Mountain, south of Russellville. For more on Governor Terral, see W. David Baird, "Thomas Jefferson Terral, 1925–1927," in Donovan, Gatewood, and Whayne, *Governors of Arkansas*, 165–71.

57. E. V. Visart to Chief, Biological Survey, February 4, 1925, box 385, General Correspondence-State Laws-Arkansas, records of the Bureau of Biological Survey.

58. "Arkansas," *American Game*, October 1925, 18; "Local Legislation May Be Abolished by Supreme Court," *Osceola Times*, February 6, 1925, 1.

59. "Who Pays for Conserving Our Game and Fish," *Baxter Bulletin*, December 18, 1925, 1.

60. "Shall It Be Changed?," *Arkansas Conservationist* 1, no. 4 (1925): 11; Arkansas Game and Fish Commission, *Fifth Bi-Ennial Report* (Fort Smith: McBride-Calvert Publishing, 1924), 19.

61. Guy Amsler, "What We Are Doing," in Arkansas Game and Fish Commission, *Fifth Bi-Ennial Report*, 10.

62. Guy Amsler, "The Outlook," in Arkansas Game and Fish Commission, *Fifth Bi-Ennial Report*, 30.

EPILOGUE

1. Nell Taylor, *Report on Income and Disbursements-March 1915 Through June 1957*, Arkansas Game and Fish Commission, Little Rock, August 18, 1957); Arkansas Game and Fish Commission, *1928* and *1930 Biennial Report*; Keith Sutton, ed. *Arkansas Wildlife: A History* (Fayetteville: University of Arkansas Press, 1998), 95, 102. The AGFC continued to establish game propagation farms during the late 1930s and early 1940s with limited success. See "New Game Refuge," *Madison County Record*, February 17, 1944.

2. Taylor, *Report on Income*; Arkansas Game and Fish Commission, *1928* and *1930 Biennial Report*.

3. Taylor, *Report on Income*.

4. S. C. Dellinger, letter to the editor, *Northwest Arkansas Times*, November 4, 1944; "Mississippi County Sportsmen Asks You To Vote for Fish and Game Amendment No. 36," *Courier News*, November 6, 1944; *Baxter Bulletin*, January 12, 1945. By 1944, Arkansas's deer population remained low. They only had two six-day-long deer seasons annually. "Deer Season in Arkansas Opens Today," *Courier News*, November 13, 1944. The AGFC reported that only eleven thousand hunters purchased deer hunting licenses. The state contained 1.77 million people. *Northwest Arkansas Times*, December 7, 1944.

5. Sutton, *Arkansas Wildlife*, 199–200; "Reported Turkey Harvests," Arkansas Game and Fish Commission, Wildlife Management Division, Little Rock, 1996. Bear numbers reached enough by 1980 to allow a limited season, and bruin numbers have continued to grow. In the 1990s, the AGFC reintroduced elk near the Buffalo River.

6. United States Census of Agriculture, vol. 1, 1950, 3; "History," Butler Center for Arkansas Studies website, accessed April 8, 2023. Currently, Arkansas contains about 14.3 million acres of agricultural land and 18.8 million of forests.

7. Sutton, *Arkansas Wildlife*, 245–46.

8. "Arkansas Feral Hog Eradication Task Force," Arkansas Department of Agriculture website, accessed April 8, 2023.

9. Paul Greenberg, "Board Chairman Runs Up Expenses," *Baxter Bulletin*, June 30, 1990; "Former Member of the AGFC Say 3 on Panel Control It, *Baxter Bulletin*, August 23, 2010; Bryan Hendricks, "Other 'Influencers' Than Internet Have AGFC's Ear," *Arkansas Democrat-Gazette*, March 12, 2023.

APPENDIX A

1. Excerpted from Arkansas Game and Fish Commission, *First Annual Report*, (Fort Smith: McBride-Calvert, 1916), 7–8.

APPENDIX B

1. Arkansas Game and Fish Commission, *First Annual Report* (Fort Smith: McBride-Calvert, 1916), 4.

2. Arkansas Game and Fish Commission, *Second Annual Report* (Fort Smith:

McBride-Calvert, 1918), 2. Notice the list contains "officers" starting in 1917 and that allows an unofficial chief game warden and another member of the commission without going over the five commissioners or eight game warden limits that the law sets.

3. Arkansas Game and Fish Commission, *Third Bi-Ennial Report* (Fort Smith: McBride-Calvert, 1920), 2.

4. Arkansas Game and Fish Commission, *Fourth B-Ennial Report* (Fort Smith: McBride-Calvert, 1922), 2.

5. Arkansas Secretary of State, *Arkansas: Biennial Report of the Secretary of State for the Years 1921–1924* (Little Rock: H. L. Pugh, 1924), 53.

6. Arkansas Secretary of State, *Arkansas Biennial Report of the Secretary of State for the Years 1925–1926* (Little Rock: H. L. Pugh, 1926), 98.

Bibliography

PRIMARY BOOKS AND PUBLICATIONS

American Game Protective and Propagation Association: The Story of its Origin, Purposes, History and Policies. New York, 1915.

American Poland-China Record Association. *American Poland-China Record.* Vol 73. Columbia, MO: E. W. Stephens, 1920.

Black, Sam. *A Soldier's Recollections of the Civil War.* Printed by the *Minco Minstrel.* Minco, OK, 1912.

Cleveland, Grover. *Fishing and Shooting Sketches.* New York: Outing, 1906.

Dickinson, Samuel, ed. *New Travels in North America by Jean-Bernard Bossu, 1770–1771.* Natchitoches: Northwestern State University, 1982.

Donaghey, George W. *Autobiography of George W. Donaghey.* Benton, AR: L. B. White, 1939.

Elliott, William. *Carolina Sport by Land and Water: Including Incidents of Devil-Fishing, Wildcat, Deer, and Bear Hunting, etc.* 1846. Reprint, Columbia: University of South Carolina Press, 1994.

Etter, Patricia A., ed. *An American Odyssey: The Autobiography of a 19th-Century Scotsman, Robert Brownlee, at the Request of His Children, Napa County, California, October 1892.* Fayetteville: University of Arkansas Press, 1986.

Feathers and Fins on the Frisco. Issued by passenger department, St. Louis & San Francisco Railroad. St. Louis, 1900.

Featherstonhaugh, George W. *Excursion through the Slave States: From Washington on the Potomac to the Frontier of Mexico.* London: J. Murray, 1844.

Federal Writers' Project. *Slave Narratives: A Folk History of Slavery in the United States from Interviews with Former Slaves.* Typewritten records prepared by the Federal Writers' Project, 1936–1938; assembled by the Library of Congress project Work Projects Administration for the District of Columbia. Vol. 2, *Arkansas Narratives.* Washington, DC: 1941.

The Field Dog Stud Book. Chicago: American Field Publishing, 1904.

The Field Dog Stud Book. Chicago: American Field Publishing, 1919.

The Field Dog Stud Book. Chicago: American Field Publishing, 1922.

Flint, Timothy. "Journal of the Rev Timothy Flint from the Red River to the

Ouachita or Washita in Louisiana in 1835," in *The Select Circulating Library*, vol. 7, part 1, 284–88. Philadelphia: Adam Waldie, 1836.

———. *Recollections of the Last Ten Years, Passed in Occasional Residences and Journeying in the Valley of the Mississippi*. 1826. Reprint, New York: Da Capo, 1968.

Forbush, Edward Howe. *A History of Gamebirds, Wildfowl, and Shore Birds*. Issued by the Massachusetts State Board of Agriculture. Boston: Wright and Potter, 1912.

Gaskins, John. *Life and Adventures of John Gaskins in the Early History of Northwest Arkansas*. Eureka Springs: Printed and for sale by the author, 1893.

Gerstäcker, Friedrich. *Wild Sports in the Far West*. Boston: Crosby, Nichols, 1859.

Harris, William, ed. *The Sportsman's Guide to the Hunting and Shooting Grounds of the United States and Canada*. New York: Angler's Publishing, 1888.

How to Insure the People of New York An Abundance of Fish and Game. Pamphlet issued by New York State Conservation Department. Albany, NY, 1912.

Hundley, Daniel Robinson. *Social Relations in Our Southern States*. New York: H. B. Price, 1860.

Huntington, Dwight W. *Game Farming for Profit and Pleasure*. Wilmington, DE: Hercules Powder, 1914.

Ideal Hunting and Fishing Grounds. Pamphlet issued by H. C. Townsend, general passenger agent, Missouri Pacific Railway Co., Iron Mountain Route. 1900.

Jacobson, Charles. *The Life Story of Jeff Davis: The Stormy Petrel of Arkansas Politics*. Little Rock: Parke-Harper, 1925.

Keefe, James F., and Lynn Morrow, eds. *The White River Chronicles of S. C. Turnbo: Man and Wildlife on the Ozarks Frontier*. Fayetteville: University of Arkansas Press, 1994.

Lankford, George E, and Federal Writers' Project. *Bearing Witness: Memories of Arkansas Slavery; Narratives from the 1930s WPA Collections*. Fayetteville: University of Arkansas Press, 2003.

Miller, James William, and Friedrich Gerstäcker, eds. *In the Arkansas Backwoods: Tales and Sketches*. Columbia: University of Missouri Press, 1990.

National Science Club. *Proceedings of the National Science Club*. Washington, DC: Judd and Detweiler, 1897.

Nichols, Charles S. *With Rod and Gun in Arkansas*. Chicago: Chicago & Eastern Railroad, 1900.

Nuttall, Thomas. *A Journal of Travels into the Arkansas Territory during the Year 1819*. Philadelphia: Thomas H. Palmer, 1821.

Rafferty, Milton D., ed. *Rude Pursuits and Rugged Peaks: School Craft's Ozark Journal, 1818–1819*. Fayetteville: University of Arkansas Press, 1996.

Rawick, George P., ed. *The American Slave: A Composite Autobiography*. 19 vols. Westport, CT: Greenwood, 1972–1979.

Reynolds, Charles B. *The Game Laws in Brief: Laws of the United States and Canada*

Relating to Game and Fish Seasons; For the Guidance of Sportsmen and Anglers. New York: Forest and Stream, 1894.

———. *The Game Laws in Brief: Laws of the United States and Canada Relating to Game and Fish Seasons; For the Guidance of Sportsmen and Anglers.* New York: Forest and Stream, 1896.

———. *The Game Laws in Brief: Laws of the United States and Canada Relating to Game and Fish Seasons; For the Guidance of Sportsmen and Anglers.* New York: Forest and Stream, 1904.

———. *The Game Laws in Brief: Laws of the United States and Canada Relating to Game and Fish Seasons; For the Guidance of Sportsmen and Anglers.* New York: Forest and Stream, 1905.

———. *The Game Laws in Brief: Laws of the United States and Canada Relating to Game and Fish Seasons; For the Guidance of Sportsmen and Anglers.* New York: Forest and Stream, 1911.

Rudd, Dan A., and Theo. Bond. *From Slavery to Wealth: The Life of Scott Bond; the Rewards of Honesty, Industry, Economy and Perseverance.* Madison, AR: Journal Printing, 1917.

Shea, John G., ed. *Discovery and Exploration of the Mississippi Valley.* Albany, NY: Joseph McDonagh, 1903.

Shewey, Arista C. *Shewey's Guide and Map to the Happy Hunting Grounds of Missouri and Arkansas: The Paradise of the Hunter and Fisherman.* St. Louis: Shewey, 1892.

The Sportsman's Paradise on the Buffalo Island Route, Arkansas. St. Louis: Woodward and Tiernan, 1897.

Stockard, Sallie W. *The History of Lawrence, Jackson, Independence, and Stone Counties of the Third Judicial District of Arkansas.* Little Rock: Arkansas Democrat, 1904.

Where to Hunt American Game. Lowell, MA: United States Cartridge Company, 1898.

Wildwood, Will. *The Sportsman's Directory and Year Book.* Milwaukee: Pond and Goldeye, 1892.

PRIMARY ARTICLES

Atkinson, J. H., and Charles McDermott. "A Memoir of Charles McDermott: A Pioneer of Southeastern Arkansas." *Arkansas Historical Quarterly* 12, no. 3 (Autumn 1953): 253–261.

Baugh, Stanley T. "James and Julia Baugh." *Arkansas Historical Quarterly* 10, no. 2 (Summer 1951): 182–209.

Billingsley, John, and Ted R. Worley, eds. "Letter from an Early Arkansas Settler." *Arkansas Historical Quarterly* 11, no. 4 (Winter 1952): 327–30.

Burnham, John B. "Association and Magazine Combine in Interest of More Game." *Bulletin of the American Game Protective Association* 6, no. 4 (October 1917): 1–8.

"Calendar of Coming Meetings." *Good Roads* 14, no. 6 (1917): 78.

Eison, James Reed, and J. E. Lindsay. "A Letter from Dardanelle to Jonesville, South Carolina." *Arkansas Historical Quarterly* 28, no. 1 (Spring 1969): 72–75.

Holland, Ray P. "The Public Shooting Ground—Refuge Bill." *Bulletin of the American Game Protective Association* 10 (July 1921): 1–5.

Huntington, John. "The Game Conservation Institute." *American Game*, April/May 1929, 48–49.

Jansma, Jerome, and Harriet H. Jansma. "George Engelmann in Arkansas Territory." *Arkansas Historical Quarterly* 50, no. 3 (Autumn 1991): 225–48.

La Harpe, Jean-Baptiste Bénard de. "Exploration of the Arkansas River by Benard de la Harpe, 1721–1722: Extracts from His Journal and Instructions." Edited and translated by Ralph A. Smith. *Arkansas Historical Quarterly* 10, no. 4 (Winter 1951): 349–60.

Shaw, John. "Shaw's Narrative." In *Collections of the State Historical Society of Wisconsin*, vol. 2, edited by Lyman C. Draper, 199–202. Madison: State Historical Society of Wisconsin, 1903.

Slaughter, Lon, and E. D. Wall. "Reminiscences of the Late Lon Slaughter as Told to Mrs. E. D. Wall, September 3, 1920." *Arkansas Historical Quarterly* 8, no. 2 (Summer 1949): 167–69.

"Soil and Climate of the American Territories: Translation of a Letter from an Enlightened French Immigrant." In *The Port-Folio*, vol. 4, no. 4 (Philadelphia: H. Maxwell, 1817).

"T. Gilbert Pearson." *Game Breeder* 18, no. 3 (December 1920): 87–88.

West, Mabel. "Jacksonport, Arkansas: Its Rise and Decline." *Arkansas Historical Quarterly* 9, no. 4 (Winter 1950): 231–58.

Wetzel, Nat. "Cunning Wiles or the Hunters and the Hunted." *Texas Game and Fish* 11 (February 1953): 8–9, 29–30.

Wolf, John Quincy, and Charles Heinrich. "Journal of Charles Heinrich, 1849–1856." *Arkansas Historical Quarterly* 24, no. 3 (Autumn 1965): 241–83.

COLLECTIONS

Brough, Charles Hillman. Papers, 1895–1935. Special Collections Department, University of Arkansas Libraries, Fayetteville, AR.

Davis, Jeff. Papers, 1849–1986. Special Collections Department, University of Arkansas Libraries, Fayetteville, AR.

Dellinger, Samuel C. Papers, 1789–1970s. Special Collections Department, University of Arkansas Libraries, Fayetteville, AR.

Donaghey, George Washington. Miscellaneous materials, 1856–1947. Special Collections Department, University of Arkansas Libraries, Fayetteville, AR.

Hays, George W. Papers, 1913–1917. Arkansas State Archives, Little Rock, AR.

Heiskell, J. N. Papers, 1829–1973. Center for Arkansas History and Culture. University of Arkansas, Little Rock, AR.

Hershey, Charley. Diary, 1883–1884. Joplin Missouri Historical Society.

Hilliard, (Mrs. Isaac H.). Diary, 1849–1860. Isaac H. Hilliard Family Papers, Louisiana and Lower Mississippi Valley Collection, Louisiana State University Libraries, Baton Rouge, LA.

Holland, Ray P. Papers, 1872–1974, Special Collections and Archives, Wesleyan University, Middletown, CT.

Hornaday, William T. Papers, 1888–1937. Wildlife Conservation Society Archives, New York, NY.

Luttig, John C. Letter, 1815. Special Collections Department, University of Arkansas Libraries, Fayetteville, AR.

McIlhenny, Edward Avery. Papers, 1911–1947. Louisiana and Lower Mississippi Valley Collections, Louisiana State University Libraries, Baton Rouge, LA.

Point Gun Club Records, 1892–1911. Center for Arkansas History and Culture, University of Arkansas, Little Rock, AR.

Robinson, Joseph Taylor. Papers, 1900–1954. Special Collections Department, University of Arkansas Libraries, Fayetteville, AR.

GOVERNMENT RESOURCES AND PUBLICATIONS

Arkansas Game and Fish Commission. Annual reports. Fort Smith: Calvert-McBride, 1916–1930.

———. *Arkansas Conservationist.* Arkansas Game and Fish Commission publication, 1924–1931.

Arkansas General Assembly. *Acts Passed at the Tenth Session of the General Assembly of the State of Arkansas.* Little Rock: Johnson and Yerkes, 1855.

———. *Acts, Resolutions, and Memorials of the General Assembly of the State of Arkansas.* Little Rock: William E. Woodruff, Jr., 1875.

———. *Acts, Resolutions, and Memorials of the General Assembly of the State of Arkansas.* Little Rock: William E. Woodruff, Jr., 1879.

———. *Acts and Resolutions of the General Assembly of the State of Arkansas.* Little Rock: Gazette Printing, 1887.

———. *Acts and Resolutions of the General Assembly of the State of Arkansas.* Little Rock: Gazette Printing, 1889.

———. *General Acts of the Forty-Third General Assembly of Arkansas.* Little Rock: Democrat Printing and Lithographing, 1921.

———. *Public and Private Acts and Joint and Concurrent Resolutions and Memorials of the General Assembly of the State of Arkansas.* Little Rock: Democrat Print, 1907.

———. *Public and Private Acts and Joint and Concurrent Resolutions and Memorials of the General Assembly of the State of Arkansas.* Little Rock: Democrat Print, 1909.

Arkansas House of Representatives. *Journal of the House of Representatives of Arkansas, Thirty-Ninth Regular Session.* Little Rock: Southern Press, 1913.

———. *Journal of the House of Representatives of Arkansas, Fortieth Regular Session.* Little Rock: Southern Press, 1915.

———. *Journal of the House of Representatives of Arkansas, Forty-First Regular Session.* Little Rock: Southern Printing, 1917.

Arkansas Secretary of State. *Arkansas: Biennial Report of the Secretary of the State for the Years 1903–1904.* Little Rock: Tunnah and Pittard, 1905.

———. *Arkansas: Biennial Report of the Secretary of State for the Years 1921–1924.* Little Rock: H. L. Pugh, 1924.

———. *Arkansas Biennial Report of the Secretary of State for the Years 1925–1926.* Little Rock: H. L. Pugh, 1926.

Arkansas Senate. *Journal for the Senate of Arkansas, Thirty-Fourth Regular Session.* Little Rock: Arkansas Democrat, 1903.

———. *Journal of the Senate of Arkansas, Thirty-Sixth Regular Session.* Hot Springs: Sentinel-Record, 1907.

———. *Journal of the Senate of Arkansas, Thirty-Seventh Session.* Little Rock: Tunnah and Pittard, 1909.

———. *Journal of the Senate of Arkansas, Thirty-Ninth Regular Session.* Little Rock: Southern Press, 1913.

———. *Journal of the Senate of Arkansas, Fortieth Regular Session.* Little Rock: Southern Press, 1915.

———. *Journal of the Senate of Arkansas, Forty-First Regular Session.* Little Rock: Southern Printing, 1917.

———. *Journal of the Senate of Arkansas, Forty-Second Regular Session.* Little Rock: Southern Printing, 1919.

———. *Journal of the Senate of Arkansas, Forty-Third Regular Session.* Little Rock: Southern Printing, 1921.

———. *Journal of the Senate of Arkansas, Forty-Fifth Regular Session.* Little Rock: Southern Printing, 1925.

Bar Association of Arkansas. *Report of the Proceedings of the Bar Association of Arkansas.* Little Rock, 1932.

Bureau of Biological Survey, United States Fish and Wildlife Service. Record group 22. National Archives, College Park, MD.

Castle, John. *Supplement to Kirby's Digest of the Statues of Arkansas.* Indianapolis: Bobbs-Merrill, 1911.

Catton, Theodore. *Life, Leisure, and Hardship Along the Buffalo.* Historic resource study of the Buffalo National River. National Park Service, Midwest Region, Omaha, NE, 2008.

City of Texarkana v. Hudgins Produce Co., 112 Ark. 17, 163–64 S.W. 737 (1914).

Colcord, Mabel, Ina L. Hawes, and Angelina Carabelli, compilers. *Check List of Publications on Entomology.* United States Department of Agriculture Library, vol. 20. Washington, DC, January 1930.

Daniels, Charlie. *Historical Report of the State of Arkansas.* Fayetteville: University of Arkansas Press, 2008.

Hempstead, Samuel H., and Elbert Hartwell English. *A Digest of the Statutes of Arkansas, Embracing All Laws of a General and Permanent Character in Force at the Close of the Session of the General Assembly of 1846: Together with Notes of the Decisions of the Supreme Court upon the Statutes.* Little Rock: Rearden and Garritt, 1848.

Holmes, George K. *Wages of Farm Labor.* United States Department of Agriculture, Bureau of Statistics, Bulletin 99. Washington, DC: Government Printing Office, 1912.

Interstate Commerce Act. 1887. Milestone Documents, National Archives (online), accessed May 10, 2022.

J. N. Adams, Jake River, and J. K. McMasters v. Jonesboro, Lake City and Eastern Railroad, No. 242, Ark. 2nd Circuit. Pleas before Honorable W. J. Driver, Judge, January 29, 1914. William H. Bowen School of Law Library, University of Arkansas, Little Rock.

Jonesboro, Lake City and Eastern Railroad v. J. N. Adams et al., 117 Ark. 54, 174 S.W. 527 (1915). Abstract and briefs, case 3306, December 3, 1914. William H. Bowen School of Law Library, University of Arkansas, Little Rock.

Lewis v. State, 110 Ark. 204, 161 S.W. 154 (1913).

Palmer, Theodore S. Circular for the Bureau of Biological Survey, United States Department of Agriculture, August 22, 1914.

———. "Federal Game Protection: A Five-year Retrospect." Reprint from the *Yearbook of Department of Agriculture for 1905*, 546–55. Washington, DC: Government Printing Office, 1905.

———. *Hunting Licenses: Their History, Objects, and Limitations.* Washington, DC: Government Printing Office, 1904.

Palmer Theodore S., and H. W. Olds. *Digest of Game Laws for 1901.* Washington, DC: Government Printing Office, 1901.

Palmer Theodore S., and H. W. Olds. *Laws Regulating the Transportation and Sale of Game.* United States Department of Agriculture Bulletin No. 14. Washington, DC: Government Printing Office, 1900.

"Reported Turkey Harvests." Arkansas Game and Fish Commission, Wildlife Management Division. Little Rock, 1996.

Sandels, Leonidas P., and Joseph M. Hill, eds. *A Digest of the Statutes of Arkansas Embracing All Laws of a General Nature in Force at the Close of the Session of the General Assembly of One Thousand Eight Hundred and Ninety-Three.* Columbia, MO: E. W. Stephens, 1894.

South Dakota Department of Game and Fish. *Annual Report.* Sioux Falls: Mark D. Scott, 1910.

Steele, J. and James M. Campbell, eds. *Laws of Arkansas Territory.* Little Rock: J. Steel, 1835.

State v. Stokes, 117 Ark. 192 (1915).

Taylor, Nell. *Report on Income and Disbursements-March 1915 Through June 1957*. Arkansas Game and Fish Commission, Little Rock, August 18, 1957.

United Mine Workers v. Coronado Coal Co., 259 U.S. 344 (1922). Transcript of record, Supreme Court of the United States, October term 1921, vol. 1.

United States Census Bureau. Eleventh Census of the United States. 1890.

———. Twelfth Census of the United States. 1900.

———. Thirteenth Census of the United States. 1910.

———. Fourteenth Census of the United States. 1920.

———. Fifteenth Census of the United States. 1930.

———. United States Census of Agriculture, vol. 1, part 23. 1950.

United States Department of Agriculture. *The Migratory Bird Treaty: The Decision of the Supreme Court of the United States Sustaining the Constitutionality of the Migratory Bird Treaty and Act of Congress to Carry It into Effect*. Department Circular 102. Washington, DC, 1920.

———. *The Official Records of the United States Department of Agriculture* 5, no. 1. Washington, DC: Government Printing Office, 1926.

United States Department of Agriculture, Bureau of Biological Survey. *Migratory-Bird Treaty-Act Regulations and Text of Federal Laws Relating to Game and Birds*. 1930.

United States Fish and Wildlife Service. National Archives Records Administration, RG 22.

———. *Officials and Organizations Concerned with Wildlife Protection*. Washington, DC: Government Printing Office, 1914.

United States House of Representatives. *Migratory Bird Refuges and Public Shooting Grounds: Hearings before the Committee on Agriculture, House of Representatives, February 16 and 17, 1922*. Washington, DC: Government Printing Office, 1922.

SECONDARY BOOKS

Addler, Dennis. *Winchester Shotguns*. New York: Simon and Schuster, 2015.

Arkansas Game and Fish Commission. *Arkansas Wildlife: A History*. Fayetteville: University of Arkansas Press, 1998.

Armitage, Kevin C. *The Nature Study Movement: The Forgotten Popularizer of America's Conservation Ethic*. Lawrence: University Press of Kansas, 2009.

Arnold, Morris S. *The Arkansas Post of Louisiana*. Fayetteville: University of Arkansas Press, 2017.

———. *The Rumble of a Distant Drum: The Quapaws and Old World Newcomers, 1673–1804*. Fayetteville: University of Arkansas Press, 2000.

———. *Unequal Laws unto a Savage Race: European Legal Traditions in Arkansas, 1686–1836*. Fayetteville: University of Arkansas Press, 1985.

Arsenault, Raymond. *The Wild Ass of the Ozarks: Jeff Davis and the Social Bases of Southern Politics*. Knoxville: University of Tennessee Press, 1988.

Bernard, Shane K. *Tabasco: An Illustrated History*. Avery Island, LA: McIlhenny, 2007.

Blassingame, John W. *The Slave Community: Plantation Life in the Antebellum South*. New York: Oxford University Press, 1979.

Blevins, Brooks. *Arkansas/Arkansaw: How Bear Hunters, Hillbillies, and Good Ol' Boys Defined a State*. Fayetteville: University of Arkansas Press, 2009.

Bolton, S. Charles. *Arkansas, 1800–1860: Remote and Restless*. Fayetteville: University of Arkansas Press, 1998.

Brock, Julia, and Daniel Vivian, eds. *Leisure, Plantations, and the Making of a New South: The Sporting Plantations of the South Carolina Lowcountry and Red Hills Region, 1900–1940*. New York: Lexington Books, 2015.

Bruce, Dickson D. *Violence and Culture in the Antebellum South*. Austin: University of Texas Press, 1997.

Callicott, Baird, ed. *Companion to a Sand County Almanac: Interpretive and Critical Essays*. Madison: University of Wisconsin Press, 1987.

Cronon, William. *Nature's Metropolis: Chicago and the Great West*. New York: W. W. Norton, 1991.

Crosby, Alfred. *The Columbian Exchange*. Westport, CT: Greenwood, 1973.

Donovan, Timothy P., Willard B. Gatewood Jr., and Jeannie M. Whayne, eds. *The Governors of Arkansas*. Fayetteville: University of Arkansas Press, 1995.

Dorman, Robert. *A Word for Nature: Four Pioneering Environmental Advocates*. Chapel Hill: University of North Carolina Press, 1998.

Dorsey, Kurkpatrick. *The Dawn of Conservation Diplomacy: U.S.-Canadian Wildlife Protection Treaties in the Progressive Era*. Seattle: University of Washington Press, 2009.

Dunlap, Thomas R. *Faith in Nature: Environmentalism as Religious Quest*. Seattle: University of Washington Press, 2004.

———. *Saving America's Wildlife: Ecology and the American Mind, 1850–1990*. Princeton, NJ: Princeton University Press, 1988.

DuVal, Kathleen. *The Native Ground: Indians and Colonists in the Heart of a Continent*. Philadelphia: University of Pennsylvania Press, 2007.

Fox, Stephen. *The American Conservation Movement: John Muir and His Legacy*. Boston: Little, Brown, 1981.

Garone, Philip. *The Fall and Rise of the Wetlands of California's Great Central Valley*. Berkeley: University of California Press, 2020.

Genovese, Eugene D. *Roll, Jordan, Roll: The World the Slaves Made*. New York: Pantheon Books, 1976.

Giltner, Scott E. *Hunting and Fishing in the New South: Black Labor and White Leisure after the Civil War*. Baltimore: Johns Hopkins University Press, 2008.

Glenn, John, Lillian Brandt, and F. Emerson Andrews. *Russell Sage Foundation 1907–1946*. New York: Russell Sage Foundation, 1948.

Gorn, Elliot, and Warren Goldstein. *A Brief History of American Sports.* New York: Hill and Wang, 1993.

Grinnell, George Bird, and Charles Sheldon, eds. *Hunting and Conservation.* New York: Arno and the *New York Times,* 1925.

Hahn, Steven. *A Nation under Our Feet: Black Political Struggles in the Rural South from Slavery to the Great Migration.* Cambridge: Harvard University Press, 2003.

Hahn, Steven, and Jonathan Prude, eds. *The Countryside in the Age of Capitalist Transformation: Essays in the Social History of Rural America.* Chapel Hill: University of North Carolina Press, 1985.

Haynes, Samuel P. *The Aquatic Frontier: Oysters and Aquaculture in the Progressive Era.* Amherst: University of Massachusetts Press, 2019.

Hays, Samuel P. *Conservation, and the Gospel of Efficiency: The Progressive Conservation Movement, 1890–1920.* Pittsburgh: University of Pittsburgh Press, 1999.

Hempstead, Fay. *Historical Review of Arkansas: Its Commerce, Industry, and Modern Affairs.* Chicago: Lewis, 1911.

Herman, Daniel. *Hunting in the American Imagination.* Washington, DC: Smithsonian Institution Press, 2001.

Herndon, Dallas Tabor. *Centennial History of Arkansas.* Chicago: S. J. Clarke, 1922.

Higgins, Billy D. *A Stranger and a Sojourner: Peter Caulder, Free Black Frontiersman in Antebellum Arkansas.* Fayetteville: University of Arkansas Press, 2005.

Hilliard, Sam B. *Hog Meat and Hoecake: Food Supply in the Old South, 1840–1860.* Carbondale: Southern Illinois University Press, 1972.

Hornady, William T. *Our Vanishing Wildlife: Its Extermination and Preservation.* New York: C. Scribner's Son, 1913.

Huddleston, Duane, Sammie Rose, and Pat Wood. *Steamboats and Ferries on the White River.* Fayetteville: University of Arkansas Press, 1998.

Jack, Zachary Michael, ed. *The Green Roosevelt: Theodore Roosevelt in Appreciation of Wilderness, Wildlife, and Wild Places.* New York: Cambria, 2010.

Jacoby, Karl. *Crimes against Nature: Squatters, Poachers, Thieves, and the Hidden History of Conservation.* Berkeley: University of California Press, 2001.

Jones, Kelly Houston. *A Weary Land: Slavery on the Ground in Arkansas.* Athens: University of Georgia Press, 2021.

Kelly, Tara K. *The Hunter Elite: Manly Sport, Hunting Narratives, and American Conservation, 1880–1925.* Lawrence: University of Kansas Press, 2018.

Kirby, Robert L. *Kirby Smith's Confederacy: The Trans-Mississippi South, 1863–1865.* Tuscaloosa: University of Alabama Press, 1972.

Kurlansky, Mark. *Cod: A Biography of the Fish that Changed the World.* New York: Walker, 1997.

Kwas, Mary L. *A Pictorial History of Arkansas's Old State House.* Fayetteville: University of Arkansas Press, 2010.

Lancaster, Guy, ed. *Bullets and Fire: Lynching and Authority in Arkansas, 1840–1950.* Fayetteville: University of Arkansas Press, 2018.

Laycock, George. *The Alien Animals: The Story of Imported Wildlife.* Garden City, NY: Natural History Press, 1970.

Ledbetter, Cal. *Carpenter from Conway: George Washington Donaghey as Governor of Arkansas, 1909–1913.* Fayetteville: University of Arkansas Press, 1993.

Marks, Stuart A. *Southern Hunting in Black and White: Nature, History, and Ritual in a Carolina Community.* Princeton, NJ: Princeton University Press, 1991.

Menna, Larry K., and Thomas L. Altherr. *Sports in North America: The Origins of Modern \Sports, 1820–1840.* Ann Arbor: University of Michigan Press, 1995.

Merchant, Carolyn. *Spare the Birds! George Bird Grinnell and the First Audubon Society.* New Haven, CT: Yale University Press, 2016.

Minard Peter M. *All Things Harmless, Useful, and Ornamental: Environmental Transformation through Species Acclimatization, from Colonial Australia to the World.* Chapel Hill: University of North Carolina Press, 2019.

Mitman, Gregg. *Landscapes of Exposure: Knowledge and Illness in Modern Environments.* Chicago: University of Chicago Press, 2004.

Morris, Christopher. *The Big Muddy: An Environmental History of the Mississippi and Its Peoples from Hernando De Soto to Hurricane Katrina.* Oxford: Oxford University Press, 2012.

Nash, Linda. *Inescapable Ecologies: A History of Environment, Disease, and Knowledge.* Berkeley: University of California Press, 2006.

Ossa, Helen. *They Saved Our Birds: The Battle Won and the War to Win.* New York: Hippocrene Books, 1973.

Ownby, Ted. *Subduing Satan: Religion, Recreation, and Manhood in the Rural South, 1865–1920.* Chapel Hill: University of North Carolina Press, 1990.

Patrick, Charles, and Joseph Mooney, eds. *The Mid-South and Its Builders: Being the Story of the Development and a Forecast of the Future of the Richest Agricultural Region in the World.* Memphis, TN: Mid-South Biographic and Historical Association, 1920.

Patterson, Ruth Polk. *The Seed of Sally Good'n: A Black Family of Arkansas, 1833–1953.* Lexington: University Press of Kentucky, 1985.

Perkins, J. Blake. *Hillbilly Hellraisers: Federal Power and Popular Defiance in the Ozarks.* Chicago: University of Illinois Press, 2017.

Petersen, David. *Heartsblood: Hunting, Spirituality, and Wildness in America.* Washington, DC: Island, 2000.

Proctor, Nicholas. *Bathed in Blood: Hunting and Mastery in the Old South.* Charlottesville: University of Virginia Press, 2002.

Ragsdale, William O. *They Sought the Land: A Settlement in the Arkansas River Valley 1840–1870.* Fayetteville: University of Arkansas Press, 1997.

Reardon, Erik. *Managing the River Commons: Fishing and New England's Rural Economy.* Amherst: University of Massachusetts Press, 2021.

Reiger, John F. *American Sportsmen and the Origins of Conservation*. Corvallis: Oregon University Press, 2001.

Rickels, Milton. *Thomas Bangs Thorpe: Humorist of the Old Southwest*. Baton Rouge: Louisiana State University Press, 1962.

Roper, James. *The Founding of Memphis, 1818–1820*. Memphis: West Tennessee Historical Society, 1970.

Rothman, Joshua D. *Flush Times and Fever Dreams: A Story of Capitalism and Slavery in the Age of Jackson*. Athens: University of Georgia Press, 2012.

Saikku, Mikko. *This Delta, This Land: An Environmental History of the Yazoo-Mississippi Floodplain*. Athens: University of Georgia Press, 2005.

Shrock, Joel. *The Gilded Age*. Westport, CT: Greenwood, 2004.

Sloan, W. David, and Lisa Mullikin Parcell, eds. *American Journalism: History, Principles, Practices*. Jefferson, NC: McFarland, 2002.

Speer, William. *Sketches of Prominent Tennesseans*. 1888. Reprint, Baltimore: Genealogical Publishing, 2010.

Spence, Mark David. *Dispossessing the Wilderness: Indian Removal and the Making of the National Parks*. New York: Oxford University Press, 1999.

Stradling, David, ed. *Conservation in the Progressive Era*. Seattle: University of Washington Press, 2004.

Sutton, Keith, ed. *Arkansas Wildlife: A History*. Fayetteville: University of Arkansas Press, 1998.

Taylor, Dorceta E. *The Rise of the American Conservation Movement: Power, Privilege, and Environmental Protection*. Durham, NC: Duke University Press, 2016.

Taylor, Orville W. *Negro Slavery in Arkansas*. Durham, NC: Duke University Press, 1958.

Thomas, E. Donnal, Jr. *How Sportsmen Saved the World: The Unsung Conservation Efforts of Hunters and Anglers*. Guilford, CT: Lyon, 2010.

Trefethen, James B. *An American Crusade for Wildlife*. New York: Winchester Press; Boone and Crockett Club, 1975.

Valencius, Conevery Bolton. *The Health of the Country: How American Settlers Understood Themselves and Their Land*. New York: Basic Books, 2002.

———. *The Lost History of the New Madrid Earthquakes*. Chicago: University of Chicago Press, 2013.

Vernon, Walter N. *Methodism in Arkansas, 1816–1976*. Madison: University of Wisconsin Press, 1976.

Vivian, Daniel J. *A New Plantation World: Sporting Estates in the South Carolina Lowcountry, 1900–1940*. Cambridge: Cambridge University Press, 2019.

Warren, Louis. *The Hunter's Game: Poachers and Conservationists in Twentieth-Century America*. New Haven, CT: Yale University Press, 1997.

Whayne, Jeannie ed. *Cultural Encounters in the Early South: Indians and Europeans in Arkansas*. Fayetteville: University of Arkansas Press, 1995.

Williams, Florence. *The Nature Fix: Why Nature Makes Us Happier, Healthier, and More Creative*. New York: W. W. Norton, 2017.

Wylder, Delbert E. *Emerson Hough*. Boston: G. K. Hall, 1981.

Young, J. P. *Standard History of Memphis, Tennessee*. Knoxville, TN: H. W. Crew, 1912.

SECONDARY ARTICLES

"Arkansas Sunday Laws." *Liberty: A Magazine of Religious Freedom* 16, no. 1 (1919): 117.

Casada, Jim. "Nash Buckingham, Southern Gentleman." Outdoor Writers Association of America website, accessed March 29, 2023.

Deaderick, William H. "A History of Arkansas Ornithology." *American Midland Naturalist* 26, no. 1 (July 1941): 207–17.

Dizard, Jan. "Why Hunt?" Center for Humans and Nature website. October 12, 2015.

Downey, Tom. "Riparian Rights and Manufacturing in Antebellum South Carolina: William Gregg and the Origins of the 'Industrial Mind.'" *Journal of Southern History* 65, no. 2 (February 1999): 77–108.

Dunlap, Thomas R. "Remaking the Land: The Acclimation Movement and Aglo Idea of Nature." *Journal of World History* 8, no. 2 (Fall 1997): 303–19.

Etheridge, Y. W. "Pioneers of Ashley County." *Arkansas Historical Quarterly* 16, no. 1 (Spring 1957): 63–77.

Hahn, Steven. "Hunting, Fishing, and Foraging: Commons Rights and Class Relations in the Postbellum South." *Radical History Review*, no. 26 (1982): 37–64.

Hansbrough, Vivian. "The Crowleys of Crowley's Ridge." *Arkansas Historical Quarterly* 13, no. 1 (Spring 1954): 55–65.

Higgins, Billy D. "The Origins and the Fate of the Marion County Free Black Society." Arkansas Historical Quarterly 54, no. 4 (Winter 1995): 427–43.

Hornady, William T. "Will You Have a Gameless Continent?" *Independent*, May 4, 1911, 946–49.

Key, Joseph P. "An Environmental History of the Quapaws, 1673–1803." *Arkansas Historical Quarterly* 79, no. 4 (Winter 2020): 297–316.

———. "Indians and Ecological Conflict in Territorial Arkansas." *Arkansas Historical Quarterly* 59, no. 2 (Summer 2000): 127–46.

———. "Masters of this Country: The Quapaws and Environmental Change." PhD diss., University of Arkansas, 2001.

"Looking Back at Shotgun History." *American Rifleman* (online), May 23, 2016, accessed February 2, 2023.

Morris, Wayne. "Traders and Factories on the Arkansas Frontier, 1805–1822." *Arkansas Historical Quarterly* 28, no. 1 (Spring 1969): 28–48.

Morrow, Lynn. "A 'Duck and Goose Shambles': Sportsmen and Market Hunters at Big Lake, Arkansas." Southeast Missouri State University Press website, accessed May 5, 2019.

———. "St. Louisans at the Knobel: Sport Tourism and Conservation in Arkansas." Southeast Missouri State University Press website, accessed August 12, 2021.

Otto, John Soloman, and Ben Wayne Banks. "The Banks Family of Yell County, Arkansas: A 'Plain Folk' of the Highlands South." *Arkansas Historical Quarterly* 41, no. 2 (Summer 1982): 146–67.

Ownby, Ted. "Manhood, Memory, and White Men's Sports in the Recent American South." *International Journal of the History of Sport* 15, no. 1 (1998): 103–18.

S & D Reflector 7, no. 1 (March 1970), 5–7.

Shea, William L. "A Semi-Savage State: The Image of Arkansas in the Civil War." *Arkansas Historical Quarterly* 48, no. 4 (Winter 1989): 309–28.

Stewart, Mart A. "From King Cane to King Cotton: Razing Cane in the Old South." *Environmental History* 12, no. 1 (2007): 59–79.

Swanson, Drew. "Growing Wild: Visions of Wildlife Management as Agricultural Science in American Forests and Fields." *Agricultural History* 97, no. 4 (2023): 177–214.

Taylor, John M. "The Spencer Pump: America's First Pump-Action Shotgun." *Outdoor Life* (online), January 20, 2022, accessed February 2, 2023.

Towns, Stuart. "Joseph T. Robinson and Arkansas Politics, 1912–1913." *Arkansas Historical Quarterly* 24, no. 4 (Winter 1965): 291–307.

Vessels, Jay. "King of the Market Hunters." *Texas Game and Fish* 10, no. 12 (1952): 11–14.

Watson, Harry. "The Common Rights of Mankind: Subsistence, Shad, and Commerce in the Early Republican South." *Journal of American History* 83, no. 1 (June 1996): 13–43.

Williams, C. Fred. "The Bear State Image: Arkansas in the Nineteenth Century." *Arkansas Historical Quarterly* 39, no. 2 (Summer 1980): 99–111.

Wintory, Blake. "William Hines Furbush: An African American, Carpetbagger, Republican, Fusionist and Democrat." *Arkansas Historical Quarterly* 63, no. 2 (Summer 2004): 107–65.

Worley, Ted R. "An Early Arkansas Sportsman: C. F. M. Noland." *Arkansas Historical Quarterly* 11, no. 1 (Spring 1952): 25–39.

Worley, Ted R., and Russell W. Benedict. "Story of an Early Settlement in Central Arkansas." *Arkansas Historical Quarterly* 10, no. 2 (Summer 1951): 117–37.

Yingling, Christopher P. "The Arkansas Game and Fish Commission." Senior thesis, SM 350, Spring 1985, Torreyson Library Archives, University of Central Arkansas, Conway, AR.

SPORTING MAGAZINES

American Field

American Register

American Sportsman

American Turf Register and Sporting Newspaper
Forest and Stream
Recreation
Rod and Gun
Shields Magazine
Spirit of the Times
Sporting Times
Turf, Field, and Farm

NEWSPAPERS

Arkansas Democrat (Little Rock, AR)
Arkansas Gazette (Arkansas Post)
Arkansas Gazette (Little Rock, AR)
Arkansas State Gazette (Little Rock, AR)
Arkansas Times and Advocate (Little Rock, AR)
Batesville Guard (Batesville, AR)
Baxter Bulletin (Mountain Home, AR)
Bureau County Tribune (Princeton, IL)
Chicago Tribune (Chicago, IL)
Courier News (Blytheville, AR)
Daily American (Nashville, TN)
Daily Arkansas Gazette (Little Rock, AR)
Fayetteville Weekly Democrat (Fayetteville, AR)
Forrest City Times (Forrest City, AR)
Fort Smith Times (Fort Smith, AR)
Gentry Journal-Advance (Gentry, AR)
Green Forest Tribune (Green Forest, AR)
Helena Weekly World (Helena, AR)
Hot Springs New Era (Hot Springs, AR)
Kansas City Times (Kansas City, KS)
Madison County Record (Huntsville, AR)
Memphis Avalanche (Memphis, TN)
Memphis Daily Appeal (Memphis, TN)
Memphis Evening Herald (Memphis, TN)
Monticellonian (Monticello, AR)
Nashville News (Nashville, AR)
New York Mirror (New York, NY)
New York Times (New York, NY)
Northwest Arkansas Times (Fayetteville, AR)
Orlando Sentinel (Orlando, FL)
Osceola Times (Osceola, AR)

Pine Bluff Daily Graphic (Pine Bluff, AR)
Pittsburg Post-Gazette (Pittsburgh, PA)
Plattsmouth Journal (Plattsmouth, NE)
Randolph Recorder (Randolph, TN)
Rich Hill Tribune (Rich Hill, MO)
St. Louis Globe-Democrat (St. Louis, MO)
St. Louis Post-Dispatch (St. Louis, MO)
St. Louis Republic (St. Louis, MO)
St. Louis Republican (St. Louis, MO)
Southern Standard (Arkadelphia, AR)
Star Progress (Berryville, AR)
Sun Herald (Biloxi, MO)
Weekly Arkansas Gazette (Little Rock, AR)

ELECTRONIC SOURCES

American Rifleman
Ancestry
Arkansas Department of Agriculture website
Arkansas Digital Archives
Arkansas Game and Fish Commission website
Arkansas Heritage
Butler Center for Arkansas Studies website
Chronicling America
Classic Fly Rod Forum
Dictionary.com
Encylopedia of Arkansas
Find a Grave
Legal Information Institute, Cornell Law School website
Lexico
Library of Congress online catalog
Newspapers.com
Oxford English Dictionary
State Parks website
United States Fish and Wildlife Service website
USLegal

Index